OXFORD SYMPOSIUM ON FOOD & COOKERY 1989

Staple Foods

Proceedings

Fishermen's wives preparing salted herring

OXFORD SYMPOSIUM ON FOOD & COOKERY 1989

Staple Foods

Proceedings

PROSPECT BOOKS
1990

Introduction

This volume of papers presented to the Oxford Symposium on Food and Cookery at Saint Antony's College, Oxford, under the joint chairmanship of Dr Theodore Zeldin and Alan Davidson in June 1989 follows the pattern of previous collections. The first three papers were given in plenary session and the others follow in alphabetical order by author.

As well as the papers presented, there were lively, interesting and valuable discussions both formally in the various sessions and informally almost anywhere where two or more people could gather together. A large variety of interesting types of bread increased the enjoyment of our Saturday lunch.

The practical help and understanding necessary in dealing with the rather eccentric group that we are, was not stinted by the Domestic Bursar and staff of Saint Antony's College, particularly in the kitchen.

To cook an excellent and interesting Swiss dinner on Saturday evening, Herr Peter Bührer came specially from Wallisen in Switzerland. We also thankfully acknowledge the contributions of Dettling AG of Brunnen, Kalenderfabrik Luzern, Plantours Réunis of Winterthur, Swissair Catering, the Restaurant Rebe of Zürich and Messrs. Paxton & Whitfield of Jermyn Street, London.

Harlan Walker
June, 1990

FRONT and BACK COVERS: Illustrations by Tomasz Siwinski to Michael Abdulla's paper.

FRONTISPIECE: *Fishermens' wives preparing salted herring* by Bert Olls, who also drew the herring at the end of Astri Riddervold's paper.

ISBN 0 907325 44 0

©1990, as a collection, Prospect Books Ltd. (but©1989, in the individual articles, rests with the authors).

Editor: Harlan Walker.
Published by Prospect Books Ltd. 45 Lamont Road, London. SW10 0HU.
Typeset by Armorel Productions, 15 Bewdley Villas, Birmingham. B18 4JX.
Printed by Kingfisher Print & Design Ltd., The Mill, Dartington, Totnes, Devon. TQ9 6TT.

CONTENTS

Page

STAPLES: SOME CONSIDERATIONS ON THE NATURE OF STAPLES ESPECIALLY IN REGARD TO ITALY *1*
    Keith Botsford

IN SEARCH OF THE STAPLE FOODS OF PREHISTORIC AND CLASSICAL GREECE *5*
    Andrew Dalby

DO PROCESSED SOCIETIES HAVE STAPLE FOODS? *24*
    Erica F. Wheeler

BULGUR — AN IMPORTANT WHEAT PRODUCT IN THE CUISINE OF CONTEMPORARY ASSYRIANS IN THE MIDDLE EAST *27*
    Michael Abdalla

PASTA — NOT ONLY ITALIAN *38*
    Massimo Alberini

BUCKWHEAT — FOOD FOR PEASANTS AND PHEASANTS *40*
    Josephine Bacon

THE SIGNIFICANCE OF WHEAT IN JUDAEO-CHRISTIAN RELIGION *42*
    Josephine Bacon

HISTORY AND PERSPECTIVES OF STAPLE FOODS IN AFRICA *44*
    Dr. Esther Balogh

THE EVER-EVOLVING STORE CUPBOARD *55*
    Suzy Benghiat

SURVIVAL KIT (16TH CENTURY SEAMEN'S FARE) *57*
    Maggie Black

THE DATE PALM: PILLAR OF SOCIETY *61*
    Holly Chase

ATOLLI: A LIQUID STAPLE *70*
    Dr. Sophie D. Coe

POLENTA — AN ITALIAN STAPLE *75*
    Anna Del Conte

STAPLE FOODS OF THE AMERICAN WEST COAST *78*
    John Doerper

DIFFERENT METHODS OF BAKING BREAD IN PRIVATE HOUSEHOLDS — A COMPARISON OF WORKING TIME, QUALITY AND COSTS *100*
    Martina Ehnle, Cornelie Pfau, Johannes Piekarski

THE METAMORPHIC POTATO: A REVOLUTIONARY ROOT *107*
    Dr. Beatrice Fink

RICE AND TRADITIONAL CEREMONY IN JAPAN *112*
    Yoshiko Hirasawa

TARHANA — FROM STEPPE TO EMPIRE Maria Johnson	*116*
BEANS OF THE SOUTHWESTERN UNITED STATES INDIANS Mary Wallace Kelsey	*119*
THE FLAVOUR OF JAPAN Max Lake	*129*
MAIZE AS A STAPLE FOOD Elisabeth Lambert Ortiz	*134*
CORN — A STAPLE FROM THE NEW WORLD Janet Laurence	*136*
A DICTIONARY OF EDIBLE AROIDS Jenny Macarthur	*142*
UNFAIR GAME Carolyn McCrum	*151*
TRADITIONAL TABLE MANNERS IN DAGESTAN Dr. Magomedkhan Magomedkhanov & Sergi Luguev	*153*
RICE, A STAPLE FOOD IN SPAIN Lourdes March	*158*
HEALTHY — OR A HEALTH HAZARD — YESTERDAY'S DIET, TODAY'S DISASTER Dr. H. Morrow Brown	*162*
THREE STAPLES OF INDONESIA: RICE, COCONUTS, TEMPEH Sri Owen with Roger Owen	*168*
COUSCOUS AND ITS COUSINS Charles Perry	*176*
WINE, WOMEN AND SONG: THE STAPLES OF LIFE Graham Pont	*179*
FISHERY AND THE UTILIZATION OF FISH PRODUCTS IN RUSSIA AND THE USSR Professor T. S. Rass	*185*
THE IMPORTANCE OF HERRING IN THE DAILY LIFE OF THE COASTAL POPULATION OF NORWAY Dr Astri Riddervold	*187*
SYLVESTER GRAHAM AND THE ORIGINS OF THE BREAKFAST FOOD INDUSTRY Elizabeth Riely	*198*
VILHJALMUR STEFANSSON AND THE ALL-MEAT DIET Gillian Riley	*202*
A MEDIEVAL STAPLE. VERJUICE IN FRANCE AND ENGLAND Brenda S. Rose	*205*

WHEAT, STAPLE FOOD FOR THE DEAD   213
    Rena Salaman

ANCIENT VEGETARIANISM — STAPLE FOODS AND CUSTOMS IN   216
AZERBAIJAN
    Dr. Emil Salmanov

FEED THE MAN MEAT   224
    Barbara Santich

ROOTS AND OTHER GARDEN VEGETABLES IN THE DIET OF LONDONERS,   228
C. 1550-1650, AND SOME RESPONSES TO HARVEST FAILURES IN THE 1590'S
    Malcolm Thick

WHEAT AND RICE RECIPES OF INDIA   236
    Kathie Webber

RYE, A DAILY BREAD AND A DAILY TREAT   240
    Joop Witteveen

RICE AND WHEAT IN MIDDLE EASTERN CULTURES   246
    Sami Zubaida

List of Participants
(as at June 1989)

Massimo Alberini, 20131 Milano, via A. Bazzini 29, ITALY.
Dr. Joan P. Alcock, 24 Queensthorpe Road, Sydenham, London. SE26 4PH.
Mrs. Anne Andrews, 243 Whetstone Lane, Aldridge, Walsall. WS9 0HH.
Josephine Bacon, 82 Stonebridge Road, London. N15 5PA.
Mrs. Priscilla Bain, 7 The Norton, Tenby, Dyfed. SA70 8AA.
Dr. Esther Balogh, Department of Food Science & Technology, Otsafemi Awololo University, Ile-Ife, NIGERIA.
Ann Barr, 36 Linton House, 11 Holland Park Avenue, London. W11 3RL.
Simon Beard, 23 Forest Hill Way, Dihden Purlieu, Southampton. S04 5AS.
Suzy Benghiat, 93-95 ave du General Leclerc, Batiment B, 75014 Paris, FRANCE.
Michelle Berriedale-Johnson, 5 Lawn Road, London. NW3 2XS.
Maggie Black, 167 Putney Bridge Road, London. SW15 2NZ.
Keith Botsford, c/o The Independent, 40 City Road, London. EC1Y 2DB.
Clare Brigstocke.
Elizabeth Carter, 11a Montague Road, London. E8 2HN.
Mollie Chadsey, 19 Finlay House, Phyllis Court Drive, Henley-on-Thames. RG9 2H.
Lisa Chaney, 40 Southfield Road, Oxford. OX4 1NZ.
Holly Chase, 8ox 3452, Groton Long Point, CT 06340, USA.
Robert Chenciner, 5a Shepherd's Market, London. W1.
Mark Cherniavski & Anne Willan, 1613 30th Street, N. W., Washington DC 2007, USA.
Janet Clarke, 3 Woodside Cottages, Freshford, Bath. BA3 6EJ.
Dr. Sophie D. Coe, 376 St. Ronan Street, New Haven, Connecticut 06511, USA.
Clare Connery, 2 Richmond Park, Belfast. BT9 5EF. Northern Ireland.
Anna del Conte, 93 Elm Bank Gardens, London. SW13 0NX.
Mrs. Merryl Cook, 13 Prince's Road, Heaton Moor, Stockport, Cheshire. SK4 3NQ.
Mr. & Mrs. Andrew Dalby, 5 Primrose Way, Linton, Cambridge. CB1 6UD.
Alan & Jane Davidson, 45 Lamont Road, London. SW10 0HU.
Caroline Davidson, 5 Queen Anne's Gardens, London. W4 1TU.
Joy Davies.
Dr. B. T. Davis, 3 Ascot Road, Birmingham. B13.
John Doerper, 111 Old Mill Village, Bellingham, WA 98226, USA.
Anne Dolamore, 101 Clapham Common North Side, London. SW4.
Christopher Driver, 6 Church Road, London. N6 4QT.
Dr. Rochelle Dubnow, PO BOX 10116, Chicago, Illinois 60610, USA.
Hugo Dunn-Meyne11, 125 Mount Street, London. W1Y 5HA.
Sarah Edington.
Martina Ehnle, Bundesforschungsanstalt fuer Ernaehrung, Garbenstrasse 13, 7000 Stuttgart 70, (Hohenheim) West Germany.
J. Audrey Ellison, 135 Stevenage Road, Fulham, London. SW6 6PB.
Becky Sue Epstein, 8315 Marmont Lane, Los Angeles, CA 90069 USA.
Sarah Jane Evans and Rachael Evans, Crescent Wood Cottage, 6 Crescent Wood Road, London. SE26 6RU.
Scott Ewing, 60 Alderbrook Road, London. SW12 8AB.
Clare Ferguson, 5 Colville Terrace, London. W11 2BE.
Dr. Beatrice Fink, 6111 Madawaska Road, Bethesda, MD 20816, USA.
Bjorn Fjellheim, Kystmuseet i sogn og fjordane, Postboks 94, Floro, Norway.
Thomas Floegel, Donnersbergstrasse 24, 6000 Frankfurt/Main 71, West Germany.
Angela Francis.
Jean Freemantle, Ongar Hill Farm, Magpie Lane, Coleshill, Amersham, Bucks.
Dr. Robert Frey, 194 Sutherland Avenue, London. W9 1RX.
Elizabeth Gabay, 25 Cottenham Drive, London. SW20 0TD.
Janine B. Gilson, Foods from Spain, 22 Manchester Square, London. W1M 5AP.
Mrs. Emily Govers, Trinity College, Cambridge. CB2 1TO.
Jane Grigson, Broad Town Farmhouse, Broad Town, Swindon, Wilts. SN4 7RG.
Catherine Haith, Department of Medieval & Later Antiquities, British Museum, Great Russell Street, London. WC1 1DG.
Nevin Halici, Konya Turizm DerneZi, Mevlana Caddesi no. 2, Konya. TURKEY.
Aileen Hall, 65 Rodney Court, Maida Vale, London. W9 1TJ.
Lyn Hall, La Petite Cuisine, 21B Queens Gate Terrace, London. SW7 5PR.
Jane A. D. Hedges, Fulscot Manor, Didcot, Oxfordshire. OX11 9AA.
Vicky Heyward, 8 Victoria Court, Cartwright Street, London. E1 8LZ.
Yoshiko Hirasawa, 111 Boulevard Saint Michel, Paris 75005, France.
Caroline Hobhouse, 19 Camden Grove, London. W8.
Martha Holmberg, La Varenne, 34 Rue Saint Dominique, 75007, Paris, France.
Dr. Annie Hood, 96 High Street, Wouldham, Rochester. ME1 3UW.
Rebecca Hubbard, 20 Nettleton Road, New Cross Gate, London. SE14.
Lynette Hunter, University of Leeds, Department of English, Leeds. LS2 9JT.

Mr. G. A. & Mrs. Barbara Inskip, 67 Beverley Crescent, Bedford. MK40 4BZ.
Rosalind Irwin, 32 Vale Court, Mallord Street, London. SW3 6AL.
Judy Jackson.
Rosemary Joekes, The Hermitage, St. Catherine, Bath. BA1 8HE.
Bruce W. Johnson.
Maria Johnson, 6 The Limes, Stratton Audley, Bicester, Oxfordshire. OX6 9DA.
Dr. Keane.
Mary Wallace Kelsey, Department of Foods & Nutrition, College of Home Economics, Oregon State University, Corvallis, OR 97331, USA.
Giana & Nicholas Kurti, 38 Blandford Avenue, Oxford. OX2 8DZ.
Dr. Max Lake, 28 The Pines, 51-53 The Crescent, Manly 2095, NSW, AUSTRALIA.
Elisabeth Lambert Ortiz, 19 Beaconsfield Road, London. W5 5JE.
Heidi Lascelles, Books for Cooks, 4 Blenheim Crescent, London. W11 1NN.
Janet Laurence, The Grooms, East Lydford, Somerton, Somerset. TA11 7HD.
Annie Hauck Lavson, 481 17th Street, #4A, Brooklyn, NY 11215, USA.
Beverly LeBlanc, Top Flat, 128 Shirland Road, London. W9 2BT.
Elbie Lebrecht, 3 Bolton Road, London. NW8 0RJ.
Charlotte Leeming, 2 Guildford Street, London. WC1N 1DR.
Paul Levy, The Observer Magazine, Chelsea Bridge House, Qeenstown Road, London. SW8 4NN.
Caroline Liddell.
Sandy Littman, 69 Harberton Road, London. N19 3JT.
Jenny Lo, 20 St. Clements Church, Arundel Square, London. N7 8BI.
Elisabeth Luard, 6 Fernshaw Road, London. SW10 0TF.
Jenny MacArthur, 13 Wavell Road, Maidenhead, Berkshire. SL6 5AB.
Carolyn McCrum, 57 Oakthorpe Road, Oxford. OX2 7BD.
Francis McKee, 5A Clarence Drive, Hyndland, Glasgow. G12 9QL.
Mike & Tessa McKirdy, Cooks Books, 34 Marine Drive, Rottingdean, Sussex. BN2 7HQ.
Jeremy McClancy, Institute of Social Anthropology, 51 Banbury Road, Oxford. OX2 6PF.
Dr. Magomedkhan Magomedkhanov, 367012 Dagestan ASSR, Makhachkala, Prospekt Kalinina 77B Kv. 19, USSR.
Lourdes March, General Castanos 7, 28004 Madrid, SPAIN.
Professor & Mrs. G. Mars, 53 Nassington Road, London. NW3 2IY.
Dr. Stephen Mennell, 7 Wheatsheaf Way, Alphington, Exeter. EX2 8QQ.
Richard C. Mieli, 4 Longfellow Place #1703, Boston, Ma 02114, USA.
Marie-Pierre Moine, 22 Arundel Gardens, London. W11.
Mrs. I. H. Montijn, Stationstraat 17, 1391 GL Abcoude, NETHERLANDS.
Janny de Moor, De Vriesweg 29. 8084 AR 't Harde, NETHERLANDS.
Sallie Morris, 12 Lawn Crescent, Kew, Richmond, Surrey. TW9 3NR.
Dr. & Mrs. H. Morrow Brown, Highfield House, Highfield Gardens, Derby. DE3 1HT.
Angela F. NaZy, H 1121 Budapest, 5 Mese U., HUNGARY.
Jill Norman, 1 Rosslyn Hill, London. NW3 5UL.
Sri & Roger Owen, 96 High Street Mews, London. SW19 7RG.
Bruce Palling, 13 Leamington Road Villas, London. W11.
Charles Perry, 12912 Eldorado Avenue, Sylmar, CA 91342, USA.
Graham Pont, Centre for Liberal & General StudiesThe University of New South Wales, P. O. Box 1, Kensington 2033, Australia.
Olive Portnoy, Oaktrees, Woodman Lane, Sewardstonebury, London. E4 7QR.
Dr. Peter Rachwall, 15 Sabine Road, London. SW11 5LN.
Professor T. S. Rass, Institute of Oceanology, Acadamy of Sciences of the USSR, 23 Krasikova, Moscow 11728, USSR.
Mrs. Catharine P. Reynolds, 6 Swan Walk, London. SW3 4JJ.
Dr. Astri Riddervold, Dyrlandsvei 9, 0875 Oslo 8, NORWAY.
Elizabeth Riely, Grove House, 60 Church Street, Kidlington, Oxford. OX5 2BB.
Dr. Peter Rietbergen, Erasmusgebouw 13. og, Erasmusplein 1, Nijmegen, NETHERLANDS.
Gillian Riley, 11 Kersley Road, Stoke Newington, London. N16 0NP.
Alicia Rios, Avenida General Peron 19 8°C, Madrid 28020, SPAIN.
Cherry Ripe, c/o ABC, 54 Portland Place, London. W1.
Jo Roberts.
Claudia Roden, 8 Wild Hatch, London. NW11 7LD.
Dr. B. S. & Mrs. Brenda S. Rose, 626 London Road, Davenham, Northwich, Cheshire. CW9 8LG.
Jeremy Round, 52 North Street, London. SW4 0HD.
Rena Salaman, 145 Tufnell Park Road, London. N7 0PU.
Dr. Emil Salmanov, Baku 370000, Azerbaidzhan SSR, Prospekt Lenina 5 Kv. 4, USSR.
Alice Wooledge Salmon, 125 Mount Street, London. W1Y 5HA.
Catherine Salzman, le Vijverstraat 16, 3052 HV Rotterdam, NETHERLANDS.

Barbara Santich, 13 King Street, Brighton, 5048 SA, AUSTRALIA.
David Scott, 21a Aigburth Drive, Liverpool. L17 4JQ.
Robert Sellick, Apple Jacks, 255 Eversholt Street, Camden, London. NW1.
Maria Jose Sevilla Taylor, 72 Brisbane Road, Ilford, Essex. IG1 4SL.
Margaret Shaida, Della Ra, Harpsden Way, Henley-on-Thames, Oxford. RG9 1NL.
Ralph & Kate Shirley, The Studio, Foston, Derby. DE6 5PX.
Yan-Kit So, 11 Gordon Place, London. W8 4JD.
Mrs. P. M. Spears, 4 Church Row, Coppice Lane, Middleton, nr. Tamworth, Staffs. B78 2AR.
Rosemary Stark, 6 Chamberlain Street, London. NW1 8X.
Jeffrey Steingarten, 29 West 17th Street, New York, NY 10011, USA.
Anne E. Sterling, 51 rue des Petites Ecuries, 75010 Paris, FRANCE.
Paul Stokes, Flat 1, Half Moon Court, 73, High Street, Bagshot, Surrey. GU19 5AH.
Malcolm Thick, 8 Buckingham Close, Didcot, Oxfordshire.
Carol Thompson, University of California Extension Center, 55 Laguna Street, San Francisco, CA 94102, USA.
Jill Tilsley-Benham, 12a Upper Mulgrave Road, Cheam, Surrey. SM2 7AZ.
Anne Trager, 1 ave. de l'Observatoire;75006 Paris, FRANCE.
Mr. C. & Mrs. R. P. Tyler, 22 Belle Avenue, Reading, Berks. RG6 2BL.
Harlan Walker, 294 Hagley Road, Birmingham. B17 8DJ.
Jennifer Walker, 10 Whitley Park Lane, Reading, Berks.
Conal Walsh, 31-39 Earlham Street, Covent Garden, London. WC2H 9LO.
Ann Watson, The Forge, High Street, Barford, Warwick. CV35 8BU.
Kathie Webber, Stoke Green Farmhouse, Stoke Green, Stoke Poges, Bucks. SL2 4HN.
Rosemary Weinstein, The Museum of London, London Wall, London. EC2Y 5HN.
Robin Weir, 104 Iffley Road, London. W6.
Erica F. Wheeler, Dept. of Human Nutrition, London School of Hygiene & Tropical Medicine, Keppel Street,
    London. WC1E 7HT.
Margaret Willes, 17 Appleby Road, London. E8 3ET.
Joop Witteveen, Tweede Oosterparkstraat 261, 1092 BN Amsterdam, NETHERLANDS.
Mary Wondrausch, The Pottery, Brickfields, Compton, Guildford, Surrey. GU3 1HZ.
Dr. Theodore Zeldin, Tumbledown House, Cumnor, Oxford. OX2 90E.
Sami Zubaida, 2 Avenue House, Belsize Park Gardens, London. NW3 4LA.

Some Considerations on the Nature of Staples, Especially in Regard to Italy

Keith Botsford

I   **The Word Itself**

Being a linguistic purist, when I received the announcement of this year's subject, staples, I - no doubt like many contributors - grappled first with the word.

The older French word for provisions, for food for man or beast, is *denrée*, from the Latin *denarius*, a coin. The United Nations' *Lexique général* (a most useful work) gives us a thoroughly rational and far more modern sense of the word. A staple is called *un aliment de base*; in industrial terms, *un produit de première nécessité*. I can live with that sense as long as I can find out how divers languages, and therefore divers cultures, came to define such a concept. What is 'first necessity' and what is a 'basic food'?

The Germans, ever thorough in word formation, bring the idea down to "the most important" item in a diet. A staple is a *Hauptnährungsmittel:* not the sort of word one wants to bandy about, and being one of those agglutinative words, no clue, etymologically, to the concept. But that is because the modern word has replaced an older one which, as we shall see, underlies the modern concept of a 'staple' - in which, buried, lies the near-homonym 'stable' and the 'stability' that is at the heart of the idea.

In the Romance languages, Italians, who obviously imported the word, waffle around the concept. In Italian, staples are the *principali generi di consumo*, the main articles of consumption. Romanians, sticking by an older connection to Latin, understand it in a more helpful sense, in that the principal meaning of the work is a *loc de antre-pozit* in which, however dimly, one can detect the French *entrepôt* (a storehouse) which I believe leads us, linguistically, nearer to the truth, for it connects with the Latin *emporium*, a market. But even in Bucharest, a staple is also, as in Italian, an *alimentul de baza*.

In the Slavic languages, by contrast, the concept of a staple is tied to the idea of fruitfulness (Russian *plod*, fruit, and *plodit*, to multiply, bears the concept of foetus: thus Czech *hlavni plodiny*, main fruits, i.e. staples). In Russian, the first meaning of staple is as in English, a market, *rynok* or *bazar*; it is only subsequently that Slavic produces its own version of *glavni produkt* or *Hauptartikel*.

The true clue to staples, linguistically, reaches us via the French, and the French got the word from Low German. Trench's *Select Glossary* brings out into the open the shift in the word's use. "A curious change," he says, "has come over this word: we should say, cotton is the great staple ... of Manchester; our ancestors would have reversed this and said, Manchester is the great staple, or established mart, of cotton."

The French original of the word (OF and MF) is *estaple*, whence *estape* and modern *étape*, a stopping off place. That in turn would seem to derive (in its LG form *Stapel*, German *Staffel*) from the modern word for the step of a ladder. The great philologist F. Muller explains the original meaning of the word, and its long evolution, showing the following progression: a prop or foundation, a support: then a heap or a stand for laying things on; then heaped wares (especially in order) or a storehouse - whence the idea of a market or entrepot or bazaar. Minsheu's *Guide to the Tongues* (1627) gives the word its then current meaning: "Staple signifieth this or that towne or citie, whether [wither] the Merchants of England, by common order or commandment, did carrie their woolles, wool-fals, cloathes, leade and tinne, and such like commodities of our land, for the viterance of them by the great." Hence the survival in market-town names, Dunstable, Barnstaple, Whitstable, etc.

The underlying metaphor, therefore, is of stability, consistency and order. A secondary level includes the idea of an array of foods in some way arranged in order, that order being originally local. Thus an agricultural market in Piedmont would be based on rice. In French, it is here that we get a reference back to the original German, for an *étal* (our 'stall') is a table on which goods are displayed in a market, hence *étaler*, to spread, to display. Finally, there is a third subjacent sense, of something that can be stored, magazined, warehoused - that, in short, need not be as fresh as produce.

One may take it, then, that staples are a substratum of culture: not necessarily prime necessities (for these would have varied even from village to village, from mountain to plain, from town to town) but foods which by tradition are thought to be necessary and therefore to be made available, even if they have to be stored. It is in that sense that the French, for instance, have always had – even in the bad old days – a system of price supports for bread, and that almost all Mediterranean countries have long historical state monopolies of salt.

It wasn't until preservation entered the food picture - curing, smoking, tinning, freezing, dehydrating, etc. - that our basic list of staples grew, and therefore it is historically quite correct to limit the idea of staples to the most ancient and basic foods, such as bread (or grains in general), rice, maize, salt and oil. These are foods without which one could not produce a meal or keep foodstuffs from day to day or from season to season. They are also foodstuffs susceptible to storage.

Nonetheless, I'm not sure it's an useful distinction for our times, in which the list of staples (in the United Kingdom) is likely to include such varied items as catsup, dry pasta, tins of baked beans and so on. We can be quite sure that some government department has, in its infinite wisdom, already prepared a list of such staples to be stocked in the underground shelters of the future.

My persistent unease with this definition of staples is that I'm not at all sure that different cultures share the same notion of necessity. The states of antiquity, we know (see Julius Caesar's propensity, at election time, to distribute food), knew that the populace may not, without disturbance, be allowed to go hungry. If we want to start from there, then we can go to the Greeks, who had only two words for food: bread and that which is eaten with bread. If there were bread riots in antiquity, and even in the recent past, similar riots would have been caused in Mexico over maize; in China with rice; in an Ireland without potatoes and so on.

But the act of not allowing people to starve is not the same as that of establishing 'staples'. A staple is one thing - we readily know what it means when we say the staple diet of much of Africa is the plantain; staples, plural, are an altogether different thing. They go beyond survival to the very roots of cultural differences.

II   The Staples of Italy

It is an unwise man who makes generalisations about food in Italy. My grandmother, after all, was born before Italy became a nation, and Bologna and Emilia in general is as different from Milan or Piedmont as Sicily is from the Veneto. I live in a part of Italy known as Sabina. It is a very ancient part of Italy, that part in fact, which founded Rome - besides, of course, getting its women raped. I live in fact right on the Via Salaria, the old road along which salt was brought to Rome.

I think, therefore, that I make no mistake in saying that there is no house in my part of the world which does not consider salt a staple. For a similar reason, Sabina producing some of the finest (though much less well-known than Tuscan or Umbrian) olive oil in Italy, I would suppose that oil falls into the same category.

But is bread a staple in my region? The villages about me do indeed make an excellent bread, less antique than the Roman roll and rather more like a much crustier version of a *pain de campagne*, and I've never seen one of my neighbours eat a meal without bread. On the other hand, I am less than convinced by its necessity. For a Sabine, to live without bread might be more possible than to live without salt or oil, and especially the former.

The reason for this is obvious: Italy as a whole has a substitute for bread in the form of pasta. Could my Sabines live without pasta? And is pasta a staple? No, it isn't, because our locals make their own, daily. A compromise, then: neither bread nor pasta is a staple, but flour, from which either may be made, is. When I think of Mirella's kitchen, or Assunta's (Assú), I see that triumvirate quite clearly. To be without any of the three would be unthinkable. Take me a hundred kilometres away and I might well eliminate oil, for instance. Emilians might think of milk as a staple, and therefore of butter and lard, rather than oil.

Thereafter, matters become more complex. Are those elements which provide the basic flavour of local cuisines to be considered staples? In a sense, it would be as inconceivable to think of Lipari islanders living without capers as it would be to consider Reggians doing without parmesan. Here again, one cannot generalise. Garlic and tomatoes, both of which are readily preserved, are quintessential ingredients of Sabine cooking, as they are for most of southern Italy - my grandmother would always refer to those living below a line drawn at Rome, as 'Africans', just as anyone north of Florence was 'German'. Garlic is omnipresent, and in mid-summer, when tomatoes are plentiful, the women of my village band together to bottle prodigious quantities of *purée* or *ragú*, tomato sauce.

If spices are to be considered as staples, then locally I would have to add basil and wild mint, *mentuccia*, as further south I would include oregano and red pepper.

But what, you may ask, has this to do with a 'staple diet'? What the Reatines eat is as simple as it is unvarying. A meal starts with pasta and is followed by meat or fish with a *contorno* of a seasonable vegetable - locally, almost invariably artichoke or *brocoletti*. The meat is, equally invariably, lamb (we are a mountainous region) or pork or veal, and sometimes rabbit; the fish generally trout. I take it that though these are part of the staple diet, they are not, being perishables, staples.

Breakfast, of course, involves another staple: coffee. And wine is every bit as much a staple as coffee.

We now have a two-tier arrangement of staples that looks something like this:

    I - Salt, Oil, Flour

    II - Tomatoes, Garlic

And propping that up is a third level: those foods or flavourings without which life would not be worth living:

    III - Basil, Mint, Coffee, Wine

The reader will notice that this forms a hierarchy based on necessity: without the first staples, man cannot live: without the second, he cannot prepare (cook) that without which he cannot live; and without the third, he wouldn't want to live.

Taxonomically, these are interesting, in that the prime necessities, so-called, are none of them 'foods' in the accepted sense. The only people I know who eat (or drink) olive oil straight are Greek merchant seamen. A diet of salt would do no man good. And similarly, flour by itself is nothing much. Even my next two tiers contain little that we would consider food.

Should we then define staples as those commodities which, while not being food, are those from which civilised man creates his daily bread? If that is so, then almost certainly the only absolutely basic staple is flour (or, in other cultures, the local grain, rice, maize, barley, rye).

But it is a curious fact of modern life - as recently as a hundred years ago, the thought was inconceivable - that most of us in the advanced nations of the world have completely lost our sense of food as being essential to survival and conceive of it only as a pleasure, albeit a necessary one. We know what hunger is by being hungry and eating: we do not know what hunger is as others know it, who perforce sometimes do without eating. Thus, when we ship sacks of rice to starving Cambodians we are offering them a staple which embodies survival, but if a typhoon strikes Houston, Texas, we are more likely to ship in commodities which, in other cultures, would seem luxuries: tinned meat and vegetables, sugar (ah, there's a doubtful staple!), buns, bars of chocolate and bottled water or soft drinks. American staples, one could almost say, consist of those foods which can be most easily prepared and most expeditiously satisfy basic appetites. That is why the American larder, as compared to my Italian, consists of things in packages, not bulk goods in sacks and bottles.

III   Conclusions

I am unwilling to be drawn into a debate about what we mean – since we all mean very different things – when we say 'staples', beyond stating the following limits:

(1)   Staples are foodstuffs that are (a) common and (b) considered necessary;

(2)   What defines a staple is conditioned by culture in the broad sense of the term.

(3)   Staples are sometimes, but far from inevitably, foods necessary to survival.

(4)   Staples relate closely to markets, to the food production of any culture, to its culinary habits and the eating preferences of its members and, as such, reflect a wide range of goods which, while thought vital and necessary, are in fact luxuries whose absence would in no way affect the physical well-being of their consumers;

(5)   A number of 'staples' in fact are 'flavours' or foodstuffs used in cooking: e.g. salt, sugar, oil, herbs, preserves, etc;

(6)   The historical development of both the word and the concept indicates that the more advanced the society, the broader the range of foodstuffs brought into the overall category of 'staples', thus showing that

(7)   Staples are more a matter of taste, convenience, accessibility and tradition than necessity.

In Search of the Staple Foods of Prehistoric and Classical Greece

by Andrew Dalby

I can remember first hearing of the term 'staple food' and being told that mine was bread. I was taken by surprise; it had not occurred to me that bread was sufficiently important in my diet, or in that of anyone around me, to be singled out in this way. That is a problem of perspective, perhaps. What is identified by some as the staple food of a society may not have had such importance in the minds of all the eaters.

That leads to a problem of sources. A study of the cookery books of modern England would not help one to identify its staple foods: we give far more variety to, and put far more effort into, the things we eat *with* our bread and potatoes. Other sources (trade statistics for example) would be a better guide. So in investigating early societies one will have to compare culinary sources with quite different ones, if they are available, to get the answer one is looking for.

I intend to concentrate on this point and to discuss the staple foods of ancient Greece indirectly, by way of asking what kind of information is available to the food historian looking at a society in the distant past. Something will be said on two kinds of information: firstly archaeology; secondly literature.

Greece is one of those countries where the diet of the inhabitants can be studied as far back as 40,000 B.C. Let us start by asking what archaeology can tell us about two very important questions: how and when wheat and barley became the staple foods of ancient Greece, and what were the staple foods even before that.

THE EARLIEST PERIOD

In the long period between 40,000 and 10,000 B.C. there was no agriculture: both animal and vegetable foods were available but as yet there is very little positive evidence that vegetable foods were used.[1] We know of animal foods because of finds of bones. It is really an assumption that the inhabitants ate the animals and discarded the bones that archaeologists find. One does not know how important meat was as compared with, say, leaves or berries or seeds.

A site which was probably continuously inhabited from about 25,000 B.C. is the Franhthi cave near Portoheli in the southern Argolid. There was an early period during which most of the bones found are of wild horse, *Equus hydruntinus*, and red deer, *Cervus elaphus*. Later, perhaps around 12,000 to 10,000 B.C., there appears to have been a greater variety of foods available: horse was still the favourite but ox or bison, goat, pig and fish were available as well as red deer.[2] At Asprohaliko, a cave in southern Epirus used from about 40,000 to about 10,000 B.C., red deer, fallow deer (*Dama dama*) and wild goat (ibex, *Capra hircus*) were the commonest animal bones.[3] There are cut-marks on the excavated bones at Franhthi, which however show no signs of having been burnt or boiled. So if we can talk of a staple food for the inhabitants of palaeolithic Franhthi and Asprohaliko and their contemporaries, it was uncooked wild meat, cut up with stone knives.

Between about 20,000 and 6000 B.C. there are signs of the development of patterns of food specialisation. Some communities were transhumant - and they chose their seasonal camps because they were near a single food source. For example, Klithi rock shelter (used about 10,500 to 8000 B.C.) is too high in the mountains of Epirus to be habitable throughout the year but is close to one of the few possible wild goat migration routes in the region. At Kastritsa cave, some way to the south and used between about 20,000 and 11,400 B.C., at least 80% of identified bones were red deer. At Sidari on Corfu there was a seasonal settlement of cockle collectors.[4] Moreover, Franhthi continued to be inhabited after 10,000 B.C. but there are signs that people at Franhthi now tended to concentrate on one food at a time: at first snails (though snails may seem to predominate around 10,700 B.C. not because they were Franhthi's flavour of the century but just because the archaeologists happened on the spot where all the snail shells, over quite a long period, were thrown away) then red deer again, then after 7000 B.C. tunny, *Thunnus thynnus*.

Some information is available about the way these foods were used. At Klithi 'the bone surface is often stained black and sometimes carries a thin calcite skin ... Many cut marks can still be identified ... Many of the bones, especially small fragments, are burnt.'[5] At Franhthi bones were burnt; even small bones were broken. It seems, therefore, that meat was roasted on the bone and that great trouble was taken to extract marrow. Since coriander (root, leaf or seed) is not very useful except as a flavouring, a find of coriander seed at Franhthi suggests culinary experimentation. Coriander seed is probably more likely to strike the archaeologist's eye than most other spices or herbs, so they may quite well have used others.

THE INTRODUCTION OF SHEEP AND WHEAT

And now a new development was catalysed by the rapid spread in Greece of sheep, *Ovis aries*, long since domesticated in the Near East, and of two kinds of cultivated wheat (emmer, *Triticum dicoccum* and einkorn, *Triticum monococcum*), all three imported, almost certainly.[6] This is not quite 'the introduction of agriculture', by the way, because people seem to have done something towards managing the goats that passed by Klithi before 8000 B.C.[7] and they took a hand in growing barley, lentils and oats at Franhthi around 7000 B.C.[8] There is nothing surprising, given the earlier history of food specialisation, in the enthusiastic adoption of a new food source that appeared to offer advantages.

It is just at this time, about 6100 B.C., that the first known settlement on an Aegean island was established, at Knossos on Crete. Where the settlers came from is unknown, but they were certainly at the leading edge of civilisation in the region. They were soon growing a new hybrid wheat, perhaps bread or durum wheat, not found so early anywhere else, as well as emmer and barley;[9] they kept sheep, goats, pigs and cattle,[10] and all four species must have been brought in by sea.

Developments are really complicated. To simplify the pattern one would suggest that as population grew, reliable barley[11] had become as important in the diet as limited meat and fish.[12] Sheep and wheat, when first introduced, appeared to offer advantages over native food sources (and it may be that they were brought by immigrants whose staple foods they already were). But in the long run, in the extremely varied and not particularly fertile terrain of Greece, and as population continued to grow,[13] each district has had to do its own experimentation and find its own balance. In particular, wheat and cattle soon became the most popular, and indeed are still popular, in the wide plains of Thessaly and Macedonia; while barley and sheep eventually won the competition in the more rugged and drier districts, central and southern Greece and the Aegean islands.[14] One suspects that this experimentation in prehistoric Greece, in the sixth, fifth and fourth millennia B.C., helped to give rise to the variety of foodstuffs, the relative absence of a single staple, that characterise classical Greek and later western cuisines.

THE CLASSICAL PERIOD

Let us now look at another source of information for the ancient food historian: literature. We must move forward four thousand years or more to classical Greece and to what ancient historians call 'literary sources'. And these come in several kinds. There are those that everybody would call 'literary' - that is, poetry, plays, fiction and memoirs in which meals are described or food is talked about. There are factual and scientific books, on history, on geography, on medicine. There are books about diet and nutrition. There used to be cookery books and at least one gastronomic traveller's book and at least one book about shopping for food. There still survives one book about the history and literature of food, drink and entertainment. It is worth while to take some examples of the information to be found in these different kinds of sources.

ILIAD AND *ODYSSEY*

Let me start by looking at the *Iliad* and the *Odyssey*, epics of the eighth century B.C. in which social life is quite fully described, though one can argue for a long time about when (if ever) daily life might really have been like that. As regards food the *Iliad* and the *Odyssey* are different from one another. In the *Iliad*[15] (in which a Greek army is shown besieging Troy, in what was already the legendary past) the staple food of that army is meat: bread is seldom mentioned. One of the few places where bread is talked of is when the food of the Trojans themselves is described,[16] but then the Trojans were at home. In the *Odyssey*[17] (which is about Odysseus on his way home from the Trojan war, and has varied settings, sometimes domestic and peaceful) there is nearly always bread at meals, sometimes as the main feature with unspecified 'relishes' to accompany it, sometimes perhaps taking second place to meat at lavish banquets. Just one meal lacks bread - the one served by the pig-farmer Eumaeus to a guest.[18] But then, if anyone can offer a meaty meal without troubling about the expense, surely a royal pig-farmer can.

The variations in meals in the *Iliad* and the *Odyssey* can probably be accepted as plausible. I have argued, in a forthcoming article, that the absence of bread among the army at Troy is because the poet of the *Iliad* could not successfully imagine the details of a real, well-planned military expedition lasting ten years. No such thing had taken place in Greece in his time. Instead he had in mind the kind of expedition that did happen in his time - piracy, brigandage, colonisation - and on that kind of expedition one foraged and did not always have the chance to grind corn and bake bread.

But to show how the staple fitted into the other foods in a well-to-do household, take this description of a meal:

> Squires [of the suitors] and busy servants were some of them mixing wine and water in bowls, some again with porous sponges washing and setting out tables, some portioning out plentiful meat ...
>
> 'Be pleased, stranger; you will be welcome among us. Having eaten a meal you shall tell us what you need,' [the host] said, and led on, and Pallas Athene followed.
>
> When they were inside the high building ... having spread linen under, he sat [his guest] on a fine expensive chair, and placed a stool under her feet. He himself put a well-made couch nearby, away from the other[s, the] suitors, lest the stranger, distressed by the noise, should not fancy his meal ...
>
> A maid bringing hand-water poured it from a fair gold jug over a silver bowl for the wash, and drew up a carved table. An old housekeeper had put out bread, and bringing it, having added many relishes, regaled them from what was there; a waiter put out bronze trays of all sorts of meats, and put out gold goblets for them; a squire often passed by, pouring wine for them.
>
> And in came the noble suitors; and next they were sitting in rows on couches and chairs, and their squires poured water over their hands and house-girls piled out bread in baskets and boys filled bowls with drink; and they set their hands to the food laid out ready.[19]

In the eighth century B.C. (or whenever the oral poets first developed that description of a meal that was incorporated into the *Odyssey*) there was already much washing of tables. So there was throughout the classical period. Tables were small, shared between a couple of diners at the most, and we are seldom told of individual bowls or plates for food - so the tables had to start out clean. It is also evident that at this period Greeks had not adopted the Eastern fashion of lying down for meals.

Milling and bread-making were jobs for women, as is shown by this short passage:

> The feasters came to the house of the godlike king, and they were driving sheep and carrying manly wine, while their wives sent bread for them.[20]

One can translate **sitos** 'bread' but one cannot say any more about this bread. The only way we know it meant 'bread' and not 'porridge' to the poet is that it was 'piled out in baskets' in the passage quoted above. The epics are closed systems: anything that is not specified in them (such as the grain used for **sitos**, the method of preparing it, the texture, colour, shape, taste and smell of it) can only be guessed at. But they tell us a great deal about the etiquette, the sociology of meals: who served the bread (usually a woman); who served the meat and the wine (nearly always a man).

GEOGRAPHY: HERODOTUS

To find out more one may move a little forward to the time when Greeks were exploring their neighbours' ways of life, making comparisons and asking questions. The first Greek prose author of any importance, Herodotus[21] (5th century B.C.), heard of an interesting experiment regarding the origin of mankind.

> Before Psammetichus ruled them the Egyptians used to think that they themselves were the oldest of all peoples. But Psammetichus decided to find out who had come first: and since then they have known that though they were older than other peoples, the Phrygians were older still.
>
> Psammetichus could discover only one way to find out the answer to the question. This is what he did. He took two newborn babies from some parents or other and gave them to a shepherd to bring up on his farm, instructing that no one should say any word in front of them, that they should be kept in an empty hut, that he should drive in goats whenever necessary to provide milk, and that he should care for them generally. Psammetichus made these arrangements because he wished to know what word the children would first speak when they had got beyond meaningless babble.
>
> And so it was done. For two years the shepherd acted in this way, until he opened the door and went in to find the children stretching out their hands and saying, **Bekos**. The first time they said it the shepherd kept quiet, but he found that they often said this word when he went in to them, so he told his master and was instructed to bring the children into his presence. When Psammetichus had heard it himself he enquired which people had a word **bekos**. He found on enquiry that the Phrygians use this word for a 'loaf of bread'. So the Egyptians accepted, after this experiment, that the Phrygians were older than themselves. I had this story from the priests of Hephaestus at Memphis.[22]

This piece of folklore not only proves that (so someone thought) the Phrygian language was the original and natural language of mankind: it also proves that the loaf of bread is the original and natural staple food of mankind, but that was so obvious to Herodotus that he does not even trouble to bring the point out!

Staple foods were of importance to Herodotus, as they indicated to him the level of barbarism or civilisation of the eaters. He recorded, for example, from hearsay the names and customs of a series of tribes living in southern Russia; the nearest lived on wheat, those more distant on milk and meat, the furthest off of all on human flesh,[23] a staple which has not yet been mentioned at this symposium.

The observations and comparisons made by Herodotus, and the more open-minded of his successors, give us clues about Greek culinary practice. He says, for example, that some of the Babylonians have fish as their staple diet. They dry it, pound it in a mortar, and eat it either kneaded like **maza** or baked like **artos**.[24] **Artos** is classical Greek for a loaf of bread; **maza** is something made out of barley, but it isn't too clear exactly what. This passage tells us, more clearly than any other, that the Greek **maza**, in the 5th century B.C., was not a baked loaf.

Among Herodotus's successors in ethnography one may cite Megasthenes, who in the early 3rd century B.C. wrote a description of India (which Greeks began to visit in Alexander the Great's time) and mentioned food. The passage is summarised by Athenaeus:

> Megasthenes in *Indian Studies* book 2 says that at dinner the table that is placed before each person is like a pot-stand, and a golden bowl is set on it, in which they first of all put rice, boiled just as one might boil **khondros**, and then many different relishes prepared according to Indian recipes.[25]

This fascinating extract not only tells us that the Indian style of eating - adding modest quantities of several highly-flavoured dishes to one's helping of rice - is more than two thousand years old; it also reminds us that the Greeks ate solid food from the table rather than from a bowl. This is why Megasthenes needed to say that the Indians had a different kind of table, one which was not eaten off directly. Again, it tells us something about **khondros**, a preparation of wheat which was not flour, not whole grain, not bread and not porridge (though some kind of bread and porridge could be made from it). **Khondros** was evidently grainy enough to be prepared by boiling: it was, surely, 'cracked wheat'. We return to it and to Athenaeus soon.

DIETETICS: THE HIPPOCRATIC CORPUS

Ancient writers on diet and nutrition are worth studying carefully; their arguments can be impressive though their conclusions are fairly often wrong. Certainly they have something to tell us about food and food customs. In the earlier period, in the 5th, 4th and 3rd centuries B.C., medical authors were anonymous: they contributed, as it were, to something that gradually built up into a miscellany (not unlike a scientific journal) containing long and short articles, supplements, disputes, corrections, case notes, that went under the name of Hippocrates. It is quite impossible to say whether the famous doctor Hippocrates wrote a single one of these texts: certainly he did not write most of them. The personalities of the many anonymous authors sometimes stand out strongly, as do their prejudices and obsessions. Not many people can have thought that barley gruel and barley water were such universal cures as did the late 5th century author of *Diet in Acute Diseases*, who apparently prescribed them for practically everything.[26] He was the founder of a school: my father thought as highly of barley water as some other people think of vitamin pills.

I will pass over barley water: my view (now as when I was a boy) is that a little barley water goes a long way. Here is the earliest text in Greek that talks about the ways of baking bread, extracted from the monograph called *Regimen* or 'Diet', which belongs to the fourth century B.C.[27]

> As for bread itself, the largest loaves are most nutritious because they are less dried out by the fire; and ordinary oven bread (**ipnites**) is more nutritious than griddle bread and spit-bread because it is less burnt by the fire. Clay-oven bread (**klibanites**) and ash-bread are the driest, the latter because of the ash, the former because the clay absorbs moisture. Bread made of good white flour (**semidalites**) is strongest of all, not counting that made of **khondros**, and is highly nutritious, but not so laxative.[28]

Translation is quite complicated here because it is on the basis of many such passages, none individually very explicit, that the meaning of the technical terms has to be worked out. Elizabeth David,[29] whose opinion is not to be lightly ignored, thought that **klibanites** meant 'mould bread' (i.e. baked in a mould placed among embers): some other texts suggest to me that the **klibanos** was bigger than that, and so, like some others, I guess it was the clay-oven (and certainly the Greeks must have had a word for the clay-oven, which was quite standard equipment for them). **Semidalites** is very difficult to translate: it really seems to mean 'bread that is best in all respects', made of fine white flour from the most suitable kind of wheat.

ATHENAEUS

The *Deipnosophists* of Athenaeus of Naucratis (end of the 2nd century A.D.) is a vast collection of source materials to which all subsequent studies of Greek food behaviour are directly or indirectly indebted.

The *Deipnosophists* takes the form of a series of dinner conversations in which every speaker was armed with quotations from Greek literature, a very high proportion of which date from before the 1st century B.C. Several hundred authors are quoted: it is clear that Athenaeus got many of his briefer quotations out of 'secondary sources' - glossaries of dialect words, studies of poets' vocabulary - because sometimes the speakers say so explicitly. But it is also certain that Athenaeus read and excerpted a great range of works himself. His favourite sources include Athenian comedy of the fifth to third centuries B.C., hexameter poetry (imitating, in a sense, the *Iliad* and the *Odyssey* and written in the third and second centuries B.C.), dietetic writers (not the anonymous ones in the Hippocratic collection, but later and more expansive authors), 'popular' and anecdotal history (written from the third century B.C. up to Athenaeus's own time, often unreliable for historical fact but excellent for social history).

The subject of conversation throughout this long book is food, dining and entertainment: there is a sequence of topics, maintained through continual digressions. Book 1: the literature of food, food and drink in Homer, wine. Books 2-3: hors d'oeuvres, bread. Book 4: the organisation of banquets, music. Book 5: lavish display and luxury. Book 6: parasites, flattery. Books 7-8: fish. Book 9: meat, poultry. Book 10: gluttony, more wine. Book 11: cups. Book 12: social behaviour. Book 13: love, women. Book 14: more music, desserts. Book 15: wreaths, perfumes. As a student of our subject Athenaeus had the advantages of wide reading, lack of prejudice and untiring enthusiasm; among his disadvantages was that he was easily confused by changes in the meaning of words. He quoted texts from the eighth century B.C. to the second century A.D., so over roughly a thousand year period. Over this period, for a variety of reasons, the written Greek language changed rather little. So it was easy for readers at the end of the period to assume that nothing had changed: but in fact, of course, there had been many changes, new words, new meanings, often linked to new patterns of behaviour. This applied to many words for food, drink and behaviour at meals, and Athenaeus did not always realise it. Another fault is that having chosen to write his book as a dialogue he did not tackle all the artistic problems that this caused. Real conversation is desultory, inconsequential, repetitive: so is the *Deipnosophists*.

The text survives in only one manuscript, copied by John the Calligrapher in Constantinople in the tenth century, a large and beautifully written manuscript which is now in the Marciana library in Venice. From this unique copy books 1- 2, the beginning of book 3, part of the final book 15 and a few other pages (perhaps about a fifth of the whole work) disappeared many centuries ago. But already before that time someone in Constantinople had made an abridged Athenaeus: there are several manuscripts of that, and they give a good deal of information about what was in the missing sections. The Greek text was first printed in 1514.

ATHENAEUS ON BREAD

The appendix below is a translation of the section of the *Deipnosophists* that deals with bread. On the basis of this text it is worth while to point out some dangers of using Athenaeus as a source. Food historians like to look for early sources: if they look for Greek sources they will find Athenaeus, and often quote him in Gulick's easily available translation in seven volumes of the Loeb Classical Library.[30] This translation is a pretty good one, better in the later volumes, and certainly the best available: but any translator from an ancient language has to make some guesses as to meanings that are simply not precisely known. Gulick made guesses. There are guesses in the translation below, some marked with asterisks. If one relies too closely on the wording of a sentence in translation, one may be relying on someone else's guess.

As to the make-up of this text in particular, it will be seen that the expression 'Athenaeus says that ...', sometimes found in histories of food, may be better avoided. Athenaeus actually vouches for no single fact in this whole passage. His speakers do make some independent assertions - such as that loaves cost an obol[31] at Alexandria - and their assertions have to be accepted with caution because, after all, Athenaeus sometimes makes them disagree with one another. But even the speakers generally limit themselves to quoting, or citing, an earlier authority. Before we accept anything said in a passage such as this, we need to know: 1. where and when the quoted author wrote, 2. whether he was writing fact or fiction, 3. whether he meant what he said literally, ironically or in some special context, 4. whether he is being quoted accurately.

Often the answers cannot be known. In the layout of this translation a special attempt has been made to indicate some problems and some solutions:

(a) If what is being quoted is a dialogue (e.g. a play) it has been marked as such with single inverted commas. But there is very often room for disagreement as to which speaker said what within the quoted passage, and so what may be fact, what may be irony, what may be a lie. The manuscript gives no help here.

(b) What appear to be verbatim quotations have been indented: indeed, most of the text is indented. But in the case of prose there may be no way of knowing whether what seems to be a quotation is really verbatim: what little punctuation is found in the manuscript is not reliable. In particular it can be hard to know where a quotation stops: see for example the final quotation from Diphilus of Siphnos and the note there. If a reader wants to follow up the statement that what I have called 'pancakes' and 'scones' (drop scones, I believe) contained oil, he cannot be altogether certain who made it, where and when.

(c) Athenaeus liked to interweave quotations. Sometimes a speaker will take one authority and keep on referring to it, inserting short quotations from other authors meanwhile. Becoming aware of this when working on an earlier section of the *Deipnosophists*, about shellfish, I saw that it also happens near the beginning of the present extract. "Pontianus" takes a string of names for bread from a scientific book by Tryphon of Alexandria: they are underlined in the translation. After an initial selection he begins to intersperse extracts from plays exemplifying each name, but the underlinings continue, because he apparently goes on using Tryphon to provide a framework.

A further point concerns the limits of Athenaeus's knowledge. "Pontianus" says that rolls "may be made of young wheat". But the passage he quotes only mentions rolls made of three-month wheat. Three-month wheat (it is clear from other sources) meant the various varieties of spring-sown wheat. "Pontianus" (and presumably Athenaeus) did not know what three-month wheat was. In other words, Athenaeus did not always understand his authorities. In this case, since he gives a verbatim quotation, one can see what went wrong. But if there are similar mistakes among all the references to dialect dictionaries, later in this same passage, it would be difficult if not impossible to recognise them and suggest corrections.

My aim, then, has not been to give a definitive survey of staple foods in ancient Greece, but some specimens of the different kinds of information on the subject that come from different sources - just a few examples of the fascination and frustration of early food history.

* * * * *

APPENDIX: ATHENAEUS ON BREAD

Cynulcus interrupted these jokes of Ulpian's.

"We need bread," he shouted. "And I don't mean King Bread, King **Artos** of the Messapians in southern Italy! There is a monograph about him by Polemon, incidentally, and he is mentioned in Thucydides book VII and by Demetrius (the comic author) in a play called *Sicily*. Demetrius says:

'And thence before a southerly wind we crossed the main to Italy, to the Messapians; and King Bread rose up and hosted us warmly.'
'A sweet man, eh?'
'Oh, a white man - a large white man.'

So that is not the **artos** we want now. We want the gifts of Demeter of the Corn, of Demeter of the Flour, as the Syracusans name her in their worship. It is Polemon, again, who gives this information in *On Morychus*, while in *To Timaeus* I he says that in Scolus in Boeotia they have put up statues of Great Wheat-Loaf and Great Barley-Cake."

The bread was soon on its way in, and a quantity of varied relishes to eat with it. Cynulcus looked it over and said:

" 'So many snares do sorry mortals lay to catch their bread,'

as Alexis puts it in *To the Well*. So, now, let us say something about bread."

"There is a classification of bread in the *Plants* of Tryphon of Alexandria, if I can bring it to mind," said Pontianus, getting in first:[32] "leavened, unleavened; flour bread, meal bread; wholemeal, more laxative, he says, than white; emmer bread, einkorn bread, millet bread. Meal bread is always made from the lesser wheats, he says: it cannot be made from barley. Then those named after the method of baking: oven-bread, mentioned by Timocles in *Honest Robbers*:

'I discovered a warm tray lying there, so I ate some of the warm oven- bread;'

*drop-scones,[33] mentioned by Antidotus in *Chorus Leader*:

'took hot scones - why not? - folded them over and dunked them in must,'

and by Crobylus in *The Suicide*:

'and taking a tray of white scones,'

while Lynceus of Samos, in his *Letter to Diagoras*, comparing the foodstuffs made in Athens with those at Rhodes, says:

Even the loaves of the market place are exalted among them. They bring them in at the beginning of dinner and in the course of it till none is left; and when the diners have finished and are full they serve up what are called 'dipped scones', a very pleasant affair so compounded of sweets and softness and so symphoniously soaked in must that a true miracle occurs - hunger recurs in the eater through the joy of eating, just in the way that sobriety frequently recurs in the drunkard;

atabyrites,[34] as Sopater in *Woman of Cnidos*:

'There was an **atabyrite** loaf, a jaw-ful!'

brockets, mentioned by Semus, in *Deliad* VIII; he writes that they are made for processions:

They are large loaves; the feast is called Great Loaves, and they say as they carry them in, 'A goat, a brocket full of suet',

pitta bread, mentioned by Aristophanes in *Age*. He depicts a baker-woman who has her loaves stolen by men who have thrown off their old age, and says:

'What's happened here?'
'Hot ones, child!'
'Are you mad?'
'Pittas, child!'
'What do you mean, pittas?'
'And very white, child,'

ember-bread, mentioned by Nicostratus in *High Priest* and by the magic gourmet Archestratus, whom I shall quote in due course; toast, as in Eubulus's *Ganymede*, [and also:]

'Hot toast!'
'What's toast?'
'Sexy bread,'

[from] Alcaeus's *Ganymede*;[35] and then *crumpets, which are thin and light, and *muffins even more so. Aristophanes mentions the former in *Assemblywomen*, saying:

'Crumpets are baking,'

while Diocles of Carystus refers to muffins in *Health* I, stating:

Muffins are lighter that crumpets.

The latter also were probably baked on charcoal, like the Athenian ember-bread, the same which at Alexandria people dedicate to Cronus and put out in the temple of Cronus for anyone who wishes. Epicharmus in *Heba's Wedding* and *Muses* - and this latter play is a reworking of the former - lists, as kinds of bread, pitta bread, *fried bread, *lardy-cake, doughnut, oil-bread, *dumpling. Sophron mentions some of these in *Mimes for Women*:

'A dinner for the priestesses, pittas and fried bread; and a dumpling for Hecate.'

I am aware, my friends, that the Athenians say **kribanon** 'clay oven' and **kribanites** 'pitta bread' with an **r** whereas Herodotus in *Histories* II said,

... in a red-hot **klibanos**,

and Sophron said:

'Who's baking lardy-cakes or **klibanitai** or dumplings?'

He also mentioned a kind of bread called **plakitas** in his *Mimes for Women*:

'... to feast me at nightfall on a **plakita** loaf,'

and a cheese-loaf in the mime called *Mother-in-Law*:

'I advise you to have a bite, because someone has sent down a cheese-loaf for the slaves.'

Nicander of Colophon in *Words* gives **daratos** as a name for unleavened bread. Plato (the comic playwright) in this passage in *Long Night* calls large loaves Cilician:

'He came with some loaves he had bought, and not little white ones but big Cilicians,'

while in *Menelaus* he says that some bread is 'of the common herd'. Whole wheat bread is mentioned by Alexis in *The Cyprian*:

'... wholly eaten a whole wheat loaf.'

In *Weeders* Phrynichus used the form **autopyritai** for these:

'... with wholemeal loaves and oily olive-cakes.'

Sophocles in *Triptolemus* mentions **orindes** bread - is this made of rice, **oryza**, or of the seed resembling sesame that grows in Africa?[36] Aristophanes in *Friers* has *rolls:

'Take a roll each!'

and:

'... or fetch me the belly of an autumn porker, and warm rolls.'

These may be made of young wheat, as Philyllius specifies in *Auge*:

'I bring the offspring of three-month wheat, hot milk-white rolls.'

Poppy-seed loaves are mentioned by Alcman in book V:

Seven couches and as many tables crowned with poppy-seed bread, with linseed bread and sesame bread and, for the girls, buckets full of **khrysokolla**,

these last being a sweetmeat made of honey and linseed. Aristophanes mentions a roll called **kollyra** both in *Peace*,

'... a great big *bap to make your knuckle-sandwich,'

and in *Freighters*,

'... and a *bap for the old men for the sake of the trophy at Marathon.'

The spit loaf, **obelias**, was so called either because it cost an obol - as bread does at Alexandria - or because it was cooked on spits, **obeloi**. Aristophanes' *Farmers*:

'... and someone happened to be baking a spit loaf;'

Pherecrates' *Absent-Minded*:

> 'To crumble a spit loaf and to give no thought to the bread;'

and those who carried them on their shoulders in processions were called spit-bearers. Socrates, in *Eponyms* book VI, says that Dionysus invented spit loaves during his campaigns.

Bean bread is also called pulse bread, so Eucrates tells us. **Panos** is bread in Messapian, and so **pania** means 'satiety' and **panios** 'sufficient' (thus Blaesus in *Worn in the Middle*, Deinolochus in *Telephus*, Rhinthon in *Amphitryon*), while **pan** is Latin for 'bread'.

A 'fruit loaf', **nastos**, is a large leavened one, Polemarchus and Artemidorus tell us; Heracleon says it is a kind of cake. Nicostratus, *Bed*:

> 'A fruit loaf as big as that, master, white, and so plump it was peeping out of the basket. And the smell, when the napkin was taken off, and a sort of atmosphere of it, mixed with honey, rose to the nostrils: yes, it was still warm ...'

There is some kind of bread called 'scraped' in Ionia, says Artemidorus of Ephesus in *Ionian Memoranda*. 'Throne' was a name for bread in Neanthes of Cyzicus, *Hellenica* VII. He writes:

> Codrus takes a slice of the bread called 'throne', and meat; and they allot it to the oldest man.

Bakkhylos, reports Nicander in *Words* II, is a name given in Elis to a loaf baked in ashes. Diphilus mentions such loaves in *She Was Wrong*:

> '... to take round ash-bread of sifted flour.'

The **apopyrias** is a kind of loaf which is baked over charcoal. Such bread has been described as leavened, as by Cratinus in *Softies*:

> 'First I have an **apopyrias**, and leavened, by Jove, not stuffed with flock!'

Here is the discussion of pearl barley and bread from Archestratus's *Gastronomy*:

> And first I shall recall the gifts of fair-haired Demeter, dear Moschus: take them to your heart. The best one can get, the finest of all, cleanly hulled from good ripe barley-ears, is from the sea-washed breast of famous Eresus in Lesbos, whiter than airborne snow. If the gods eat pearl barley, this is where Hermes goes shopping for it. It is passable at seven-gated Thebes and at Thasos and at some other cities too, but just like grape-pips as against Lesbian: accept that as certain.

> Take a Thessalian *bun, a circling whirl of dough well kneaded under hand; they call it 'crumble' there, meal-bread as others say. I also commend to you a child of flour, the ember-bread of Tegea. Fair is the loaf that famous Athens sells to mortals in her market-place; those from the clay ovens of vinous Erythrae, white and blooming with the gentle seasons, are a joy with dinner.

After all this, greedy Archestratus recommends that one's baker should be a Syrian or a Lydian: it was not yet recognised that bakers from Cappadocia are the best. This is how he puts it:

> Have a Phoenician or a Lydian in your household, a man proficient in the daily making of all the kinds of bread you may order.

Antiphanes in *Omphale* agrees that Athenian bread is out of the ordinary:

'How should a gentleman ever leave this roof who may see these fair- skinned loaves conquering the kitchen with their massed assaults; may see them shape-changing in the clay ovens, a magic show made in Athens, taught to the citizenry by Thearion?'

This Thearion is the baker mentioned in Plato's *Gorgias* (coupling him with Mithaecus, Plato writes:

'... "Which men have been, or are, best at caring for our bodies?" And you would tell me, absolutely seriously, "Thearion the baker and Mithaecus who wrote the book on Sicilian cookery and Sarambus the wine merchant," - that these men have been outstanding at caring for our bodies, the first providing outstanding bread, the second, cuisine, and the third, wine,')

and in Aristophanes' *Gerytades*, and in this passage in his *Aeolus in the Kitchen*:

'I come to you from Thearion's bakery, home of the clay ovens.'

Eubulus in *Orthannes* mentions the bread of Cyprus as out of the ordinary:

'It is a damned hard thing to ride away from Cypriot loaves: they draw hungry men to them like a magnet.'

Ephippus in *Artemis* mentions bun-shaped loaves, which will be the same as rolls:

'... from Alexander, up there in Thessaly the bun-eating, an oven full of bread;'

and so does Aristophanes in *Acharnians*:

'Greetings, my little bun-eating Boeotian!'"

After this one of the literary critics present, called Arrian, said:

"This food, colleagues, has passed its best.

We welcome not pearl barley when the city's full of bread!

nor do we welcome the cataloguing of all those breads; because I happened to come across another book, by Chrysippus of Tyana, called *Baking*, and I have tried the kinds named here by our friends. Now I am going to tell you something about kinds of bread:

The bread called **artoptikios** 'mould bread' differs from **klibanikios** 'pitta bread' and from **phournakios** 'oven bread'. If made with dryish dough it will be white and good to eat on its own; if with slack dough it will be nice and light but not white. Pitta bread and oven bread are better from softer dough.

In Greece there is a bread called 'soft', made with a little milk and oil and a proportion of salt. The dough must be slack. This bread is called Cappadocian, because that is where soft bread is mostly made. The Syrians call such bread **lakhmá,** and it is found very serviceable in Syria because it is eaten very warm; and it is like a flower.

The bread called 'boletus' is shaped like a boletus mushroom. The kneading trough is greased and sprinkled with poppy seed on to which the dough is put, and while proving it does not stick to the trough. When it is to be placed in the oven some wheat meal is sprinkled on the baking tray and then the loaf put on that; and it takes on a very beautiful colour, like smoked cheese.

Streptikios 'plaited' bread includes a little milk, and oil is added, or, if none, suet. The so-called **artolaganon** 'crumpet bread' calls for a little new wine, pepper, milk and a little oil or suet. **Kapyria**, or **trakta** as they are known, require the same mixture as ordinary bread.'"

When the Roman pedant had thus exemplified his critical creed, Cynulcus said:

"Demeter, what learning! No wonder Amazing Fishface[37] here has got umpteen students and makes more money out of his fine learning than Gorgias and Protagoras. Goddesses! I'm hard put to it to say whether he's blind, or whether the whole lot of those who signed up as his students have only one eye between them and can't see for the crowd. Happy are they in life - no, perhaps already in a happier state than life - if their teachers teach them such stuff."

Magnus, who liked his food, spoke up on the poor critic's side:

"'With unwashed feet, with earth for a bed, with sky for a roof ...'

(said the comic author Eubulus),

'with filthy tongues, and eating other men's wealth for your dinner ...'

and didn't Diogenes, forefather of you all, once gobble up a whole cake at a dinner, and, when questioned, say Yes, it was excellent bread? You Cynics –

'close companions of the white tuna steaks ...'

to quote the poet Eubulus again - you never give way to others. You talk and talk till someone throws you bread or bones as if to a dog. So how should you know that 'dice' (and not the kind you are always playing with)

... are square loaves, flavoured with dill, cheese and oil,

as Heracleides says in *Cookery*? One that Fishface overlooked! as he did the **thargelos**, called **thalysios** 'harvest-home' by some (Crates in *Attic Dialect* says that **thargelos** is the name of the first bread of the new harvest) and the **sesamites**; and he did not notice the so-called **anastatos**, baked for the symbol-bearers of Athena Polias. There is a **pyramous** loaf as well, baked with sesame and so possibly the same as **sesamites**. These are all listed by Tryphon in *Plants* I; so are the **thiagones**, loaves baked for the gods in Aetolia. Among the Athamanes certain loaves are called **dramikes** and **araxeis**.

The lexicographers list some names of loaves: Seleucus gives **dramis**, a Macedonian name, or **daraton**, Thessalian. **Etnitas**, he says, is a bean bread; **erikitas** one made of cracked, unsifted, coarse wheat. Amerias calls whole wheat bread **xeropyritas**, 'dry-wheat': so does Timachidas. Nicander says that the Aetolians call loaves made for the gods **thiagones**. The Egyptians call their sourish bread **kyllastis**; Aristophanes mentions it in *Daughters of Danaus*:

'Cry **kyllastis**, cry Petosiris!'

so do Hecataeus and Herodotus and also Phanodemus in *Attic History* VII. Nicander (the poet of Thyateira) says that it is bread made of barley that the Egyptians call **kyllastis**. Brown loaves are called 'grey' by Alexis in *The Cyprian*:

'So which way did you come?'
'It was hard, but I did get some ready-baked.'
'Oh, damn him! - Well, how many have you?'
'Sixteen.'

'Bring them over here ...'
'Eight white ones and eight grey.'

Blema 'missile' is a name for warm bread dipped in wine, according to Seleucus. Philemon in *Various Sacrifices* I says that bread from unsifted wheat, containing the whole grain, is called **pyrnon**; **blomiaioi** 'breakers' are loaves with cuts in the crust (in Latin these are called **kodratoi** 'quartered'); and that **brattime** is a name for the bran loaf called **eukonon** by Amerias and Timachidas. Philitas in *Unruly* gives the name **spoleus** to a kind of bread eaten only by relatives [at funerals].

Barley cakes can be found written up by a great many authors, including Tryphon. In the Athenian **physte** 'puff' the meal was not too well ground; there are also **kardamale** 'cress cake', **berex**, **tolypes** 'balls', **akhilleion** (probably the latter was made from akhilleion barley), **thridakinai** 'lettuces', **oinoutta** 'wine cake', **melitoutta** 'honey cake' and **krinon** (name of a figure in group dancing in Apollophanes' *Fool*). What Alcman calls **thridakiskai** are the same as Athenian **thridakinai**. Alcman writes:

... dripping with lettuce-cakes and **kribana**,

on which Sosibius comments in *On Alcman* III that **kribana** meant a kind of breast-shaped cake. 'Health' is the name of the barley cake they give one to taste at sacrifices. Hesiod describes some kind of barley cake as **amolgaia**:

A creamy barley cake and the milk of goats now running dry ...

meaning rustic and healthy, because **amolgos** is 'prime'. Allow me not to list - since, sadly, memory fails me - all the sacrificial cakes and sweets given by Aristomenes of Athens in *Religious Requisites* III. I myself, when young, knew this author as an old man, an actor in classical comedy and a freedman of that very cultured monarch Hadrian, who used to call him 'Athenian Partridge'."

"Freedman!" said Ulpian. "Did any classical author use that word?"

Someone said that *Apeleutheroi*, *Freedmen*, is the title of one of Phrynichus's plays, and that Menander uses 'freedwoman' in his *Girl Who Was Beaten*.

Ulpian added a rider: "And what is the difference between **apeleutheros** and **exeleutheros**?"

But we agreed to leave that problem for the present, and were about to get to grips with the bread when Galen said:

"We shan't start dinner before we have told you what the medical fraternity[38] have to say about bread and cakes and gruel. Now in *Diet in Sickness and Health* Diphilus of Siphnos says:

> Wheat bread is more nutritive, more assimilable and altogether better than barley, whether it be from fine bread-wheat flour (better than other wheats) or whole bread from unsifted meal: this latter is considered the most nutritive of all...

(Philistion of Locri[39] says that bread-wheat flour bread is more invigorating than meal bread. He puts this in second place, then bread from the flour of other wheats.)

> Bread of spring-wheat flour[40] is less emulsifying and less nutritive. Warm bread of any kind is more assimilable, more nutritive and more emulsifying than bread which has cooled: it is also stimulant and quickly metabolised. Cold bread is filling and slowly assimilated. Really old, dry bread is less nutritive and is astringent and not emulsifying. Ember-bread is heavy and slowly assimilated because it cooks unevenly. Oven bread and stove bread is indigestible and slowly assimilated. Scones and *pancakes are less constipating because they contain oil, but less

wholesome because of the steam. Pitta bread is superior in all these qualities: it is emulsifying, wholesome, digestible and quickly metabolised; it is neither astringent nor laxative.

The physician Andreas says that there is a mulberry bread made in Syria, which causes baldness. Mnesitheus says that wheat bread is more digestible than barley cake, and that even einkorn bread is quite sufficiently nutritive, being digested without much trouble. Bread from the emmers, he says, eaten in quantity, is heavy and indigestible: hence those who eat it are sickly. And you must know that grain that is not parched or separated produces wind, heaviness, stomach-ache and headache."

After all which we decided it was time to start eating ...

BIBLIOGRAPHY

Bailey 1983
Hunter-gatherer economy in prehistory ed. G. Bailey, Cambridge 1983. (Cambridge University Press)[Includes G. Bailey and others, 'Epirus revisited: seasonality and inter-site variations in the Upper Palaeolithic of north-west Greece', pp. 64-78.]

Bailey and others 1983
G. N. Bailey and others, 'Asprochaliko and Kastritsa: further investigations of Palaeolithic settlement and economy in Epirus (north-west Greece)' in *Proceedings of the Prehistoric Society* vol. 49 (1983) pp. 15-42.

Bailey and others 1986
G. N. Bailey and others, 'Palaeolithic investigations at Klithi: preliminary results of the 1984 and 1985 field seasons' in *Annual of the British School at Athens* vol. 81 (1986) pp. 7-35.

Bökönyi 1969
S. Bökönyi, 'Archaeological problems and methods of recognizing animal domestication' in *The domestication and exploitation of plants and animals* ed. P. J. Ucko and G. W. Dimbleby, London 1969, pp. 219- 229.

Butcher and Lang 1879
The Odyssey of Homer tr. S. H. Butcher and A. Lang, London 1879. (Macmillan)

Caughley 1977
G. Caughley, *Analysis of vertebrate populations*, Chichester 1977. (Wiley)

Dakaris, Higgs and others 1964-7
S. I. Dakaris, E. S. Higgs and others, 'The climate, environment and industries of Stone Age Greece' in *Proceedings of the Prehistoric Society* vols 30-33 (1964-7).

Dalby forthcoming
A. Dalby, 'Greeks abroad: food and social organisation among the Ten Thousand', forthcoming.

David 1977
E. David, *English bread and yeast cookery*, London 1977. (Allen Lane)

De Sélincourt 1954
Herodotus, *The histories* tr. A. de Sélincourt, Harmondsworth 1954. (Penguin)

Evans 1964
J. D. Evans, 'Excavations in the Neolithic settlement of Knossos 1957-60. Part 1' in *Annual of the British School at Athens* vol. 59 (1964) pp. 132 ff.

Garnsey 1988
P. Garnsey, *Famine and food supply in the Graeco-Roman world*, Cambridge 1988. (Cambridge University Press)

Gulick 1927-41
Athenaeus, *The Deipnosophists* ed. and tr. C. B. Gulick. 7 vols, London 1927-41. (Heinemann/Harvard University Press)

Halstead 1981
P. Halstead, 'Counting sheep in Neolithic and Bronze Age Greece' in *Patterns of the past: studies in honour of David Clarke* ed. I. Hodder and others, Cambridge 1981, pp. 307-339.

Halstead 1987
P. Halstead, 'Man and other animals in later Greek prehistory' in *Annual of the British School at Athens* vol. 82 (1987) pp. 71-83.

Hammond 1987
Homer, *The Iliad* tr. M. Hammond, Harmondsworth 1987. (Penguin)

Hawkes 1969
J. G. Hawkes, 'The ecological background of plant domestication' in *The domestication and exploitation of plants and animals* ed. P. J. Ucko and G. W. Dimbleby, London 1969, pp. 17-29.

Hort 1916-26
Theophrastus, *Enquiry into plants* ed. and tr. A. Hort, Cambridge, Mass. 1916-26. (Harvard University Press/Heinemann)

Jacobsen 1969
T. W. Jacobsen, 'Explorations at Porto Cheli and vicinity. Preliminary report II: the Franchthi cave 1967-1968' in *Hesperia* vol. 38 (1969) pp. 343-381.

Jacobsen 1973
T. W. Jacobsen, 'Excavation[s] in the Franchthi cave, 1969-1971' in *Hesperia* vol. 42 (1973) pp. 45-88, 253-283.

Jacobsen 1981
T. W. Jacobsen, 'Franchthi cave and the beginning of settled village life in Greece' in *Hesperia* vol. 50 (1981) pp. 303-319.

Jarman and Jarman 1968
M. R. Jarman and H. N. Jarman, 'The fauna and economy of early Neolithic Knossos' in *Annual of the British School at Athens* vol. 63 (1968) pp. 241-264.

Joly 1967
Hippocrate, *Du régime* ed. and tr. R. Joly, Paris 1967. (Les Belles Lettres)

Jones 1931
Hippocrates ed. and tr. W. H. S. Jones. Vol. 4, London 1931. (Heinemann/Harvard University Press)

Lloyd 1978
Hippocratic writings ed. G. E. R. Lloyd. Revised ed., Harmondsworth 1978. (Penguin)

Payne 1973
S. Payne, 'Kill-off patterns in sheep and goats: the mandibles from Asvan Kale' in *Anatolian studies* vol. 23 (1973) pp. 281-303.

Payne 1985
S. Payne, 'Zoo-archaeology in Greece: a reader's guide' in *Contributions to Aegean archaeology: studies in honor of William A. McDonald* ed. N. C. Wilkie and W. D. E. Coulson, Minneapolis 1985, pp. 211-244.

Renfrew 1969
J. M. Renfrew, 'The archaeological evidence for the domestication of plants: methods and problems' in *The domestication and exploitation of plants and animals* ed. P. J. Ucko and G. W. Dimbleby, London 1969, pp. 149-172.

Renfrew 1972
C. Renfrew, *The emergence of civilisation: the Cyclades and the Aegean in the third millennium B.C.*, London 1972. (Methuen)

Renfrew 1973
J. M. Renfrew, *Palaeoethnobotany: the prehistoric food plants of the Near East and Europe*, London 1973. (Methuen)

Sordinas 1969
A. Sordinas, 'Investigations of the prehistory of Corfu during 1964-1966' in *Balkan studies* vol. 10 (1969) pp. 393-424.

Weinberg 1970
S. S. Weinberg, 'The stone age in the Aegean' in *The Cambridge ancient history* vol. 1 part 1. 3rd ed. by I. E. S. Edwards and others, Cambridge 1970, pp. 557-618. [Originally published separately, Cambridge 1965.]

Zohary 1969
D. Zohary, 'The progenitors of wheat and barley in relation to domestication and agricultural dispersal in the Old World' in *The domestication and exploitation of plants and animals* ed. P. J. Ucko and G. W. Dimbleby, London 1969, pp. 47-66.

NOTES

1. Jane Renfrew in Jacobsen 1973 p. 68.
2. Jacobsen 1969, 1973, 1981.
3. On Asprohaliko (and Kastritsa, excavated earlier) see Bailey 1983, esp. p. 69; Bailey and others 1983; Dakaris, Higgs and others 1964-7.
4. Sordinas 1969.
5. Bailey and others 1986 p. 18.
6. Halstead 1987 p. 74.
7. Bailey and others 1986 p. 20: 'most animals were killed as they were approaching maturity', a kind of selection that hunters of wild animals do not regularly make (Caughley 1977 p. 189) but that herdsmen typically do (Payne 1973). But this is quite speculative: see further Bökönyi 1969 p. 222, Bailey 1983 pp. 9-10 and the references given there.
8. Jacobsen 1981, Payne 1985. On the probable nature of the earliest steps towards cultivation see Hawkes 1969 pp. 22-3.
9. Evans 1964 p. 140; Renfrew 1969 pp. 160, 163-4; Renfrew 1973 pp. 47, 200- 203. On the continuity of the settlement at Knossos cf. Weinberg 1970 pp. 608- 618.
10. Jarman and Jarman 1968.
11. Barley, if not native to Greece, appears from the Franhthi evidence to have been introduced well before wheat: see J. Renfrew in Jacobsen 1973 p. 68. On its native habitat see Zohary 1969. Garnsey 1988 pp. 10-14 observes that in modern Attica the wheat crop failed more than one year out of four, the barley crop about one year out of twenty.
12. On the change from meat to vegetable staple food see Halstead 1981 p. 314.
13. On population density in Neolithic and Bronze Age Greece see Renfrew 1972 pp. 225-264.
14. Renfrew 1972 pp. 270-280.
15. The new translation by Hammond 1987 is reliable enough for most purposes.
16. *Iliad* 8.547; but also the meal in Achilles' tent, *Iliad* 9.216.
17. There is an accurate but old-fashioned translation by Butcher and Lang 1879.
18. *Odyssey* 14.45-114.
19. *Odyssey* 1.109-160.
20. *Odyssey* 4.620-624. The word *sitos* is a sort of Greek equivalent of 'staple food'. Here it means 'bread'; elsewhere in the epics it sometimes means 'food' in general; in later Greek it means 'grain', 'cereal', and often an army's 'food supply'.
21. The translation of de Sélincourt 1954 is convenient.
22. Herodotus 2.2.
23. Herodotus 4.16-18.
24. Herodotus 1.200.
25. Athenaeus 153d-e. Kathie Webber points out that the custom described is south Indian. I suppose that the food presentation of north India, using bread, very much as the Greeks did, was not thought by Megasthenes to be worth commenting on.
26. There is an English translation by J. Chadwick and W. N. Mann in Lloyd 1978 pp. 186-205.
27. There is an English translation in Jones 1931; it is worth consulting the French translation by Joly 1967.
28. *Regimen* 2.42.2.
29. David 1977 pp. 209-210.
30. Gulick 1927-41.
31. In classical Greece an obol had been a tiny silver coin, one sixth of a drachma. In imperial Egypt bronze money provided the small change.
32. Terms apparently taken from Tryphon's *Plants* are underlined.
33. An asterisk marks words whose meaning should not be taken too literally. They are approximately right but there is not enough evidence to pin down the meaning of the Greek precisely.

34. Unknown: possibly a variant of **autopyrites**, 'wholemeal bread', discussed below.
35. It is not clear whether the quotation comes from Eubulus, Alcaeus or both. Pollux attributes the same fragment to Eubulus, hence the text added in square brackets. I translate *dipyros* 'toast', but it might equally mean 'biscuit' or 'biscotte'. The word was still in use in Byzantine times, and was translated by Latin *panis biscottus*, 'twice-cooked bread'.
36. Described by Theophrastus, *Study of Plants* 4.18.4 and identified by Hort 1916-26 with *Corchorus trilocularis*. This sentence is mistranslated by Gulick.
37. The nickname Blepsias appears to liken the unfortunate Arrian to a kind of grey mullet.
38. Literally 'the sons of the Asclepiads', Asclepius being the god of healing.
39. Here Galen appears to interrupt his quotation from Diphilus.
40. It appears from the style and technical vocabulary that Galen now reverts to quoting Diphilus.

Do Processed Societies Have Staple Foods?

By Erica F. Wheeler

Nutritionists define a staple food as the one which generally contributes most energy to the diet of an individual or a group. The insertion of "generally" is important here, for on any one day and for any one person the staple might be pushed into second or third place. But when evaluating the results of large scale surveys, or of national food balance sheets, they would expect to find a staple food contributing between 30% and 70% of the total energy. For example, in 1981, rice contributed 64% to the diet of Thailand and maize, 41% to that of Honduras. Typical staple foods would be cereals such as rice and wheat, root crops such as cassava and potatoes, and other high-starch foods such as sago and plantain. It is not uncommon to find a country with a two-staple system, varying seasonally or regionally. Brazil gets 38% of its energy from all cereals, and 30% from rice and wheat.

The value of this classification has been that it enables the nutritionist to make some rough but generally useful predictions about the nature of the diet and likely problems associated with it. For example, protein deficiency (if it exists at all) would never be expected to occur where a cereal is the staple food: only where a low-protein tuber such as cassava is predominant. The cereals are poor sources of vitamins A, C and D, so these must come from other foods if the diet is to be adequate. Some staple foods are notorious for lacking a particular nutrient: thus the combination of an unavailable vitamin (niacin) and low levels of an amino acid (tryptophan) in maize (corn) accounts for the risk of pellagra, a niacin deficiency disease, in peoples who eat large amounts of maize. Knowledge of the staple food of a community or nation, then, has been regarded as a short cut to some knowledge of the nutrient pattern of the diet.

However, all these predictions must be hedged with qualifiers about the other foods being eaten with the staple, and there are numerous examples of traditional dietary patterns, and methods of preparing foods, which valuably supplement the staple. The Latin American practice of soaking maize in lime water has the effect of releasing the unavailable niacin, and enhances the value of the staple. The Sri Lankan practice of cooking rice in coconut milk and eating it with fish adds valuable fat, protein and vitamins.

The staple food concept appears to fit in with another beloved of nutritionists: that is, "food groups". In health and nutrition education and home economics programmes, there are found simple classifications of foods into categories such as "energy foods", "protective foods" and "body building foods"; or "cereals", "dairy products", "green vegetables", and so on. The healthy, desirable, or "balanced" diet, is then defined as one which contains some of each food group, every day. This approach has been criticised for over-simplification, in that (for example) the cereals are both "energy" and "body-building" foods, and beans could be classed as both "legumes" (when eaten dried) and "green vegetables" (when eaten at the immature stage). Be that as it may, the food group schema has been used very widely, and thousands of school and college students have absorbed it into their thinking about food and diet.

Thus the simple concept of the staple food, and the "food group" approach to the healthy diet, are built into the teaching and practice of nutrition and home economics, in both industrialised and "developing" countries. In asking whether an individual has a "good" or "healthy" or "balanced" diet, the nutritionist is employing a mental set which involves these concepts and which assumes that the objective of good dietary practice is to achieve a varied diet which contains elements of all the simple food groups.

These concepts, however, have been overtaken by the massive rise of food technology and of the food industry in industrialised countries. It may still be realistic to describe a villager in a "developing" country as consuming a diet made up of basic foodstuffs such as a cereal, an oil or fat, some vegetables, some dairy products, and possibly some meat or fish. Given the addition of flavourings and drinks, this is a dietary pattern recognisable in poor rural communities from India to Mexico. But in towns and cities, and throughout the highly urbanised industrial countries, the picture is quite different. Some people maintain the "rural" pattern, buying basic foodstuffs in markets and preparing them at home. But even these are almost certain to use some processed foods: bread, soft drinks, and snacks cooked and sold from market stalls. The more industrialised the country, the more two trends will be apparent: women "go out" to paid work, and a wide range of processed foods is available and used. It is obvious that these trends are linked, since women in full-time waged employment have little time for preparing meals from basic foodstuffs.

The increased use of processed foods has two effects on the structure of human diets. First, it tends to blur the distinction between food groups. A characteristic of most processed foods is that they are more like cooked dishes than single foods. A range of complexity of foods can be constructed:

Scale of complexity of foodstuffs

COMPLEXITY LEVEL

single		preserved	other	dishes
raw	modified		foods	combined
food		cooked	added	together

TERMS USED

food ··
staple   snack ··
              drink ··
raw ·······················cooked···································
fresh·······················preserved ····························
                            processed ·························
                                        dish/recipe ·····
                                                    course
                                                    meal
                                                    diet

On this scale, processed foods are usually somewhat more complex than the home-prepared or cooked-from-raw equivalent. Preservatives, flavourings and colourings are added, and these substances may be "foods" in themselves. Salt and sugar are obvious examples. Texture-modifying additives, also, may be foods such as cornflour or gelatin. Thus a processed food such as corned beef will contain a significant amount of carbohydrate, whereas the original beef contained none. Secondly, the food industry has had a good deal of success in diversifying products. "Bread" may be anything from a wholegrain loaf, replete with all its original bran and vitamins, to a highly refined white roll containing milk solids, salt, and fat as well as the wheat flour from which about 30% of the original material has been removed. "Cream" may contain anything from 20 to 50% fat, and milk may be high-fat, semi-skimmed, or skimmed; and may or may not be enriched with extra calcium and vitamins. The boundaries of "food" are changing, and the supermarkets are redefining terms such as "bread" and "sugar". This means that the old nutritional categorisations have broken down. For example, the development of products like yoghurt with muesli, and cheese with embedded nuts, means that dairy products can no longer be regarded as fibre-free. A host of exotic composite semi-snack foods such as samosas, pizzas, and bhajis have joined the "traditional" British composite foods like sausages and pies, to make the definition of a "high protein food" or "low fat meal" increasingly meaningless.

All this makes the life of the simple-minded nutritionist fairly difficult. The evaluation of individuals' diets, given the enormous range of commercial food products available, is increasingly a matter of speculating as to whether Sainsbury's cream-filled biscuits have the same chemical composition as Tesco's. And what of the concept of a staple food? The British staples, traditionally, have been regarded as wheat and potatoes. But even in 1962 potatoes contributed only 5% to total energy in the average British diet, being outstripped by wheat products, sugar and sweets, milk and butter. The contribution of bread, and of all wheat products, has fallen steadily from 20 and 26% respectively in 1962 to 15 and 19% in 1987, at which time sugar and sweets contributed 13% to energy. Wheat presents great problems, because of the number of foods classified as "meat" or "vegetable" products, which do contain wheat flour. In my opinion, it is no longer possible to state the amount of wheat flour eaten by a UK citizen, without subjecting her diet to chemical analysis. On the one hand, the consumption of "plain bread" is declining in some parts of the UK, as rice, pastas and pizzas become ever more popular. On the other, many complex processed foods, from biscuits to tomato sauce, contain wheat products.

All in all, it would appear that the simple concept of a staple food, the staff of life, the "real" food which underpins the rest of the diet, is obsolete in industrialised countries: as obsolete as concepts like "the village" and "the agricultural labourer". Technology has moved on, and in doing so has made it necessary to redefine "diet", "food", and "meal". It can be argued that simple, basic, raw foodstuffs are of decreasing importance to the majority of the population. They are the raw materials of a major industry. They are of interest to a relatively small army of people with specialist interest in food: chefs, gastronomes, enthusiastic amateur cooks, dietitians, health and organic food consumers. They are of especial importance, too, to ethnic minority groups who retain their own characteristic foodways and cooking methods: although the supermarkets are busily targeting these groups. But to the mainline British population there is no food which supplies the majority of its energy intake in a simple, minimally processed form. There is, instead, a combination of foods, many of which contain wheat, but which cannot simply be categorised as "wheat foods". No longer are wheat and potatoes the staple foods of the UK. The potato has been displaced long since, and wheat is disguising itself in so many ways as to have become virtually unquantifiable.

An alternative view of "a staple food" may, of course be presented: this is, the food which most people perceive to be an important one, which forms the basis of their main meal. This concept may lag well behind reality. I suspect that many people in the UK would mention wheat bread if asked to name such a food. This in itself is an interesting speculation, as it implies that the staple is itself a "convenience" food which does not need cooking.

The demise of the staple food, and the move to complex processed foods, also means the demise of the simple food group concept. Most composite processed foods overlap the boundaries of any food groups system. Is a deep-fried meat samosa a protein, fatty, or carbohydrate food? Do we categorise it as providing energy or protein? Interestingly, this problem brings to light an essential weakness which has always been present in the food group system: that although people may buy foods singly, they nearly always cook and eat them in combination. Dietary advice and analysis has never been easy for this reason.

If modern food technology has exposed the folly of attempting to force human eating patterns into unrealistic categories, it will have performed an unexpected service.

Bibliography

Ministry of Agriculture, Fisheries and Food, "National Food Survey" series, HMSO.

Fig. 9

Bulgur -
An Important Wheat Product in the Cuisine of Contemporary Assyrians in the Middle East

by Michael Abdalla

Drawings by Tomasz Siwinski

Bread is the basic food for Assyrians. But it is also the main food for the majority of world nations though the quantity eaten decreases as civilization develops. For this reason I am going to discuss a different wheat product which, after bread, takes the second place on the Assyrian scale of importance and quantity consumed. It is known under different, though similar, names. I have adopted the name which is used in the literature. It is the most important preserved product, stored at homes as part of winter reserves.[1] Bulgur is known all over the Middle East.

Among traditional semi-finished wheat products, such as couscous (northern Africa) and frumenties (Scotland), bulgur is significant for its long history. The oldest record of this product was found on Assyrian cuneiform tablets. In documents of Ashurnasirpal II, the king of Assyria in the 9th c. B.C., there is a description of preparations for a great feast of consecration of a newly-built palace in the town of Kalhu (Nimrud). Included is a text which mentions the kind of food served. One of the dishes is called "gubibate". Archaeologists not familiar with Assyrian cuisine erroneously associated it with Turkish-Arab "kebab".[2,3] However, tourist guides to Iraq, written by the Iraqis themselves[4] and foreigners, (for instance Poles[5]), in parts which discuss the most sophisticated national foods of Iraq, the historical character of a dish known as "kubba" is emphasized (in Syria it is also known as "kibbe"). The roots of the dish go back to the feast described above and the dish itself is associated with the name of king

Ashurnasirpal II. It is true that today the dish is made of bulgur and that its most original and delicious variant is called "mosul" (after the town of Mosul, formerly Nineveh, the capital of Assyrians). However, I think it more likely that the ancient "gubibate" is nothing else but "kbebat"[6], a dish made of bulgur and of a similar name, widely known among Arabic-speaking Assyrians.

More information about bulgur can be found in the Bible. According to Fellers[7] and Haley and Pence[8], the Hebrew word "arisah" used in the book of Ezekiel (Ezek. 44:30) and Nehemiah (Neh. 10:37), initially translated as dough and later as coarse flour, denoted bulgur. It very likely that these ancient chroniclers knew bulgur. Both spent some time in Mesopotamia, the motherland of bulgur. Ezekiel, having been apprehended and taken to Babylonia by Nabokadnezar in the year 597 B.C., lived near Nippur (Ezek. 1:1). Likewise, Nehemiah lived in the court of Persian kings after they had conquered Babylonia.[9]

Different phonetic variants of the name used to denote the semi-finished wheat product in question can be written as follows: bulgur, bulgor, burghol, burghul, byrghel[10], gurgur[10]. The first of the variants is widespread in Turkey from where it entered Europe; in most scientific publications, including - which is worth stressing - a pioneer experimental (and not ethnographic!) work, published in 1953[11], it is known under the name of bulgur. The three other names are used in Arab, and mainly Asiatic, countries. The last two names (byrghel and gurgur) are used by Assyrians.

On the basis of information obtained from Bulgarians it follows that bulgur has been known for a long time in this Balkan country (although its usage range is narrow); presently it is used during sacral ceremonies in the Bulgarian Orthodox church. This role of bulgur helps us suppose that the product entered Bulgaria not through Muslim Turks who conquered the country at the end of the 14th c. but through emigrants from ancient Syria[12] who, driven away by the emperors of Byzantium in the 8th c., found shelter in Bulgaria.[13] In addition to the name,"bulgur" there are also two other names in Bulgarian dialects: "bongur" and "bangur".[14] Likewise, the inhabitants of western parts of the Soviet Union near the Polish border eat a Christmas Eve supper dish known as "kutia" whose basic ingredient is wheat grain. Below I will discuss an Assyrian dish with a similar name "kutle".

References to bulgur being used in bartering were made as early as three centuries ago. They can be found in manuscripts of monks from an Assyrian monastery Mor Matay (near Mosul in Iraq). The monks, living on their work, used to prepare every year large quantities of bulgur from wheat grain grown on land belonging to the monastery. Subsequently, they traded their bulgur to the Kurds through followers of the same faith for other commodities which the monastery needed.[15] At that time monasteries served as inns free of charge. Hence the concern of their hosts (monks) to provide them with reserves of food.

Until the end of the 19th century bulgur in the Middle East was made within households in quantities sufficient to provide for the needs of one family. Information on a factory-scale production of bulgur comes from Jordan where, beginning in 1963, 150 to 200 tons of this semi-finished product were made annually.[16]

Emigration of the inhabitants of the Middle East to, among other places, the United States of America, and sabbaticals of Western scientists and scholars in some countries of this region (for instance, at the American University in Beirut), helped forward research on the economic, culinary and nutritious values of bulgur. This contributed to its popularization in even the most distant of countries. Bulgur became the subject of interest at research centers in the USA. Experiments have been conducted to improve the technology of its production and its adaptation to the requirements of a new consumer and his/her living conditions. As early as 1892 the Armenian Grains Company, active in the United States, started the production of bulgur and sold it to Americans who had come originally from the Middle East.[7,17] Beginning at the time of World War II, increasingly larger quantities of bulgur were produced and the surplus was exported to such Asiatic countries like Korea, India and to the Middle East.[8,18,19,20] In 1963 bulgur was used in the nutrition programme for US school children as a dish served for lunch.[17,21] According to 1971 data there were seven plants of bulgur in the United States with a total production output of 454 thousand tons.[7,22] Ten years later the export of bulgur amounted to 268 thousand tons, and that of soya-enriched bulgur - 71 thousand tons.[23] Attempts were also made to increase the sales of bulgur - new varieties of ready-made foods were prepared such as canned bulgur,[24,25] bulgur in wafers[26] and instant bulgur.[27,28]

Figs. 1 - 4

The traditional way of obtaining bulgur in an Assyrian village

In Assyrian villages bulgur is produced at home in July and August. Each family prepares 100 - 200 kg of recently harvested wheat for this purpose. First the grain is carefully cleaned: it is sieved manually, sorted on large round trays (every odd grain is removed), rinsed in water several times and dried in the sun (Fig. 1 - 4). In addition to family members neighbors also take part in these activities. The activities last a few weeks and are done in leisure time, usually in the afternoons of long summer days.

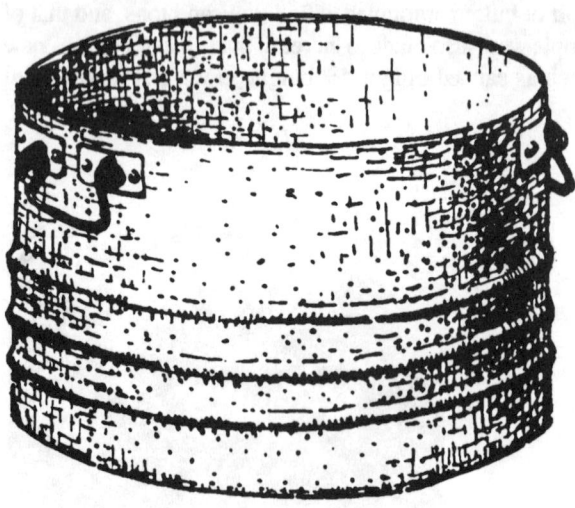

Fig. 5

Perfectly cleaned grain is then boiled in large metal tubs. Basically two types of tubs are used. One called "qaqwo" is made of thin sheet metal and shaped as a cylinder with a diameter of about 100-120 cm and height of 60-80 cm (Fig. 5). The other, called "margella", is artistically made of tin-plated copper and shaped as a "flattened" barrel with thick walls (Fig. 6). Usually such a tub belongs to or is loaned by the village administrator or a single old woman. It is loaned to the peasants on security, which is usually a golden ring. The tub "travels" from one household to another, staying in one household one day only, until it has completed the circle and served all the inhabitants. Often a few neighboring households use the tub which is placed conveniently. The atmosphere accompanying such work helps synchronize the stages of home preparation of the reserves and is conducive to mutual assistance.

The tub is placed on three stone bases or on a few clay bricks stacked together (used to build houses), at a height of about 30 - 40 cm above the ground. Sometimes a small hole is dug in the ground under the tub to increase the capacity of the hearth and to protect the fire against wind. Dried animal dung, wood, cloths and recently also car tires are used as fuel. Often a permanent place is allocated for the tub in the village or urban district to allow for the preparation of wheat for bulgur.

The tub is half filled with water and the fire is lit. The grains are poured in when the water starts to boil. This is supposed to shorten the boiling time.[29] The surface of the water should be about 20 cm above the grain. During boiling the grain is slightly stirred with a wooden or metal shovel so that the absorption of water is uniform. The process ends when the grain has become sufficiently soft. This is tested by the smell, taste, and feel of the grain. The boiled wheat is called "danoke" or "shleeqa". It has become a tradition to offer this wheat to children who have gathered around, each with his/her own plate. The children take the shleeqa home and eat it together with their parents and brothers and sisters along with cream and sugar or, less frequently, salt. Hence, the host, when planning the quantity of bulgur, takes this treat into consideration.

Fig. 6

"Shleeqa" is then taken to terrace roofs, less frequently to front or backyards and put on jute mattresses or straw mats or a previously cleaned concrete surface where it is thinly spread. The drying, depending on the weather, lasts from 1 to 3 days. During this time the grain is stirred twice a day, in the morning and in the evening. The moisture of the dried grain is very low, between 8 - 10%. The grain darkens and contracts and looks like wrinkled peas.

At the second stage of bulgur preparation the grain is shelled. To make taking off the shell easier, the grain is sprinkled with water to make it more flexible. The shelling is done manually in a stone hole called a "gurno", resembling a mortar. A "gurno" is hand made and fixed permanently to the ground in the yard. At other seasons it is used as a tank from which chickens drink water. The grain sprinkled with water, in ratios of 10 kg, is poured into the "gurno" and struck with wooden hammers. This is done by a few men who stand opposite one another. A woman sitting next to them stirs the grains quickly after each stroke(front cover). In some towns grain shelling resembles typical group work; dried grain belonging to a few or even several families is mixed and shelled. Nobody suffers any loss because one variety of wheat, a high protein wheat, is usually used to make bulgur. In the latter case, however, the "gurno" is shaped like a long stone gully around which a few or several men sit. Their wives sit between them and their task is to stir the grain. During the work, men often want to show off in front of their wives and strike the grain with all their might. The host's daughters prepare meals for their hard-working parents and at leisure time encourage their fathers. This stage of bulgur preparation is very colorful and truly folkloristic. It is often accompanied by folk songs sung by the children. One of these songs has even been recorded; the record was released in 1972 in Aleppo in Syria. Its lyrics are as follows:

> Himma u haye, himma u ha,[30]
> Come girls, come ha!
> Come and sing, sing, sing!
> All say, say, say:
> Himma u haye, strike ha!
> Strike haye, strike ha!
> The girls have run and have come!
> They gathered around byrghel in silence.
> They rolled up their sleeves and stood ready.
> They looked more beautiful to the eyes of the boys.
> Himma u haye, strike ha,
> Strike haye, strike ha!
> The girls as beautiful as doves,
> Shapely figures with plaited hair.
> Lord be praised for their creation!
> He molded them in the honey of the earth.
> Himma u haye, strike ha,
> Strike haye, strike ha!
> The boys leaned down and were happy,
> Their eyes made wider from looking.
> The girls' hands got wounded,
> And rested after byrghel shelling.
> Himma u haye, himma u ha,
> Strike haye, strike ha!

In towns a different method of grain shelling is known. It is more efficient and less arduous than the other two methods. A single-horse treadmill, called a "dang", is used for this purpose. It is found at fixed locations and is operational seasonally, not only to shell grain for bulgur but also to shell unboiled grain to be used for the preparation of another semi-finished wheat product called "garso". Grain sprinkled with water is put on a round cylindrical platform made of stone or on a hardened surface and next it is shelled by the weight of a vertically placed stone rotated around its axis by the horse(back cover). Such a treadmill is commonly used also in Asia and Northern Africa to squash vegetable oil.

Recently, mechanical shelling machines have appeared which, towed by donkeys or mules, travel from village to village. Most farmers who tried this highly simplified and quick method have not commissioned it any more. It appears that the machines damage the grains by making them round and consequently deteriorating their taste in dishes.

After shelling, the grain mass is spread for a while in the hot sun. This is supposed to facilitate the separation of dried husks from the grain and emphasize the quality of a nut-tasting dish. In the end the husk is blown away from the grain mass by the wind and the process of separation is completed. Small cupfuls of the grain are taken and poured down from a height of about two meters. The bran "pyrto" is blown away by the wind and pure grain falls straight onto the ground.

In the next stage of bulgur preparation the grain is ground. A quern "goresto" is used for this purpose. The quern consists of two flattened stone cylinders of which the bottom one is fixed permanently to the ground and the top one is moved manually around a joint wooden axis. The grain is poured into the quern by hand through the middle opening(Fig. 9, page nn). Depending on the grinding surface of the cylinders, and the mass of the top stone and the speed of rotations, the grain is ground into several particles of a different size. The bruised grain is sieved through sieves "'arbole"; three fractions, each of different granulation and taste, are obtained. This is due to the fact that during sieving different anatomical parts of the grain reach each fraction. Each of the fractions is used to make specific dishes. The thickest of them (constituting about 60 - 70%) is called "resa d gurgur", the middle (about 20 - 30%) - "perda" or "samdo" and the smallest (about 10%) "esta d gurgur". The percentage of particular fractions in the entire mass that is sieved depends on the intensity of grinding in the quern and the mesh size. Some housewives often make their own sieves. Their flange is made of wood and the bottom of leather. Sieves are probably the only household utensils which are reluctantly lent to others. The taste of particular bulgur fractions depends on the sieve quality.

Grain grinding is often done at night. A few women take part in this activity and a folk song accompanies it. It is sung in chorus to prevent the women from falling asleep. Its lyrics express the pain and fatigue of their tiredness, their love for a sleeping child, a great wish to return the magnificent times of the ancestors, their sorrow about an existing injustice, hope for better times of life in peace and longing for a distant husband:[31]

> My quern, oh my quern!
> You took my strength!
> Even my piece of food!
> Yet you are not capable of feeding my baby!
> I am carrying your heavy weight.
>
> I work on hot days.
> I stay awake at nights,
> Till my arms are tired.
> Of the entire camel only its ear belongs to me.
> I cannot satisfy my hunger.
>
> Turn oh quern smoothly!
> Like water current,
> Which treats some and tires others.
> Some have selected rest,
> But I cannot quench my thirst.
>
> Turn oh quern, make no noise!
> Dream has overcome the eyes of my baby.
> Do not let him wake up!
> If he cries, what shall I feed him with?
> My food has finished.

> Turn oh stone, so that time could pass!
> And the ancestors could be with us,
> And days could become sweet.
> The sister could see her brother,
> And I could see my baby grow into an adult.
>
> The riches of my earth are abundant,
> Stored in the warehouses of Emir.
> Each measure of wheat for one lyre -
> A poor man does not have it.
> I do not even have bran to eat.
>
> My quern, the sun will rise!
> Peace will be made on the Earth!
> After our dream - prosperity!
> After our pain - will be created!
> My husband will return, I will sing to him.

Sometimes bulgur is obtained from partly ripened wheat (at the stage of wax ripeness) in whose grains the starch has not been structured completely. Harvested ears of wheat are bound into bundles, singed over fire and the chaff is removed manually. These grains, called "forehto" ("freeke" in Arabic) are the material from which, after drying, grinding and fractioning ready bulgur is obtained. In this case neither boiling of the grain nor its shelling are necessary. Moreover, "forehto" is a delicacy for children and teenagers.

The initial processes applied to the raw wheat grain in the preparation of bulgur, such as rinsing, boiling and drying contribute to the removal of damage of the grain microflora, denaturation of proteins (including enzymes) and gruelling of starch. The low moisture of bulgur prevents unfavorable processes in the raw grain. A secondary contamination of bulgur is more difficult owing to the relatively low moisture of the air which prevails in the regions of the Middle East inhabited by Assyrians. Consumers of bulgur in the Middle East think that it never gets spoiled. Its particular fractions are stored separately in jute sacks, usually for one year till the next harvest.

Bulgur in the kitchen

Among Assyrians, as well as all over the Middle East, dishes made of bulgur have for centuries been greatly acclaimed. They are prepared at homes on ordinary days and on various special occasions as well as being served in snack bars and luxurious restaurants. They are an important item on the menu of every Assyrian family, both with respect to their quantity and quality. However, as yet these dishes have not been sufficiently documented.

Bulgur is treated as a universal food. It is prepared with vegetables, meat, milk, eggs and leguminous plants. Thus, many dishes of a wide range of taste and high nutritive value can be prepared. It is boiled much faster (about 10 minutes) and considerably easier to boil than loose rice.

From among culinary products made with bulgur we can especially note "tyrhayno" (Arab "kishk" or "kushuk", Turkish "tarhana"). Bulgur is mixed with yogurt and left for a few days to ferment. The resulting pulp, in which particles of bulgur disintegrate, is molded into small balls which are subsequently dried in the sun and stored at home for one year. After diluting and heating these balls are served as soup for breakfast on holidays. On fasting days "tyrhayno d maye" - "water tyrhayno" is used, without the addition of yogurt fat. The fermentation begins when yeast is added as leaven. The dish tastes a little sour.

Bulgur itself is very aromatic and can absorb smells and tastes of the ingredients with which it is mixed. Its thicker fraction is mixed with another wheat product, called "sheiraye", in a 10:1 ratio. It is a home-made "spaghetti type" macaroni. Most of the dishes are prepared from this mixture. The simplest and most basic of them is bulgur boiled in a proportional quantity of water. Then very hot milk fats are spread on the surface of boiled bulgur. Bulgur prepared in this way is molded into a pyramid on a plate. An egg, fried in butter (byrghel u be'e) or pieces of boiled or roasted meat or poultry (byrghel u basro) are put on top. The dish is usually served with bread (!) and tomato and onion salad seasoned with lemon juice and oil.

The fraction of bulgur, however without "sheiraye" but with minced meat, is used to stuff tomatoes, cucurbits and eggplants (these dishes are called "mehshi"), the leaves of fresh grape-vines ("aprahe") and thin ram's intestines ("hishayat"). Recently, however, most people have begun to use rice instead of bulgur for this purpose. Rice is a more prestigious raw material because it is imported! Moreover, the white colour of rice in these dishes is more attractive and its softer structure after boiling is more desired. This type of dish is served with garlic and citric acid.

"Falafel" is an important combination of bulgur, broad beans and chic-peas. It has become very popular in many countries. The cereals of leguminous plants are soaked in water for a day. The processes taking place at this time, similar to fermentation, improve the culinary values of the dish. When all three ingredients are mixed and the pulp is molded into meat balls, the latter are fried in oil. "Falafel" is the most frequent "sandwich" served in snack bars, in outdoor food stalls and at schools all over the Middle East. It has also become an international dish, like Barb kuskus. "Falafel" is factory made in, for example, Israel and Australia.

A dish similar in appearance to "Falafel" but with meat is called "kibbe". Different varieties of it are made, each being differently prepared. I think that it is more popular among Middle East Christians than among Muslims. Bedouins, for instance, whose cuisine is extremely poor, have not known of bulgur until recently.

"Kibbe" is prepared in the following way: the middle fraction of bulgur is soaked in water until it gets swelled. Even then the bulgur is ideally loose. One of the varieties of "kibbe" is eaten raw. The soaked bulgur is mixed with the same quantity of beat sheep veal. This dish is similar to Polish "tatar" in which the yellow of an egg is replaced by bulgur! In other varieties the bulgur mass is mixed with meat and seasoned with cinnamon and baked.

An elaborate ritual is connected with the preparation of "kibbe". Among some Assyrians it is one of the first dishes eaten after the "fast of Niniva" (a strict two and a half day long fast). It is believed that the dish will prevent headaches and dizziness.

"Ballo'" is another dish. It consists of a set of different quantities of bulgur and lentils. First whole grain lentils are boiled in salted water until soft. Bulgur is soaked in this hot concoction and after it has swelled it is mixed with boiled lentils. "Ballo" is a typical fasting dish as no animal fat is used to prepare it.

"Kutle" is an equally popular dish. It is also called "kbebat . Fine and middle fractions of bulgur are mixed and soaked in warm water until they join. Next they are hand-molded into balls the size of hens eggs. The balls are then hand flattened into 0.5 cm thick cakes. Lightly fried meat stuffing with onions and parsley is placed on the cakes. The cakes are then molded into different shapes, closed and boiled in water and sometimes also fried in fat. Sometimes fried "kutle" is dipped into stirred yolk of an egg. "Kutle" is a holiday dish, prepared for Sunday dinner. Some Assyrians eat it on Easter.

During family picnics, enjoyed enormously by Assyrians in spring evenings, a salad called "tabbule" tastes especially good. It is made of a small quantity of bulgur and many fresh vegetables such as tomatoes, lettuce, spring onion, onion and a lot of parsley and lemon juice. Aromatic "tabbule" is also seasoned with oil.

Bulgur is also a raw material from which a cake bread, about 1 cm thick and 10 cm in diameter, called "tyhrata d gurgur", is made. A fine fraction of bulgur is soaked in warm water until it swells and then it is mixed with fine groats of raw wheat, similar to macaroni semolina. The dough thus obtained (without fermentation!) is molded into cakes which are then baked in an oven or on a hot metal plate. The shape of bulgur is only slightly changed in this process.

This discussion on the origin of bulgur and an overview of its culinary values in the cuisine of contemporary Assyrians (though not only Assyrians) could be expanded. This requires investigating written sources or historical memory of different nations in the Middle East. Differences in the forms of bulgur dishes and the methods of making them more palatable could also be mentioned. Yet, the accomplishment of this task would require intensive research, observations and interviews, done in the Middle East. An extremely abundant amount of Assyrian literature, both that of pagans and Christians, would be of great assistance here. A researcher of this problem, however, might come across certain difficulties resulting from the fact that culinary issues are treated there as a marginal field, not worthy of scientific interest.

NOTES

1. Preparation of reserves known as "mozuno" or "moune" is a history- and climate-dependent tradition in the rural household of Assyria. In the past the reserves helped sustain long periods of siege by Muslims in the Middle East.
   Reserves are prepared during summer. A folk proverb related to this says: "uha d domeh bu qayto, kfoyesh bu satwo kafino" - "he who sleeps in summer goes hungry in winter".
2. Mierzejewski A.: *Tajemnice glinianych tabliczek* [The Secrets of Cuneiform Tablets]. Warsaw, Iskry 1981, pp. 277 - 281.
3. Zawadzki S.: *Podstawy gospodarki nowoasyryjskiej swiatyni* [The Bases of the Management of a New Assyrian Temple]. Poznan, Adam Mickiewicz University Press. *Series Historia* [History], No 94, 1981, p. 23.
4. Anonymous: *Iraq - a Tourist's Guide*. Baghdad. dateless. pp. 114 - 115.
5. Nielubowicz B.: *Irak - maly przewodnik turystyczny* [Iraq - a Small Tourist Guide]. Warsaw, KAW 1979. pp. 22 and 116.
6. In Arabic there is no consonant "g". In Assyrian it is the third letter of the alphabet.
7. Fellers D.A.: 'U.S. Wheat Products in World Feeding Programs'. (in:) *Wheat - Production and Utilization*. Edited by Inglett G.E., AVI Pul. Co., Westport 1974, pp. 467 - 474.
8. Haley W.L., Pence .W.: 'Bulgur, an Ancient Wheat Food'. *Cereal Sci. Today* 1960, vol. 5. No 7. pp. 203 - 204.
9. Tyloch W.: *Dzieje ksiag starego testamentu* [The History of the Books of the Old Testament]. Warsaw, KiW 1981, pp. 192 - 201, 356 - 360.
10. The verb "barghel" derived from "byrghel" means in Assyrian "to make loose", "to separate individual particles", whereas the verb "gargar" derived from "gurgur" denotes "to strike against one another" which probably , refers to grain shelling.
11. Saracoglu S.: 'The Thiamine Content of Turkish Wheats and Corresponding Bulgurs'. *Cereal Chemistry* 1953. vol. 30. No 3, pp. 323 - 327.

12. Michal Gawlikowski, a Professor of the Warsaw University, specialist in the art of the Middle East, claims that the name Syria comes from Greek and is abbreviated from Assyria. It initially denoted coastal provinces of the Assyrian empire. Greek sailors discovered it in the 7th c. B.C. *Sztuka syrii* [The Art of Syria]. Warsaw, WAiF 1976, p. 6.
13. Syrian sects, known in Bulgarian literature as "Paulicans", exerted a certain influence on certain Bulgarian circles. Their traces are detectable in the life of Orthodox Bulgarians back to the 16th century. Christo Vakarelski: *Etnografia bulgarii* [Bulgarian Ethnography]. PTL, Wroclaw 1965. p. 2. Tadeusz Wasilewski claims that those Syrians were advocates of the so called Paulican heresy. They were displaced from northern Syria, precisely from the vicinity of Theodoziopul and Melitana, which the Byzantinians took from the Arabs, to Bulgaria after 746. *Historia bulgarii* [The History of Bulgaria]. Ossolinieum 1988, p. 44.
14. Christo Vakarelski, ibid., p. 61. The famous French voyager, Alphonse de Lamartine, who in the years 1832 - 1833 visited the cities of the Middle East: Beirut, Damascus, Constantinople, describing all the things he had seen and heard, did not mention bulgur anywhere (Lamartine: *Podróz na Wschód* [Voyage to the Orient]. Warsaw, PIW 1986). So one can draw a conclusion that: either he did not have opportunity to be the witness of preparing bulgur nor eat it, or bulgur had not yet been known thereabouts.
    Written and handed down orally the accounts of Assyrian refugees confirm the first suggestion. For the first time they reached those cities - carrying bulgur with them - in the year 1895, the effect of massacres made of them in the mountains of Mesopotamia (Issak Saka: *Kanisati As-syrianiya* [My Assyrian Church] (in Arabic). Damascus 1985). When they settled they had to begin the production of bulgur by themselves. Such phenomena can also be observed today. Assyrian Students in Poland bring most of all bulgur with them and ask for it of every man who goes to their homeplace.
15. Hanna Q.: 'Qaryat Telesqof bayna almadi walhader' - [The village of Telesqof Yesterday and Today] (in Arabic). (in:) *Min turatina Hash-sh'bi fi qura Naynawa* [On our rural folklore in the vicinity of Mosul], published by the Society of Speakers of the Assyrian language, Iraq 1981, p. 111. See also Shamaya S.: 'Lawha folkloriya' [The folklore scene] (in Arabic). *Qala Suryaya* 1976 (No 12), pp. 49 - 52.
16. Shetty M.S., Amla B.L.: 'Bulgur Wheat'. *Journal of Food Sci. and Techn.* 1972, vol. 12, No 9. pp. 163 - 165.
17. Aykroyd W.R., Doughty J.: 'Wheat in Human Nutrition'. *FAO National Studies* No 27. Rome 1970, pp. 64 - 66.
18. Pence J.W., Ferrel R.E.: 'The Composition of Commercial Bulgur'. *Cer. Sci. Tod.* 1965. vol. 11, No 10. pp. 587 - 589.
19. Shepherd A.D., Ferrel R.E., Bellard N., Pence J.W.: 'Nutrient Composition of Bulgur and Lye Peeled Bulgur'. *Cer. Sci. Tod.* 1965, vol. 11, No 10, pp. 590 - 592.
20. Smith S.G., Barta E.J., Lazar M.E.: 'Bulgur Production by Continuous Atmospheric Pressure Process'. *Food Techn.* 1964, vol. 18, No 1, pp. 89 - 92.
21. Nuefeld C.H., Weinstein N.E., Mecham D.K.: 'Studies on the Preparation and Keeping Quality of Bulgar'. *Cer. Chem.* 1957, vol. 34, No 5, pp. 360 - 370.
22. Anderson R.H., Moran D.H., Huntley T.E., Holahan J.L.: 'Responses of Cereal to Antioxidants.' *Food Techn.* 1963, vol. 12. No 11, pp. 1587 - 1592.
23. Bookwalter G.H.: 'World Feeding Strategies of Utilization Cereals and other Commodities'. *Cer. Sci. Tod.* 1983 vol. 28, No 9, pp. 507 - 511.
24. Ferrel R.E., Heid M., Pence J.W.: 'Canned Products From Bulgur and Wheat'. *Food Techn.* 1963, vol. 17, No 8, pp. 56 - 58.
25. Cooley M.J., Ferrel R.E., Pence J.W.: *U.S. Patent* No 3 228 771. 1966.
26. Ferrel R.E., Pence J.W., Ulson R.L., Taylor E.G.: 'Food for Fallout Shelter. IV - A Special Shelter Ration Based on a Cereal Wafer'. *Food Techn.* 1962. vol. 16, No 9, pp. 45 - 49.
27. Ferrel R.E., Pence J.W.: 'Effect of Processing on Dry Wheat Expansion of Wheat Bulgur'. *Cer.*

Chem. 1963. vol. 40. No 2, pp. 175 - 182.
28. Ferrel R.E., Shepherd A.D., Theiling R.H., Pence J.W.: 'Gun-Puffing Wheat and Bulgur'. *Cer. Chem.* 1966, vol. 43, No 5, pp. 529 - 537.
29. The advantages of such boiling, also in order to retain the nutritive values of products, have been scientifically proved.
30. "Himma u haye, himma u ha" are characteristic phrases describing the sounds made when striking bulgur with wooden hammers. They are uttered by men who want to encourage each other and maintain the pace and rhythm of strokes. Smiths, for example, use a shorter version, namely, he, he...
31. Shamaya S.: op.cit., pp. 49 - 52.

Pasta: Not Only Italian

by Massimo Alberini

There are valid reasons, historical, geographical and linked to natural eating habits, why pasta keeps a mainly Italian image. Even the names of some of the shapes, vermicelli, spaghetti, maccheroni (although changed to macaroni) used in many countries, contribute to its Italian physiognomy, though in fact this food is also present in other cuisines. The same thing occurs in music, where the expressions "adagio", "allegretto", "presto" figure in scores that are totally foreign to Italy.

The Italians had always been convinced that they were the creators of their national dish, even though they did not know its exact origins. Then, some years ago, everything changed. This was when people started to say that pasta was Chinese and that it was Marco Polo who made it known in Italy. Apart from the fact that there are documents, in particular in Genoa and Parma, showing that lasagne and maccheroni were eaten before Marco Polo's return to Venice, the definite proof is in fact to be found in his book *Il Milione*. In chapter 166 of the version called *L'Ottimo* [The Best] - when describing the customs of the Population of Sumatra (therefore not China) Marco Polo clearly says that a flour, the present-day sago, produced from the so-called bread plant, is used to make "many good foods of pasta, and I too have eaten them many times." Nothing more.

As far as I can make out, the "Chinese" story originates from an article entitled "A Saga of Catai" that appeared in the American magazine *Macaroni Journal* in 1929. There it was written that a sailor in Marco Polo's expedition had seen a Chinese girl preparing long strands of pasta, and that the sailor's name was Spaghetti. Obviously an unlikely tale.

In fact pasta is one of the three kinds of product in which cereal flour is used. The other two are the baked kind, bread and focaccia, and the porridge kind, polenta in Italian. The discovery of the tombs in Cerveteri with bas-reliefs illustrating the tools to make tagliatelle and lasagne made people think that pasta had Etruscan origins. But the writings of Cicero and Horace show that those lagane, the Latin name, were not cooked in water but fried, the way crescentine are cooked to this day in Emilia. Foreign visitors to Naples and its province in the 19th century brought back prints of maccheronari - pasta vendors - and pasta-eaters (using their fingers) as souvenirs, and these convinced many people that pasta was the typical food of the southern peasants and eaten almost exclusively by them. This is true up to a point: Pulcinella is always hungry for maccheroni and, for this reason, a new pasta factory, Voiello, has chosen him as the symbolic character for their packaging.

In 1835, however, while he was touring Campania, Alexandre Dumas wrote that the common people, the so-called lazzaroni, were living mainly on melons and pizzas, and he went on "Today maccheroni is a European food which has travelled like civilisation, and is to be found everywhere." Years before him, in 1747, Hannah Glass was giving the recipe for vermicella pudding (spelled with a final a instead of an i), without mentioning the Italian origins of the dish.

Singing the praises of pasta as a food of the south ignores the fact that, many centuries before the arrival in the kitchen of the tomato, this dish had a particular importance in the diet of northern Italians thanks to one factor, Parmesan cheese. The proofs of this alliance are many. In 1284 the monk and chronicler from Parma, Salimbene da Adam wrote the following about Brother Giovanni da Ravenna: "I have never seen a man that ate a plate of lasagne with cheese with such relish." A century later Giovanni Boccaccio describes the famous mountain of grated Parmesan, in the village of Bengodi, which is used to dress maccheroni and ravioli. The union of these two elements is considered so perfect as to have originated a proverb quoted by the philosopher Giordano Bruno in 1534: "Come il cacio sui maccheroni" - like cheese on maccheroni. This perfect "entente" will continue, one could say for ever in Italy and outside. In 1851 the Neapolitan, Chevalier Delbono, writes that his fellow citizens, however fond they are of pasta with tomato, "honour maccheroni with Parmesan." And Queen Victoria's Chef, Charles Almé Francatelli, completes the recipe for "Maccaroni with cream" with "Parmesan cheese".

This twinning of the two ingredients can explain why today, with the transition from pasta made on a small scale to industrially made pasta, the largest pasta factory, Barilla, is located in Parma, a geographical coming together that is explained by reasons of history.

It is natural to ask what now remains of the ancient tradition either in Italy or abroad. There is no doubt that Italian pasta is now produced only with durum-wheat semolina, by virtue of a regulation that sets Italian producers apart from those who would like to see the use of soft wheat flour, thus producing inferior pasta. As a result, Italian pasta is now far better than it was in the old days, when it was often badly dried.

Even foreigners know that pasta must be cooked to the right point and not too much. After a period in which pasta tended to be dressed excessively with sauces called "alla carbonara" or "alla carrettiera", there is today a return to simpler preparations which identify pasta with good homely flavours. Pasta is, after all, the symbol of home cooking. After a century and a half, the rule set out by Gioacchino Rossini in a letter to a friend is still valid. "In order to produce a dish of appetising maccheroni, you need good pasta, the best butter, excellent tomato sauce and Parmesan. And an intelligent cook who dresses the pasta and brings it to the table."

Buckwheat - Food for Peasants and Pheasants

By Josephine Bacon

When I was at school, we were taught that there were five types of food grain - wheat, barley, oats, buckwheat and rye. In that order. Yet, strictly speaking, buckwheat, *Fagopyrum esculentum*, is not a grain at all, although it was inserted before rye in the school list.

Buckwheat is the fruit of a herbaceous plant which originated in central Asia, and is native to Siberia. It has the advantage of flourishing in a variety of cool climates, like the grains, but, unlike them, it can grow on very poor soil and requires little fertiliser. It is thus grown mainly where the soil is poor, but thrives in a continental climate - harsh winters and hot summers. It is hardly known at all in England, since our summers are not hot enough for it, though according to André Simon, it is grown here by gamekeepers as food for pheasants.

Buckwheat was brought into Russia by the invading Tartars in the sixteenth century, along with tea. However, while tea was a drink for the wealthy, since it could only be grown in the warm south, every serf could afford to cultivate a little plot of buckwheat. It thus came to be called by the generic name for all types of gruel or porridge - kasha. In fact, the correct name for buckwheat in Russian is "grechikha" (гречиха). The full name of the porridge should therefore be "grechikhaya kasha" (гречихая каша). There are many Russian proverbs that attest to its importance in the diet, of which the most basic is "Cabbage soup and kasha are our fare" ("щи и каша, пиша наша"). Other proverbs involving kasha are "He has kasha in his head" (instead of brains), "The kasha was cooked" (similar to "the goose was cooked") and " to cook kasha with someone" (similar to "talk turkey with someone").

The English term "buckwheat" is a corruption of "beech-wheat", the term by which the seeds are also known, since buckwheat grains resemble beech-mast. The pale brown grains have four triangular faces, a very unusual shape to occur in nature. Buckwheat has a strong flavour, similar to that of rye or wild rice but more pungent. The flavour is strengthened by dry roasting.

Buckwheat has never been popular in England, where the climate is not really suitable for its growth, but it is grown extensively in Canada and the northern U.S.A. In Canada, it is eaten by Canadians of Russian, Polish and Ukrainian origin in dishes they have brought with them. In the U.S., buckwheat is eaten as grains by people of the same origins, and is also included in breakfast cereals, such as Grape-Nuts, but is mainly consumed in the form of a flour.

The same is true of France, where significantly buckwheat is called "sarrasin", an indication that it was brought to France by the invading Saracens. In fact, a little-known name for buckwheat in English is Saracen Corn. The Saracens never actually penetrated as far north as northern Britanny, yet that is where buckwheat has entered the national cuisine, in the form of pancakes (galettes de sarrasin). Galettes de sarrasin are best eaten with sausages, such as the local andouillettes. In fact, buckwheat is most popular in pancakes, on its own or mixed with wheat, as in blinys. Blinys are Russian pancakes in which the buckwheat flour is mixed with an equal quantity of wheat flour and the whole raised with yeast. Buckwheat flour has the same effect as potato in lightening a yeast dough. In Finland, a bread called Tatarleipa is made with the addition of buckwheat flour. Again, the name is an indication of the origin of the flour. The Finns also make blinys which they call blinut.

Although buckwheat seeds can be boiled with water to produce a thick porridge like a millet porridge in consistency, they are far more delicious when first mixed with a beaten egg and dry-roasted before boiling. This makes the grains separate but fluffy. The combination of buckwheat grains with wild mushrooms, particularly ceps *(Boletus edulis)* is a particularly felicitous one. The custom of eating the whole grains as a staple accompaniment to a savoury dish, in the same way as polenta, rice or bulghur, spread across Asia to European Russia, Byelorussia and Poland. It finds its westernmost expression in Sterz, eaten in Styria, in southern Austria. Unlike most grain substitutes, buckwheat combines well mainly with savoury foods. Mention is made in the *Atlas of Food* of a sweet buckwheat dish eaten in Russia, but

I find this most unlikely and have no evidence of it in any Russian cookery book I possess. André Simon claims the Americans eat it with maple syrup, and it is true they eat buckwheat pancakes with maple syrup, but since they also pour the syrup over sausage and bacon this can hardly be called a sweet in the sense of dessert. I would be interested to know of any sweet dessert made with buckwheat.

Buckwheat has really come into its own in England now that there is a fashion for "whole grains" and it is to be found in every health food shop, in both light (unroasted) and dark (roasted) forms. Green buckwheat tea is also believed to have curative powers. Buckwheat flour is also easily available for pancakes and breadmaking. The seeds are imported mainly from Belgium and Holland.

A nutritional analysis of 98 grams of buckwheat, ground into flour, made by the U.S. Department of Agriculture gives the following results:-

Water content	12%
Calories	340
Protein	6.3 g
Fat	1.2 g
Carbohydrate	77.9 g
Calcium	11 mg
Phosphorus (in the form of phytic acid)	86 mg
Iron	1.0 mg
Sodium	0
Potassium	314 mg
Vitamin A	0
Thiamin	.08 mg
Riboflavin	.04
Niacin	.4
Ascorbic Acid	0

These are values for flour made from unroasted buckwheat flour, or light as opposed to dark flour. Roasting brings out the flavour and so darker flour is often more popular. The values will naturally vary depending on where the flour is grown and the potassium is lost in the roasting process, but as can be seen from the above, buckwheat is nutritionally far poorer than other grains since it contains no B vitamins to speak of. It has about the same calorific value as wheat.

All the above proves that buckwheat is a staple at least in some parts of the world. If further proof were needed, Savella Stechishin claims in her *Traditional Ukrainian Cookery* that kasha is to Ukrainians what oats and porridge are to the Scots. What better proof?

Sources

Nutritive Value of American Foods in Common Units, Agriculture Handbook No. 456, USDA, November 1975.
The Complete Russian Cookbook by Lynn Visson, Ardis, 1982.
Traditional Ukrainian Cookery by Savella Stechishin., Trident Press, 1976,
A Concise Encyclopaedia of Gastronomy by André Simon, Collins, 1952.

The Significance Of Wheat In Judeao-Christian Religion

By Josephine Bacon

Wheat, especially in the form of bread, has a central role in religious symbolism in the Judaeo-Christian tradition, and a consequent importance greater than that of any food. This is reflected, not only in religious practice, but in slang phrases and expressions current throughout the Christian world. The ritual significance accorded to wheat may even account for the strenuous - and often ruinous - efforts to grow it in unsuitable climates, such as in Scandinavia. According to palaeo-botanists, grains are not the preferred food of Homo sapiens, nor do they provide the best diet. In fact according to Lee[1], agriculture itself does not do a lot to improve the human diet or lifestyle, and hunter-gatherers have a much easier time of it, providing the food sources are adequate. In other words, if there is enough wild food available, there is no need to grow it. In any case, the preferred foods of humans would be, in the following order, meat, preferably from a large animal, then tuberous roots, and only lastly grain. The theory is that grain is a last resort when other foods have run out.

The value of agriculture lies in the opportunity it affords "to grow and harvest more food from a unit of space in a unit of time"[2].

Unquestionably, wheat is a good source of nourishment. In its wild state, it is high in protein and B vitamins, though a large amount needs to be gathered to make a meal. Compare the earliest cultivated varieties of wheat, some of which are still grown today, such as emmer and einkorn, with red or winter wheat. They contain one-quarter the grain. The only part of the world where large stands of wheat grew wild was the Middle East.

The earliest substantial settlement is to be found at Jericho, in the Jordan Valley, where archeologists have unearthed remains dating from 7,500 BC, the Neolithic Age. Jericho is an oasis in the Judaean Desert, lying below sea level in the Jordan Valley at a convenient crossing point, and thus on a major trade route. It has its own spring, so water is plentiful, and archaeologists have recently discovered that large stands of wild wheat were to be found nearby. It was the most natural thing for these earliest farmers to store seed from these wild plants and sow it close to their houses. The earliest inhabitants of Jericho were Semites, who worshipped the ancient Semitic gods. They practised crop rotation and irrigation, reared livestock and lived in round, brick-built houses. This civilisation unaccountably disappeared in about 4,000 BC, by which time the city-state of Akkad was becoming powerful in Babylonia. The original Akkadians, who were also Semites, had settled in the area around 8,000 BC. Other Semitic tribes learned how to cultivate the wild emmer and einkorn wheats, as well as other crops, especially barley. Grain was cultivated because it was handy to have it near at hand, and the more of it that was harvested, the further away one had to go to find more. There is much evidence in prehistory - to say nothing of more recent times - that no attempts at conservation were made. Witness the total extinction of silphium, thanks to overconsumption by the Romans. Yet, almost as soon as it began to be cultivated, efforts were being made to improve wheat. For instance, emmer and einkorn have the disadvantage of having heads which fall off too easily when ripe. Wheat that was not only better yielding but also more robust was needed.

Unquestionably, the importance of wheat over other grains is its ability to make a delicious food when fermented. In that climate, natural fermentation (sourdough) is easy and quick. The discovery of yeast must therefore have been as natural and spontaneous as the discovery of fire or the wheel. The Egyptians made barley bread to brew yeast, a more sophisticated process, since barley contains less gluten and therefore rises less easily. The advantage of wheat is thus obvious.

However what the Semitic and Egyptian civilisations had in common was their original need to resort to increasingly sophisticated methods of cultivation to ward off the inevitable famine that would otherwise have accompanied their rising populations. Wheat, made into its most economical form of bread, is thus the obvious way to ward off starvation. Hence its immense importance in the context of Middle Eastern religions.

The first religious festival mentioned in the Old Testament is Passover. Regardless of the evidence for the Exodus from Egypt - which is scant - and the story of the Hebrews baking their bread on their backs, there is evidence of a previously existing festival called the Feast of the Unleavened Bread. Why would the bread have been unleavened? Possibly, because at that time of year (early spring), unripe wheat had to be used to make bread. Flour made from unripe wheat will not rise when mixed with yeast. This is how the flatbreads of Scandinavia originated.

Unleavened bread, being the first bread of the year, thus had a sacrificial quality. The origin of the sacrificial wafer in the Communion ceremony is certainly that of an unleavened bread, and it may go back to a pre-Jewish origin.

The symbolic importance of the unleavened bread has been perpetuated in Judaism right from the days of the Temple. The sacrifices brought to the Temple included the shew-bread or hala. Leviticus 24, 5:7 states:-

"And thou shalt take fine flour, and bake twelve cakes thereof: two tenth-parts of an ephah shall be in one cake. And thou shalt set them in two rows, six in a row, upon the pure table before the Lord. And thou shalt put pure frankincense with each row, that it may be to the bread for a memorial-part, even an offering made by fire unto the Lord".

The Talmud claims that these wheat flour breads were unleavened. They probably resembled rotis or chapattis, and were laid in a tray of golden moulds which were raised slightly on supports so that they could be easily visible, when arranged on an altar. The moulds were shown in two stacks of six, representing the twelve tribes. A red canopy hung over the breads and above that a canopy of animal skin. Legend has it that this animal only lived in the days of the Temple and is now extinct. The bread was baked by bakers from the Garmu family. Every Sabbath morning the breads were replaced. The old breads would be placed on another golden table which was first covered by a blue and purple cloth, symbolizing the sky and the sea respectively. The old breads were then shared among the priests who constantly guarded the shew-bread table.

This devotion to bread as the symbol of all food has been retained in Judaism in the form of the hallah, eaten on the Sabbath. The Talmud prescribes that the bread must be baked from a wheat intended specifically for bread-making as opposed to cakes, but it can be from wheat, barley, spelt, oats or rye flour. Although anyone who has seen a challah will immediately associate the plaited shape with this traditional bread, in fact no shape was prescribed for the bread. In communities outside Europe, the Sabbath bread is the usual shape, i.e. a large round loaf. However, it only dates from the 15th century, and is modelled on the festive breads baked in Eastern Europe, and especially the Ukraine, for important occasions. In German-speaking communities, the challah is known as barches, a corruption of beracha.

In countries where other staples - rice, maize, millet, tuberous roots - grow far better than wheat, the symbolism of the bread is meaningless in religious terms, and has been replaced by the local staple. However, there is no question that the spread of wheat into colder climates and areas where it was never previously known (south and central America) must have facilitated the spread of Christianity.

References

1   Lee, Richard B., *Man, the Hunter*, Aldine Press, Chicago, 1968.
2   Cohen, Mark Nathan, *The Food Crisis in Prehistory*, Yale University, 1977.

History and Perspectives of Staple Foods in Africa

by Esther Balogh

> Iyan ni ohje, okà li ögun, airi arara l'a nje eko. (Pounded yam is real food, eba tastes as good as a medicine, and when there is no other food we eat eko.) Yoruba proverb.

It is impossible to write about staple foods in Africa without compassion. Staple foods nourish hundreds of millions of Africans from the cradle (figuratively speaking) to the grave. In case of recurrent famines (in the Sahel, Ethiopia, Sudan, Angola) (Iyer, 1984; Tworse, 1984; Harris, 1986) the bare fact of survival depends on them.

The traditional African attitude to food, that it is to extinguish the pangs of hunger, is now unfortunately returning. My own students, and junior colleagues living on meagre rations, constantly express the views classically formulated by Audrey Richards in 1932 (quoted in Mintz, 1985):

> "To the Bemba (Southern Bantu people in Zimbabwe) each meal, to be satisfactory, must be composed of two constituents: a thick porridge (ubwali) made of millet and the relish (umunani) of vegetables, meat or fish, which is eaten with it ... Ubwali is commonly translated by 'porridge' but this is misleading. The hot water and meal are mixed in proportion of 3 to 2 to make ubwali and this produces a solid mass of the consistence of plasticine and quite unlike what we know as porridge. Ubwali is eaten in hunks torn off in the hand, rolled into balls, dipped in relish, and bolted whole ... To the Bemba, millet porridge is not only necessary, but it is the only constituent of his diet which actually ranks as food ... But the native, while he declares he cannot live without ubwali is equally emphatic that he cannot eat porridge without a relish, umunani ... The functions of the relish are two: first to make ubwali easier to swallow, and second to give it taste. A lump of porridge is glutenous and also gritty - the latter not only owing to the flour of which it is made, but to the extraneous matter mixed in on the grindstone. It needs a coating of something slippery to make it slide down the throat ... The Bemba himself explains that the sauce is not food. It prevents the food 'coming back'."

This is exactly the attitude so succinctly expressed by the Yoruba proverb - pounded yam is real food. Staple foods for us in Africa are not just 'food in history', but main contributors to the web of history. I am fully aware of the complexity of historical analysis (Marwick, 1973; Walsh, 1967; Carr, 1961) and the danger of taking one sided view-points, but in spite of this, I should like to risk the statement, that the future of Africa, in the critical period of the next hundred years, depends on its supply of staple food.

One of the leading cultural historians of our time (Braudel, 1981), analysing the history of everyday life, compared the growth curves of populations in Europe and Africa (Table I). With the use of a study (Harrison, 1984) commissioned by the FAO, and other organisations of the United Nations, we may extend the estimates for Africa till 2100 (Table II). As a brief explanation I shall have to add, that this table is composed from two sets of UN sponsored estimates, and evidently their predictive value is decreasing with time, for example for 2000 the low estimate for Africa is 756 million, the high one 886 million, while the respective numbers for 2100 are 1354 million and 4444 million (I have taken the middle estimates).

This table shows that Nigeria is not only the most populous country of Africa, but early in the next century may become the third most populous country of the Earth, after China and India. I can confirm these estimates from Nigerian sources, for example the late Professor Igun of the University of Ife, one

Table I

Estimated populations of Europe and Africa from 1650 'til 1950 (in millions) *(Braudel, 1981)*

	1650	1750	1800	1850	1900	1950
Africa	100	100	100	100	120	199
Europe*	100	140	187	266	401	594

* (European Russia included)

Table II

Estimated populations of Nigeria and Africa 'til the end of the 21st Century (in millions)
(Ingun, 1981)

	1931	1975	2000	2025	2050	2075	2100
Nigeria	19	65	150	285	-	-	-
Africa	-	380(401)	780(835)*	1542	2166	2507	2591

* (including Union of South Africa)

of the most distinguished demographers of the country, in his inaugural lecture (Igun, 1975) estimated, that our population will lie between 129 and 150 million by 1994. According to his opinion there is no reason for optimism, i.e. that a demographic transition will set in and decrease the rate of increase from annual 2.5% in 1952. On the contrary he expects that, because of the structure of the evolving population, this will increase to 4.0% by 1995. Another professor (Marciniak), this time from the Ahmadu Bello University, Zaria, in an indepedent analysis forecasted 154 million Nigerians for 2000 (after considering a low variant of 140 million, and a high one of 171 million). Professor R. B. Davidson (University of Calabar) puts his estimate for 2000 as 215 million.

I could go on for a while quoting different population estimates but, instead, I should like to point out that the census data and population estimates in Nigeria constitute a most sensitive issue, because, of course, the voters' list of elections and representation in popular assemblies must relate somehow to them. The census results of 1973 were officially rejected as grossly falsified and this led to the fall of General Gowon's government. The (alleged) failure of the last elections (that is manipulation of voters' lists) toppled Alhaji Shagari's rule.

In essence, what I am saying about the Nigerian population is relevant to most of Black Africa, or even most of Africa, including the Arab-speaking north. Among other states, Kenya, Ethiopia, Burundi and Egypt have the most serious population problems. As far as I know, the governments are not doing much to temper the explosive growth of their populations, maybe because this is beyond their power and influence. One prominent educator and politician, the first Vice-Chancellor of the University of Ife, the late Professor H. A. Oluwasanmi, once told us in private, when we talked about the population issue, that an African politician cannot advocate population control, because that would finish him and his party at the next election. This leaves the government with only one possible measure - to try to develop (staple) food production to match the population curve.

Indeed there is no lack of agricultural projects, and it can be said that at least, in principle, most of them are in a good direction. Are they effective in increasing the production of staple foods? This is a most difficult and most delicate question to answer, because of course there is no reliable data base for assessment of the level of agricultural production. It is true that national and international organizations (among them, the FAO) publish agricultural statistics, but everybody inside Africa knows that at the grassroot level there is no data collection. In most of the African countries, small-holders, subsistence farmers, constitute the majority of the population. The farmers provide food for themselves, and maybe some to the local or roadside markets, where small dealers buy for the large towns. All that is possible to say is whether this was a good year (from the point of view of rain, the prevalence of pests and plant diseases) or bad, but without quantitative parameters. We have of course data on exports and imports (from the ports); and, from the escalating prices of commodities when governments restrict imports, we have to conclude that the local production does not satisfy demands. Starvation starts not when all food is exhausted but when the poor in the towns are not able to buy enough food.

The FAO report I mentioned (Harrison, 1984) is the most pessimistic. It emphasizes that, already in 1975, 54 of the so-called developing countries (and territories) could not support their population from their own lands using traditional agricultural techniques. By the year 2025 the developing countries considered in the report will have two and a half times higher population (a total of 5.1 thousand million) and the food supply could be critical in 65 countries. From the 5 regions considered (Africa, South West Asia, South East Asia, South America and Central America) the "prospect of feeding its future population seems most clouded for Africa". The food supply in 1975 was already critical in 22 countries (of the 51 countries in the region) with a population of 188 million (which was half of the African total at that time). By the year 2000 the situation is expected to become more serious, as the total African population (with a growth rate of 2.9% a year) is expected to reach 780 million. At that time the number of countries with critical food supply will probably be 29 with a total population of 466 million.

The FAO report also states that potentially higher yields could be reached by more advanced agricultural methods; however the ratio of irrigated land is not only lowest in Africa, but also the potential for the extension is the lowest here. (In 1980 only 1.5% of the total of arable area was irrigated.) Mechanization of agriculture, extensive use of fertilizers, pesticides and other agricultural chemicals, and the use of high yielding varieties, depending on these, are advocated, but is this realistic? Can Africa absorb the high level of agricultural technology elaborated in North America, Europe and the Far East?

As you know well, in most African countries the governments frequently change, and whenever a country struggles back to an elected civilian government, the military again takes over within a few years, usually by corruption charges (Gutteridge, 1965). Concerning agricultural policies, this means that the policy changes every few years, perhaps before it could have achieved something. For example, in Nigeria the 'National Accelerated Food Production Programme', 'Operation Feed the Nation' and the 'Green Revolution' followed each other in quick succession (Aribasala, 1983; Boszormenyi, 1980, 1983). The present administration initiated the 'Accelerated Wheat Production Programme' and set up a Raw Materials Directorate.

The international and particularly the North American literature (as we know from VOA broadcasts) is highly critical of the inefficiency of the aid programmes of the 1970s and 1980s. African countries at the same time accuse the West of exploiting Africa under the disguise of aid. I do not recall reading a single approving sentence in the Nigerian press which supported the policies of IMF, World Bank, USA or EEC. The burden of interest on loans has increased to the extent that it totally cripples the economy of the African countries (including Nigeria) and alienates the people. I am not an economist, and I should not like to attempt an analysis of the complex factors which lead to the failure of these different aid programmes. Subjectively, I believe that the enormous inefficiency of the state bureaucracy and all of the African institutions should be blamed (Harman, 1982, reprinted and distributed in President Shagari's re-election campaign, Oji, 1982). I also cannot suggest remedies. No question, that African countries need more aid, but different from that given in the two previous decades.

Meanwhile, the food situation is constantly deteriorating. Achievements of the previous period of food science and technology now seem to be luxuries (Aboyade, 1985) which even (or particularly) the educated middle class cannot anymore afford. It is a time for stock-taking.

Africa is a very large continent - the cradle of mankind (in the true sense), but in my short historical analysis of the origin of its staple foods, I cannot and need not go back to millions of years. As you know well, food production (both agriculture and pastoralism) started in the so-called 'fertile crescent' of the Middle East some 8 - 10,000 years ago. We now realize that the 'fertile crescent' was not particularly fertile, but ancestors of domesticated animals (goat, sheep, cattle, pig) freely roamed there, and wild cereals (wheat, barley) constituted natural meadows, from which their seed could be easily collected by stone-toothed sickles. For the next step, agriculture moved to the river valleys of Tigris, Euphrates and the Nile around 4000 B.C., where the high civilizations of the Sumerians and Egyptians emerged. In this sense both of these classical civilizations were based on 'imported' crops and animals - this of course does not detract anything from their greatness.

We have good reasons to believe that besides this important Middle Eastern centre of the origin of cultivated plants, there was another centre (Ethiopian highlands) and a more diffuse zone (the West African savanna) in Africa (Vavilov, 1926; Harlan, 1975; Oliver and Fagan, 1975). The Sahara, particularly its mountainous ranges, was quite wet about 10,000 years ago, and since then gradually dried up - with two short reverses (McIntosh and McIntosh, 1983). The conditions there about 4000 - 2000 B.C. were quite similar to those in the 'fertile crescent' at 8000 B.C., that is a gradual drying up forced the hunter-gatherers to change to pastoralism and agriculture. The most important staple crops which emerged from the savanna zone are sorghum (Guinea corn) and pearl millet *(Pennisetum typhoides)*. We have archeological evidence only after 2000 B.C. for the cultivation of pearl millet and Guinea millet *(Brachiaria deflexa)* from Karkarichinkat on the bend of the Niger river, and for pearl millet from Tichett (in the Mauretanian Sahel) from about 1000 - 900 B.C. Pearl millet, sorghum and African rice *(Oryza glaberrima)* from the third century B.C. were recovered from the famous town Jenne-jeno (McIntosh and McIntosh, 1982) in the inner delta of the Niger river, but their cultivation must have started earlier. Botanists generally place the origin of sorghum in the Eastern Sahel-Sudan zone, or even in Ethiopia. The finger millet *(Eleusine coracana)* is probably also of Eastern Sahel-Sudan or Ethiopian origin (Simmonds, 1976), while the fonio and black fonio *(Digitaria exilis* and *D. ibura)* are definitely West African (Harlan, 1975). The fonio, with the local name acha, is still grown on considerable scale on the Jos plateau in Nigeria (Anthonio and Isoun, 1982). The most important Ethiopian cereal is the tef *(Eragrostis tef)* (Ullendorff, 1960; Mesfin, 1987), but we shall have to mention that early forms of wheat and barley reached Ethiopia in pre-historic times (and are still produced there as I personally observed). The Ethiopian oat *(Avena abyssinica)* probably developed from a weed of barley fields.

The most remarkable feature of the distribution of cereals in the northern half of Africa, is that neither the wheat nor the barley reached the West African Sahel-Sudan zone before modern times. In the same way, the cereals domesticated here are not recorded from classical Egypt or later from Roman North Africa (Barton, 1972). It is not that there was a lack of trade contacts or that these crops are climatically unsuitable for production South-North of Sahara. They were also not immobile, for we have archeological and linguistic evidence, that pearl millet, finger millet and sorghum reached India around 2000 BC. (Simmonds, 1976).

Before going further on this line, I should like to discuss one other aspect of the early West African agriculture. Some agricultural historians, notably Harlan (1975), argued that besides the savanna-born agriculture, another parallel development occurred on the savanna/forest border at the same time (or even earlier). There is no solid archeological evidence for this theory, but it seems probable on anthropological/cultural grounds. Digging sticks and hoes are more suitable for cultivation of yam and black eyed bean than for the small grained cereals. The West African yams were earlier described as two species: yellow yam *(Dioscorea cayenensis)* and white yam *(D. rotundata)* but now considered to belong to the same species complex, the yellow yams perhaps a little closer to the wild form (Kay, 1987). The yams (this and some other species) were originally collected and gradually became cultivated. Unfortunately, we cannot affix any time scale to the process.

Theories of the origin and distribution of the black-eyed bean (cowpea, *Vigna unguiculata*) closely parallel the history of sorghum. According to Stanton (1966) and Harlan (1975), it was drawn into cultivation in West Africa, but Steele (in Simmonds, 1976), argued for an Ethiopian centre (5000 B.C.) from which it reached West Africa after 3000 B.C., and was introduced to India (with sorghum and finger millet) around 1500 B.C. and to China and the Far East somewhat later. India can be considered as a secondary gene centre for many cultivars earlier described as species *(cylindrica, sesquipedalis, sinesis)*. There are several other *Vigna*, still accepted as good species, which were independently domesticated in the Far East (Kay, 1979). Black-eyed beans reached Europe from India before 300 B.C., but the main European bean in Antiquity was the Mediterranean broad bean *(Vicia faba)* which also spread along the Nile to Ethiopia.

Let me now return to the African distribution of sorghum and pearl millet, to which black-eyed beans could be added. This is attributed to an important historical event: the Bantu migration. This theory was first advocated by Greenberg (1966) and was later modified by Guthrie (1967) and others (Oliver and Fagan, 1975; Phillipson, 1977). According to the theory linguistic evidence shows that the so-called Bantu languages of eastern and southern Africa (several hundreds of them) are all closely related and could not have arisen much more than 2000 years ago from the West (Central) African Congo-Kordofian family of languages (Hickerson, 1980). According to Greenberg the pre-Bantu homeland should have been on the grasslands of the Benue river valley (the present Eastern Nigeria and Cameroon) while Guthrie places it to the south east of the Congo Basin, and others to the neighbourhood of Lake Victoria, where the northern and southern savanna belts meet. One school of thought considers the spread of Bantu languages as mass movement of peoples who carried with them a full cultural assemblage: cattle (Middle Eastern-European type), grains and pulses (sorghum, millet, beans) characteristic (cooking) pots and the technology of iron making. Iron first appeared in Africa south of the Sahara in the Nok culture (centered on the Jos plateau in Nigeria) around 300 - 500 B.C. (Shaw, 1981; Gillon, 1986). Recent evidence shows some earlier data close to Lake Victoria, which could be as old as 700 B.C. (van Noten and Raymaekers, 1988).

Of course it is possible that words of languages and traits of the cultural assembly moved more by diffusion than by migration, because some other cultural traits involving crops and staple foods made from them moved across Africa not from the North West to the South East, but from the East to the West. I already mentioned the case of sorghum. Similarly, the important oil plant sesame is also probably East African (Ethiopian) but spread both west- and eastward and reached the Middle East and India before 2000 B.C. Black-eyed beans followed a similar trend.

Bananas and plantains originated in the Indonesian archipelago and probably reached Madagascar and the East African coast in the first millennium B.C., then diffused through the Congo basin to West Africa before the first Portuguese reached there. (Good evidence for the diffusion process is the impoverished gene-pool of the West-African bananas.) Coconuts, also from the eastern Indonesian islands, crossed Central Africa overland before the arrival of the European traders (the Arab traders of the East African coast never sailed into the Atlantic) (Horton, 1987; Boorstin, 1983).

Most of the edible aroids (important tropical root crops) are of South East Asian origin. Taro ('old coco yam') reached Egypt only around 500 B.C. and supposedly moved up the Nile valley and across Sudan to West Africa about 2000 years ago. (Tannia, 'new cocoyam', was introduced only after Columbus's travel to the New World, and never became popular, at least in West Africa.)

Rice *(Oryza sativa)* became cultivated in the Yangze delta around 5000 B.C. by Austro-Asiatic and Malayo-Polynesian people (Ramseyer, 1989) but its agriculture was much improved when the Chinese culture reached the area towards the first millennium B.C. Rice probably came to the east coast of Africa and Madagascar earlier than to the Nile valley (only after 700 A.D.). As far as I know, rice reached West Africa only in modern times, becoming an important crop just in these days.

In the 'Age of Discovery' (15th/16th centuries A.D.) almost all of the Far Eastern and American crops, which can be grown in tropical and subtropical climates, were introduced to Africa. At the same time many African cultivated plants reached the West Indies, Central America, Brazil and even the Far East (e.g. bambara groundnut, *Voandseia subterranea*, to Indonesia). The process was as fast as the introduction of new crops to Europe, but it cannot be as well documented. Among the new staple crops which came from Asia, the water yam *(Dioscorea alata)* and the breadfruit *(Artocarpus communis)*, should be mentioned. The American maize, cassava, sweet potato and groundnut became important staple crops. The potato is becoming an important crop on the Jos plateau in our time. But in my view, no other crops more profoundly influenced African cooking than the American vegetables and spices (capsicum peppers, tomato, squash, and leafy vegetables such as *Amaranthus* and *Celosia* spp.).

Summarizing the history of African staple plants, we can say that these could be divided into three groups: (a) originally African, (b) pre-historic introductions from the Middle East and Asia and (c) late Medieval introductions from Asia and America. I should like to emphasize that this is only the classification of the historians and plant geographers. The common people do not make this distinction and are totally incredulous if we claim that some of their favourite foods were introduced only a few hundred years ago or even more recently. But does that really matter? After all, what we like to eat in Europe or in the USA mostly came from other territories or continents. This sharing of the resources of the earth, which was a spontaneous historical process, is the best evidence of the common 'human nature'.

Africa's sharing the staple crops with the world makes it practically impossible for me to discuss all of these crops across innumerable tribes (minorities) distributed in more than 50 nations; but see examples of regional cooking in Cuzen (1983), Mars and Tooleye (1984), Mesfin (1987), Murphy (1981). I shall concentrate on the utilization of yams by the Yoruba of Nigeria (and Republic of Benin). Peculiarly, most of the ethnographers and anthropologists were rather indifferent about the details of cooking procedures and very few of them actually gave recipes (and variants). Bascom's two articles (1951a, b) are notable exceptions.

First, I should like to consider briefly the utensils and methods (that is technology in the kitchen), then elaborate on some recent industrial developments.

The Yoruba 'kitchen' of the recent past was of the utmost simplicity: a bare ground on which round-bottomed pots (quite similar to those excavated from 500 B.C. campsites of the Bantu migration) could be set on three stones on an open fire. There were of course other kinds of pots, for example, for carrying water or keeping palm oil. There was no oven of any kind and therefore no baking or roasting. As far as I know, palm oil was never used for deep frying. For me, it is rather peculiar that no spit was used for meat, perhaps because the game meat is too tough and dry. Simple querns - not much more than a slate slab of suitable size (Afolabi Ojo, 1966) - are used for leafy vegetables, capsicum peppers and tomato. The more elaborate wooden mortars are principally for yams, but also can deal with other tubers and plantain. Mortars (and pestles) for dehulling and partially pulverizing grains are of a little different construction, and mostly used by the Northern Yorubas on the border of savanna, where grains were more easily available.

The only staple crops (at least, in the last hundred years) which reached the market in processed form were cassava (as gari) and dried yam slices, which could be pounded to 'elubo' flour. The rural processing of gari and its utilization for different staple foods is well covered in literature (see Ngoddy, 1976). Mechanization of the gari production was also attempted, and it is the best example of the so-called 'appropriate technology' in Nigeria. We are, however, very worried because of the recent upsurge of gari consumption as it is nutritionally very deficient and causes chronic poisoning and contributes to the widespread goitre. This is also widely discussed (Delange, 1974; Imevbore and Boszormenyi, 1983;

Madagwu, 1979; Nestel and MacIntyre, 1973; Olarewaju and Boszormenyi, 1975; Olurin, 1975) - so let me concentrate on yams.

'Real' yams, as I already discussed, belong to the *cayenensis/rotundata* complex. Growing of these yams requires much better soils than cassava, and requires arduous work and costly operations (staking, similar to the production of hops in Europe) so the yam is increasingly becoming a much favoured, but luxury crop. The growing season is 8 - 12 months; the harvest of the new yam starts in the second half of June and reaches its peak in mid-September. Yam tubers are stored in traditional open air 'yam barns', which are vertical wooden frameworks on which the large tubers are tied individually by means of string or raffia. After 2 - 4 months the weight loss from desiccation and rot caused by a combination of microbial and insect attacks amounts to 30 - 40% of the tubers stored in that way. This large storage loss is sharply reflected in the price, which doubles or triples from the harvest till June.

The most favoured form of yams is the pounded yam (iyan), in the form of yam 'loaf' arranged on a separate dish and served with different stews ('soups'). This is a white, slightly elastic (a criterion of freshness!) solid mass, which could be sliced, but of course it is torn and rolled into a mouthful-ball by hand and eaten dipped into the stew. I can say 'objectively' (as an observer coming from a different culinary tradition) that 'iyan' is really good, much better than the kind of 'loafs' made from cassava, cocoyams, sweet potato, breadfruit, plantain or maize.

Pounded yam is made by 15 - 20 minutes of heavy manual work. The yam slices after being boiled in unsalted water, are put into a wooden mortar with just a little of the hot cooking water and pounded with a long heavy pestle. Ideally, it is done by two females, alternately letting the pestles fall into the cooked yam pieces, until they get a perfectly smooth, steaming mash. The time of mashing depends on the quantity and quality of yam, and the expertise of the pounders. When considered done the mash is smoothed into 'loafs' (of about $1/2$kg size), placed into covered bowls and served immediately. No person, who is offered food in a house, is offended by waiting for the pounding exercise, and is actually pleased if time is taken to provide a well pounded yam. Needless to say there are jokes about other human factors contributing to the enjoyment, when the weather is hot and pretty girls tend to do the exhausting job without much clothing. (For the complete truth, I must say that the pounding females are mostly of grand-mother age in African terms). In spite of the esthetic enjoyment of the consumers (not the makers), in the 1970's efforts were made to lighten the work and adapt it to the small kitchens in multistorey apartment houses.

The first yam pounding machine was developed by Professor Makanjuola of Ife University (Makanjuola, 1974). The machine is large and heavy, as it makes pounded yam for eight by using a 1.75 horse-power (H.P.) electric motor. The motor drives by a belt a long cage-like rotor (beater), the leading edges of which are sharpened to a knife edge, in a tall open cylinder. Professor Makanjuola's experimental model was developed further and marketed by Addis Engineering Co. (Lagos). It makes about 1kg processed yam in one run with a power consumption of 1.25 H.P. The knife-blade is attached to the cover and the rather flat, closed container can be lifted and engaged with the shaft of the motor (this arrangement is for safety purposes).

Finally, I should like to mention that Matsushita Electric Trading Co., a well-known Japanese firm, also marketed in Nigeria a special food processing machine for steam cooking and pounding yam. The cooking and pounding is done in the same container in two steps. Instead of knives, a paddle throws the yam pieces to the wall of the container to smash them but (with imported motor) the Japanese machines are rather expensive and I do not think many of them were sold and used. Mrs Anthonio (Anthonio and Isoun, 1982) says that 'with the most powerful mixers' (blenders), pounded yam can be done 'quite successfully, but since very little water is used and the final product is very firm, do not try it with standard mixers.' My mixers are only standard ones, so I never tried it myself, but I have some doubts. The problem is not the strength of the force, but its application, and remixing of stuff attached to the wall of the vessel and avoiding further treatment. I do not know any published study which compared the quality of pounded yam made by the different machines and by the traditional process, but I would not be surprised if machine-made products are less satisfying.

As I already mentioned, pounded yam is a prestige, luxury food particularly in the off season from May to August. The heavy storage losses might have initiated the native technology of yam preservation, which converts it at first to a dried chip, which is then pounded to a flour. This flour ('elubo') can be reconstituted with boiling water to a kind of soft pudding ('amala') to substitute for pounded yam. Unfortunately, 'amala' has a rather repulsive, greyish-brown colour, and is not elastic but somewhat sticky. I confess I cannot eat it; and, even among the native Nigerians, there are many who have reservations.

The rural technology of 'elubo' making consists of washing and peeling the tubers, then slicing thinly (0.5 - 1cm) and sun drying on mats or trays (or on the ground on abandoned roads). Interestingly, already the first Yoruba cookery book (Mars and Tooleye, 1934) recommends 'to wash the slices thoroughly and leave to soak in a bowl of boiling water until the water is cold, before drying under the sun' - this amounts to a kind of parboiling, which could reduce the excessive enzymatic browning (Onuoha, 1981), and in my view it may reflect external influence.

In our attempt at the Department of Food Science and Technology, University of Ife, led by Professor P.O. Ngoddy (Ngoddy, 1979; and Onuoha, 1983) we developed a method, a modern industrial process of eliminating the storage loss, and providing a convenient food ready within minutes, and comparable in organoleptic properties with the traditional pounded yam. I should like to outlike briefly the industrial process. (Full details are available in a feasibility study and can be provided on request.)

On arrival at the factory site, the tubers, after weighing, are unloaded from the lorry. The tubers then may be stored in the traditional barn or in platform storage structures. The tubers are fed to the plant in the order by which they arrive, that is, according to the 'first in, first out' plan.

From the storage, the tubers are taken to a roller inspection table, where they are sorted into identical size (weight) classes for the soaker/washer in which they are cleaned by jets of water and rotating brushes. For a very small plant using 1 ton fresh tubers per hour the peeling could be done manually by employing 15 - 20 operatives (about 50kg/hour rate) but for the suggested larger size plants this has to be done by a caustic/steam peeling system. The surface of the tubers is heated up by jet injection of steam in a vigorously agitated dip tank before they are transferred to a lye applicator. The softened up peel is removed in a following spin peeler/washer unit. The tubers are then transferred to a conveyor belt to remove by hand any rot, sun-burn or other blemish. Then comes a fully automatic cutting operation in two steps (precutting and dicing) which convert the tubers into regular-sized pieces (e.g. 1cm^3 cubes). This is necessary to provide exact conditions for the next two steps of the operation: the sulfite dip and the blanching, steam-cooking, which are the most critical ones in the process. These steps in a small scale operation can be executed in batches, but for a larger plant an automatic continuous system is necessary.

The properly blanched and pre-cooked yam dices are dried in a two step process. The first of these could be a continuous belt drying, but because of economic considerations, semi-continous tray or track drying was proposed (and used). In both cases the dehydration is finished in so-called bin dryers. The small, white, uniform yam cubes (which look much like cube-sugar) are obtained with about 5% water content. In this form the yam is entirely stable, and if kept free from moisture and insects, it can be stored almost indefinitely. This is important because it was envisaged that the first stage of the process would run continuously (two shifts plus maintenance) during the yam season (of about 6 months) but the milling (by a hammer mill) would run through the year (but with a much smaller mill). The product of the mill is a free-flowing, fine flour (110 - 120 mesh) which is then packed into good quality, thick polythene bags.

The reconstitution of the yam flour by the consumer is very simple: by pouring the flour into a pot and adding twice its volume (measured in cups) of boiling water, and stirring with a wooden spoon for about 3 minutes; and then forming into a yam loaf and serving in the traditional way. The instant pounded yam was favourably assessed by taste panels on several occasions and our department was producing it for years for experimental purposes (selling the surplus, which never satisfied the demand). Then we went to the Kaduna Trade Fair in February 1979, set up a small production line (producing 20kg yam flour per hour), a small buffet tasting site (providing stew to go with the yam), a projection room with a slide show with tape-recorded narrative explaining the process, the use and advantages of the instant yam. Without prejudice, I may say, that our performance (Professor Ngoddy, Dr Onayemi and myself with technicians and junior staff members) was a success.

After the performance at the exhibition (sponsored by the National Root Crops Production Co., a state-owned company), a very detailed feasibility study was submitted in 1979. Then nothing happened; as far as I know, none of the proposed 'federal factories' (altogether 16 to satisfy the different states of the federation competing for investment funds) were established. I never saw a published analysis of this 'failure' and did not have the opportunity to discuss it with Professor Ngoddy (who meanwhile left to go to the University of Nigeria, Nsukka and went into the higher administration there). Regretfully, as Hickson (1986, 1987) correctly says, there is no analysis of the African decision making processes, and perhaps it would not be feasible under military regimes. The provisional explanation, which I offer, is my own, and it could be wrong.

I believe that the National Root Crops Production Co., which at that time was interested mainly in cassava products, overexpanded its resources, trying to establish too many factories at too many sites (mostly not chosen by strict economic considerations). The Federal Government changed (twice) and internal credits were exhausted and the external credits of Nigeria dried up. A quality product like 'instant pounded yam' needs imported chemicals (lye, sulfite, fumigants), plastic bags and of course machinery (and spare parts). Professor Ngoddy is an outspoken practitioner of 'intermediate technology' (a variant of 'appropriate technology') that is, whenever it is possible he would like to promote locally manufactured machinery (Ngoddy, 1988). Only a few people realize the limitations of this approach; importation of different units, instruments, tools, specific raw materials (as stainless steel) are still necessary. 'Intermediate technology' as we understand it, is not a rule, which can be automatically applied; the extent of self-reliance depends on the type of product for which it is applied, and on the scale of application. In general terms it means more employees and less automatic machinery. (It was envisaged in the yam feasibility study, that eventually 2% of the total Nigerian yam production - that is about 140,000 metric tons - would be converted into yam flour.) At the time when the feasibility study was made the biggest proposed production unit (Model III) would have required about 2 million Naira total investment. (The Naira at that time stood at about $0.6 US, which was of course artificial and overvalued.) From this 2 million about $1/2$ million would have been for processing and auxiliary equipment. The majority of this equipment would have come from abroad, and an American company was recommended as contractor (or partner in development). Of course international loans would have been required to finance the equipment (particularly if it would have been for 16 factories).

In my view industrial establishments in Nigeria inevitably need imports, even if the raw material that they use is completely produced in Nigeria, as in our case the yams. That means, that they have to be capable of generating some foreign currency, selling some of the products on the international market. The product therefore must be of a good quality, capable of competing on the international market. This requires high quality standards (that is, more imported machinery, chemicals, packing materials, quality control laboratories). A vicious circle from which there is no easy escape for the Third World countries.

I believe that the quality of our instant yam reached international standard, but I do not think that it would have been able to break into, to a significant extent, the North American and EEC 'instant potato' market. Probably some could be sold in West Europe and the USA as a 'speciality food' particularly at the present Naira/Dollar rate. However, the economic situation in Africa (and particularly in Nigeria) changed so much in the last decade, that it would require a new feasibility study to answer this question with taste panels and marketing research overseas.

REFERENCES

Aboyade, O.: *Administering food production prices in Africa: lesson from international experiences.* Internatl. Food Policy Res. Inst., Washington, 1985.

Afolabi Ojo, G. J.: *Yoruba culture.* Univ. Ife Press, Ile-Ife, 1966.

Anthonio, H. O. and M. Isoun: *Nigerian cookbook.* Macmillan, London, 1982.

Aribasala, T. S. B.: *Nigeria's Green Revolution. Achievements, problems and prospects.* Green Revolution Natl. Committee, Ibadan, 1983. Mimeogr. 82 pp.

Barton, I. M.: *Africa in the Roman empire.* Ghana Univ. Press, Accra, 1972.

Bascom, W. R.: Yoruba food. *Africa* 21: 41-53, 1951.

Bascom W. R. Yoruba cooking. *Africa* 21: 125-137, 1951.

Boorstin, D. J.: *The discoverers.* Penguin, Harmondsworth (1983), 1986.

Boszormenyi, Z.: *What is Green Revolution?* 13th Ann. Conf. Nutr. Soc. of Nigeria, Ibadan, 28th Nov. 1980. Mimeogr. 37 pp.

Boszormenyi, Z.: *The real Green Revolution.* Lecture for the Inst. of Ecology, Univ. Ife, 7th July 1983. Mimeogr. 55 pp.

Braudel, F.: *Civilization and capitalism.* Vols. I - III. Collins, London, 1981-84.

Carr, E. H.: *What is history?* Penguin, Harmondsworth (1961), 1964.

Cuzen, P. (Ed.): *Cooking in Botswana.* Printing and Publ. Company Botswana, Gaborone (1980), 1983.

Davidson, R. B.: How many Nigerians? *West Africa*, Aug. 20, 1979.

Delange, F.: *Endemic goitre and thyroid function in Central Africa.* Karger, Basel, 1974.

Delano, I. O.: *Yoruba proverbs - their meaning and usage.* Oxford Univ. Press, Ibadan, 1966.

Douglas, M.: Population control in primitive groups. *British J. Sociol.* 17: 263-273, 1966.

Douglas, M.: *Exciting breakthrough: the UNIFE process for the manufacture of instant pounded yam flour.* Promotion leaflet for the Kaduna Trade Fair, February 1979 4pp.

Gillon, W.: *A short history of African art.* Penguin, Harmondsworth (1984), 1986.

Greenberg J.: *The languages of Africa.* Mouton, The Hague, 1966.

Guthrie, M.: *The classification of Bantu languages.* Internatl. Afr. Inst., London, 1967.

Gutteridge, W. E.: *Military regimes in Africa.* Methuen, London, 1975.

Harlan, J. R.: *Crops and man.* Amer. Soc. Agron., Madison, 1975.

Harman, N.: The most African country. *Economist*, Jan. 23, 1982.

Harris, M.: *Breakfast in hell. A doctor's experiences of the Ethiopian famine.* PAN Books, London, 1986.

Harrison, P.(Ed.): *Land, food and people.* FAO, Rome, 1984.

Hickerson, N. P.: *Linguistic anthropology.* Holt, Rinehart & Winston, New York, 1980.

Hickson, D. J. et al.: *Top decisions.* Blackwell, Oxford, 1986.

Hickson, D. J.: Decision-making at the top of organizations. *Ann. Rev. Sociol.* 13: 165-192, 1987.

Horton, M.: The Swahili corridor. *Sci. Amer.*, 25(3): 76-84, 1987.

Igun, A. A.: *Population - the primordial and universal denominator* (Inaugural lecture). Univ. Ife Press, Ile-Ife, 1975.

Imevbore, A. M. A. and Z. Boszormenyi: *Iodine supply and goiter in Nigeria.* 2nd Natl. Workshop on the Internatl. Drinking Water and Sanitation Decade, Owerri, 24th Feb., 1983. Mimeogr. 19 pp.

Iyer, P.: Ethiopia - the land of the dead. *Time*, Nov. 26, 1984.

Iyer, P.: Just one day later, this boy was dead. *Daily Telegr.*, Oct. 5, 1988.

Kay, D. E. (Ed.): *Food legumes.* Trop. Products Inst., London, 1979.

Kay, D. E. (revised by E. G. B Gooding): *Root crops.* Trop. Devel. & Res. Inst., London (1973), 1987.

Kay, D. E.: *Lessons to be learned : drought and famine in Ethiopia.* Oxfam, Oxford, 1984.

Madagwu, E. N.: Cyanide content of gari. *Toxicology Letters* 3: 21-24, 1979.

Makanjuola, G. A.: A machine for preparing pounded yam and similar foods in Nigeria. *Appropriate Technol.* 1(4): 9-10, 1974.

Marcaniak, A.: Forecasting Nigeria's population. *Daily Times*, Feb. 15, 1979.

Marwick, A.: *The nature of history.* Macmillan, London (1970), 1973.

Mars, J. A. and E. M. Tooleye: *The Kudeti book of Yoruba cookery.* CSS Bookshops, Lagos (1934), 1979.
McIntosh, S. and R. McIntosh: The secrets of ancient Jenne. *Topic* No. 144, 20-31, 1982.
McIntosh, S. K. and R. J. McIntosh: Current directions in West African prehistory. *Ann. Rev. Anthropol.,* 12: 215-258, 1983.
Mesfin, D. J.: *Exotic Ethiopian cooking.* Ethiopian Cookbook Enterprise, Falls Church, 1987.
Mintz, S. W.: *Sweetness and power.* Penguin, Harmondsworth (1985), 1986.
Murphy, P. K.: The food and cooking in rural Gambia. In A. Davidson (Ed.): *National and regional styles of cookery.* Prospect Books, London, 1981.
Nestel, B. and R. MacIntyre (Eds.): *Chronic cassava toxicity.* IDRC, Ottawa, 1973.
Ngoddy, P. O.: Gari mechanization in Nigeria: the competition between intermediate and modern technology. In N. Jequier (Ed.): *Appropriate technology. Problems and promises.* OECD, Paris, 1976.
Ngoddy, P. O.: *The industrial manufacture of instant pounded yam flour. A feasibility study.* National Root Crops Production Co., Enugu, 1979, 103 pp.
Ngoddy, P. O.: *Technological issues and strategies in the development of Nigeria's food industries.* Keynote address. 12th Annual Conf. Nigerian Inst. Food Sci. and Technol., Maiduguri, 18th Oct. 1988. Mimeogr. 14 pp.
Ngoddy, P. O. and C. C. Onuoha: *Selected problems in yam processing.* Symp. on Yam Biochemistry. Anambra State Univ. of Technol., Enugu, May 3, 1983, Mimeogr. 34 pp.
Oji, M. K. (Ed.): *The Nigerian ethical revolution 1981-2000 A.D.* Van Boekhoven Bosch, Utrecht, 1982.
Olarewaju, O. C. and Z. Boszormenyi: The process of detoxification and residual cyanide content of commercial garri samples. *West Afr. J. Biol. and Appl. Chem.,* 18: 7-14, 1975.
Oliver, R. and B. M. Fagan: *Africa in the Iron Age, 500 B.C. to A.D. 1400.* Cambridge Univ. Press, Cambridge, 1975.
Olurin, E. O.: *The fire of life - the thyroid gland* (Inaugural lecture). Ibadan Univ. Press, Ibadan, 1975.
Onuoha, C. C.: *A study of enzymes in Nigerian yam species.* Ph.D. Thesis, Dept. of Food Science, Leeds University, Leeds, 1981.
Phillipson, D. W.: The spread of the Bantu language. *Sci. Amer.,* 236(4): 106-114, 1977.
Phillipson, D. W.: *Pounded yam machine by ADDIS.* Addis Engineering, Lagos, Promotion leaflet, 2 pp.
Ramseyer, U.: The origin and history of rice: from shifting agriculture to wet rice cultivation. *Swissair Gazette* 1989 (2): 17-18.
Shaw, T.: The Nok sculptures in Nigeria. *Sci. Amer.,* 244(2): 114-123, 1981.
Simmonds, N. W. (Ed.): *Evolution of crop plants.* Longman, London, 1976.
Stanton, W. R.: *Grain legumes in Africa.* FAO, Rome, 1966.
Twose, W. R.: *Cultivating hunger.* Oxfam, Oxford, 1984.
Ullendorff, E.: *The Ethiopians. An introduction to country and people.* Oxford Univ. Press, Oxford (1960), 1973.
van Noten, F. and J. Raymaekers: Early iron smelting in Central Africa. *Sci. Amer.* 258(6): 84-91, 1988.
Vavilov, N. I.: *Studies on the origin of cultivated plants.* Inst. Applied Botany and Plant Breeding, Leningrad, 1926.
Walsh, W. H.: *An introduction to philosophy of history.* Hutchinson, London (1951), 1967.

The Ever-evolving Store-cupboard

By Suzy Benghiat

From the suggestions made in the November 1988 circular, I chose the wider approach as it was set out in the following sentence: "It has also been suggested that we could include work on substances which, while not the basic source of nourishment...are essential ingredients of a particular cuisine, for instance something that everyone would have in their store-cupboard" but my paper is about the 'evolving store-cupboard'. It fits in with a subject I am particularly interested in, i.e. the evolution of eating habits and the ever-increasing speed at which this evolution occurs. This paper is based on experience, my own and that which I have been observing in the past few years.

Let us examine first the criteria that induce us to modify what our own essential ingredients are even within 'a particular cuisine'. These depend on a number of objective and subjective factors.

Objective factors: 'the store cupboard' itself. It could be a cellar, a larder, a kitchen unit, a refrigerator, or a freezer. The size of the storage space is also important; so is the way supplies are purchased, from a mobile unit, a corner shop, a supermarket, or a hypermarket. We must also consider means of transport, budget, time for shopping, and time for cooking.

What you eat is so directly linked with the kind of life you lead that, even within a defined type of cuisine (and by that I do not mean defined by nationality but by your own style of cooking and eating), the essentials vary as your life evolves. The cupboard of a single person is not the same as that of a childless couple, a couple with babies, with growing children, with young adults at home, or away from home. These living conditions are so important that it is quite difficult to adjust when there is a sudden change in your situation. It takes time to re-organise your life around that change and it shows in your store-cupboard - it is either too full or not full enough; and it often contains the wrong items.

Some subjective factors: do you like cooking; is yours an open house; or do you entertain only by invitation or not at all.

Furthermore, one must make a distinction between two types of staples:

1) the staples you need for your day to day cooking. In addition to the main ingredients, these would include herbs, spices, condiments. But also some convenience items. In France, for instance, there is quite a tradition of using bottled or tinned items for convenience, but it is also often in order to give a luxury touch to ordinary food. For example a rich consommé or a bisque de homard to add to a sauce or a soup.

2) emergency stores: to fall back on when your usual supplies have been exhausted or when you have unexpected guests. It will contain a larger proportion of convenience foods. These are what I call the improvisation side of the cupboard - extremely useful when adapting to changing circumstances.

In the last couple of decades the increasing availability of home freezers, the supply of pre-cooked foods, and the introduction of micro-wave ovens have revolutionised even the concept of 'essential food items to store'. Of course, the timing and pace of these changes vary according to where they are taking place. In the last couple of years, the pace in France makes you quite dizzy with, on the one hand the big offensive of the junk food brigade - by that I do not mean only the junk you see, but all that goes into manufactured food items - and on the other hand, the famous chefs rushing to cook for frozen and vacuum-packed food manufacturers. The food industry is attacking the French with a pair of pincers: on one side the junk, on the other the luxury, as if to catch up on wasted time.

Incidentally I have recently noticed a striking new development, which will - if it goes on as it has - reduce the contents of the store cupboard to the bare minimum. It is the feverish consumption of snack biscuits, bars, etc. It is interesting to see the enormous advertising campaign (especially on television) aimed at the young and successful urging them to crack, or pop, or snack, a bar of this or that to become even younger and more successful! This is relatively new in France but, judging from the displays in most shops and supermarkets it must be pretty successful.

But this is another story - or is it?

Survival Kit
(16th Century Seamen's Fare)

by Maggie Black

Although, surprisingly, there is no mention of food as a staple in either my S.O.E.D. (1936) or Webster's New Collegiate Dictionary (1965), there seem to be two ways of defining a staple food. The first (S.O.E.D.) is as a commodity having the chief place among the articles produced or consumed by a given group, implying that these processes are matters of choice, usually that the food is produced in return for cash or is purchased. Thus we can say that the 17th-century English upper classes bought and ate meat as a staple of their diet. On the same tack, Webster states that a staple is a commodity in constant demand, something used, needed or enjoyed by many individuals. The Old French and medieval Latin origins of the word (which mean a purchasing place or its main support) uphold the concept that a staple is a vital negotiable product, the source from which a group gets its living, whether or not they eat it themselves.

There is, however, another meaning of *staple* in which the element of choice is missing. Instead, climate-plus-terrain or poverty through deprivation determines the principal, i.e. the staple foodstuff which a group eats. Mealie-meal (ground maize) is the staple foodstuff of most South African Blacks for instance but not from choice. It only became the basis of their diet in the 19th century when the spread of White control and White farms reminiscent of 14th-15th century enclosures in Britain denied them the hunting areas and much of the grazing land of their mixed economy.

Climate has a more radical and abiding effect on what folk use as staple foods than most conquerors[1]. Seasonal rains or, in temperate zones, the warming of the earth in Spring dictate what quickens. Since the earliest recorded times, European master and man alike have practised rituals to bribe or bully the forces of Nature to 'do their stuff', and any peasant in the pre-industrial past watched passionately at winter's end for the first greening of the fields, seeing the small stock in his grain-pit shrink day by day. The priorities of a Third World mother who does not know where her children's next meal will come from if her store runs out or moulders, are still as narrowly focussed. All she seeks is a filling food which will not decay before it is needed and which - more important - is reproducible. This and water are what she deems the essential constituents of her family's diet. Its staples. Its survival kit.

These restatements of the obvious suggest that other historical modern parallels than the examples I have given would be worth exploring. But my purpose here is a very modest one. I have always been interested in travellers' food, and this year's topic has given me a chance to start writing about it - because if ever there were times in the past when men were reduced to what Danny Kaye called 'the simple bare necessities of life', it was as travellers. The soldier in unfriendly country, dependent on what he carried in his back-pack or saddle-bags; the wandering pilgrim or leper - or, worse off still, the branded, exiled murderer taken from sanctuary - with nothing but his scrip and perhaps a begging bowl; a small ship's company in a tiny Tudor vessel, far from land and beyond reach of succour if supplies ran out or rotted.

These last knew well that their supplies were all they had to live on until they made landfall, or - since many English ships doubled as itinerant merchantmen[2] and privateers - until they found a well-stocked foreign prize. Under Elizabeth, there was always a fair chance of it on the West Indies run; the known routes and Caribbean anchorages were thronged all summer by the 1570s. But they might be out of luck, be forced to seek water and rummage the ship on an empty shore or even be driven to winter there by storms or shipwreck.

They might well, in any case, winter in the islands, perhaps trading or looting, living 'off the land' to conserve their remaining stores against need.

What food supplies would a little ship of Elizabeth's time carry for its crew on a Caribbean escapade of this kind?

A lot depended on the contract which the ship's captain made with the men he recruited. He provided the victuals and recouped himself from the profits of the enterprise before the share-out; and he might be mean or open-handed by nature, alert or easily cheated by the contractors who supplied him. Their dishonesty was notorious, but a good and vigilant master would spot and refuse short measures, rancid or mouldy goods, unseasoned and therefore leaky barrels[3], and sour beer. He might even add a few medicines to the basic stores, whereas a meaner man would skimp his purchases.

Sheer space limited even the basic supplies. Bear in mind that tackle, tools, and pitch, sails, arms from arrow-heads to cannon-balls, firearms and fuel all had to be carried as well as sale goods, gifts and trinkets as bribes. Most of the privateers' ships were very small - Drake raided the Spanish Treasure Trains in 1572 with one ship of 70 tonnes and another of 25 tonnes[4]. Where they stowed their gear is a mystery. However, for comparison, a naval ship of 100 tons carrying 200 men needed 83 tons of space for 2 months' provisions. In another case, the stores for naval personnel being fed at a contract rate of 5d a day when at sea, were deployed thus: half the available space was given to the beer, a quarter to wood and water, and the rest to the solid food.

Beer was by far the most important staple in the English seaman's diet. It was his only drink, and he held strong views on it. Lord Howard of Effingham complained when recruiting seamen to sail against the Armada, 'I know not which way to deal with the mariners to make them rest content with sour beer.' The discontent was understandable. All fresh water, even the purest, was mucky, not deemed a fit drink even then - and it was in any case needed for washing the clothes and cookware, and, vitally, for soaking the heavily salted protein foods and hard dry pulses and biscuit. So men at sea, in the salty air drank- and needed to drink - quantities of of the mild 'small beer'[5] which was the daily issue.

As for their solid foods, our most coherent accounts of them come from the lists of stores bought and issued to both the English and Spanish Armada fleets, which are so similar that we can take them as being a more or less common international diet, even though the precise products and quantities vary. The daily English issue per man, as recorded by Lord Burghley in February 1588, was:

Every Day
1 lb biscuit
1 gallon beer

Sundays, Tuesdays, Thursdays
2 lb salt beef

Mondays
1 lb bacon
1 pint pease

Wednesdays, Fridays, Saturdays
¼ stockfish or ⅛ salt ling
4 oz cheese
2 oz butter

The Duke of Medina Sidonia recorded the basic Spanish issue as:

Every Day
1½ lb biscuit or 2 lb fresh bread
1⅓ pints wine or 1 pint Candy wine (Cretan Malmsey reputed the best)

Sundays, Thursdays
6 oz bacon
2 oz rice

Mondays
  6 oz cheese
  3 oz beans or chickpeas
Wednesdays, Fridays, Saturdays
  6 oz salt fish (tunny or cod, squid or 5 sardines)
  1½ oz oil (olive)
  ¼ pint vinegar
  3 oz beans or chickpeas
Wednesdays only:
  6 oz cheese

Wine and oil rather than beer and butter are no surprise, but it is sad that no flavourings were admitted as necessaries by either side. They were certainly carried. We know that an English whaler in 1575 carried mustard seed (for pickling fish?) and that the Armada itself shipped near 50,000 strings of garlic.

The foods deemed essential to the European seaman's diet thus consisted only of heavily salted meat and fish, hard cheese, oil or very salt butter, pulses and hard-baked, ground grain (biscuit). It is no wonder that scurvy was a killer, although some shrewd captains already had an inkling that fresh foods might relieve it. What is strange is that the men themselves accepted it, like dysentery and typhus, as a fact of shipboard life, attributing it to salt food, sea air and - yes - sour beer, against the obvious evidence that their land-based fellows (especially prisoners whose diet was almost identical with theirs) were also in at least a pre-scorbutic condition by the time winter (or their sentence) ended[6].

The sorry story of years spent and lives lost in searching for a fail-safe cheap antiscorbutic after 1600, when Captain James Lancaster tried lemon juice, is well known; even after 18th-century scholars especially James Lind by his experiments had proven its worth by mid-century, the only change in the 1588 Navy rations by 1745 was that 1 lb of pork replaced beef on Sundays and Thursdays, beef or cheese replaced fish, and oatmeal was added to the (halved) ration of peas. The sailors who manned the First Fleet which went to Australia in 1780 got the same diet plus vinegar as an antiscorbutic.

The overall quantity of food per man was equivalent to 5000 calories a day on paper, but it was in fact much reduced because the quality was so appalling. Nonetheless it was nowhere near the irreducible minimum. At a long-term encampment or settlement, people's diet only fell below that level when they had neither hope of future supplies from seed-corn or breeding stock nor of supplies from home[7]. Such was the plight of the first colony planted in Virginia in 1585. Privateers were more resourceful - and more optimistic. Drake, sailing off a hostile shore with two small boats in a storm, and with only a gammon of bacon and 40 lb of biscuit for his 24 men nonetheless told them to trust in God and sail on. At other times, with brash and sometimes rash insouciance, he and his fellows set about trying new foods; and although they lost some good men, it was remarkable how many survived. It may have been due to the gambler's or fighting instinct which had made them take ship in the first place - the instinct to beat the system if possible in a life which was as brutish and short at home as it was half a world away. They may just have felt that in seeking a survival diet, it was a case of 'nothing venture, nothing win'. Or they may have believed, as Captain Drake did himself, that they were in the hand of a benevolent, Protestant, pro-British Almighty who would not let them fall.

Notes

1. Conquerors or others can beat climate by creating a dust-bowl as in the rain-forests of South America.
2. The famous Sir John Hawkins, Treasurer of the Navy and third in command of the Armada was, earlier, a well-known merchant and slave-trader, for which activity the Queen lent a ship.
3. Drake's raids along the Spanish coast in 1587, in which he destroyed many small cargo boats carrying seasoned barrel staves for the Armada was a punishing blow to its chances. Often, the leaky casks used proved to contain only few inches of scummy water.
4. Compare these with the Mary Rose (700 tonnes).
5. Made from the last strainings of the mash-vat, 'small beer' contained 150 calories per pint (modern ¾ pint).
6. *The Englishman's Food* by J.C. Drummond and Anne Wilbraham (Cape 1958) compares the value of the diets of sailors, paupers and prisoners (Appendix A).
7. Reproductive capacity in a staple foodstuff was only important for those with a life expectancy of at least one growing season in the same place. It was then vital; and the irreducible minimum of the staple was the least one could manage on without consuming one's stock for reproduction. Once that was eaten, one had to abandon camp or die (assuming there was no suitable local food to use as a substitute).

Some Books Consulted:

The Englishman's Food. Drummond and Wilbraham. Cape 1958.
Administration of the Royal Navy. M. Oppenheim. John Lane/The Bodley Head 1896
Drake's Raids on the Treasure Trains 1572-3. Ed. J. and J. Hampden. Folio Society. 1954.
Voyages and Discoveries. R. Hakluyt. Ed. Beeching. Penguin Classics. 1987.
The First Colonists. Hakluyt et al. Illus John White. Folio Society 1986.
History of the Royal William Yard. L.W. Stephens. nd.
The Spanish Armada. Martin and Parker. Hamish Hamilton. 1988.

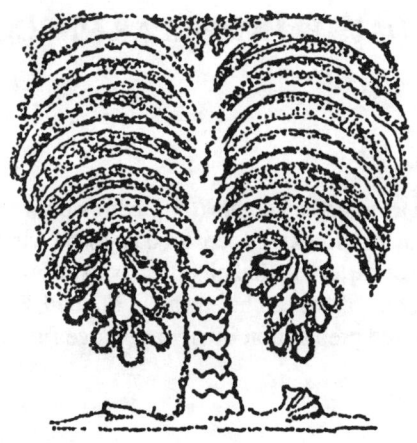

DATE-PALM MOTIF

circa 520 AD

Author's sketch of a pier capital excavated in Istanbul at Saraçhane, the site of the Byzantine church of St. Polyeuktos, built under the patronage of Anicia Juliana.

The Date-palm: Pillar of Society

By Holly Chase

"And did you not love Ishullanu, the gardener of your father's palm grove? He brought you baskets filled with dates without end; every day he loaded your table." - THE EPIC OF GILGAMESH [1]

Found during excavations at neolithic settlements throughout the Mideast, date pits attest to human consumption of the fruit at least seven thousand years ago. Precisely when and where man and the date-palm began their symbiotic relationship is unclear, but by the third millenium B.C., the epic of Gilgamesh paid tribute to the date-palm in Sumerian, a language that possessed nearly "one hundred and fifty different words for the various kinds of [date] palms and their different parts." [2]

Under the Code of Hammurabi (Babylonian legislation based on earlier Sumerian law), date-growing, share-cropping, and the sale and cutting of palms were subject to official regulation. [3] For ancient Mesopotamians dependent on grains and pulses for the sheer bulk of their diet, dates were also a staple. A high sugar content, pleasant flavor, and prolonged resistance to spoilage would have made dates an appealing food, but farmers also valued the palm for its other products and properties.

Along with the olive, the date is a botanical marker of the boundaries of the geographical and cultural world known as 'the Mediterranean'. Although (with a productive life of approximately 75 years), the date cannot outlast the olive, it can, in its shorter life, bear more fruit, supply more calories, and facilitate the growing of secondary crops in a fashion unsurpassed by any other commonly cultivated tree. Every part of the date-palm can be utilized for human benefit. To this day, rural communities in North Africa and the Middle East are sustained by palms. The intense human efforts necessary for their successful cultivation have helped to shape the societies in which the the date-palm is, without hyperbole, the tree of life.

Well into this century, two very different Mideastern societies have been dependent on the date to the point that a lack of dates constituted famine. The settled peasantry, engaged in the labor-intensive cultivation of date gardens, harvested the fruit and other parts of the palm for their own consumption.

Surplus dates could be bartered to nomadic herders (the Bedu) who would supply the peasants with milk products and, perhaps wool, from their flocks. Except for dairy products and what meat they obtained from their herds, the Bedu had to buy or barter for all their other foodstuffs - flour for bread, rice, sugar, coffee, tea, and dates. Of those staples, only dates can be eaten out of hand. Except for sugar (used with tea and coffee), all the other foods required preparation involving scarce fuel and water .

***

The date-palm's botanical appellation, *Phoenix dactylifera*, is suggestive of fable. In ancient Greek, *phoinix*, 'date-palm', was also the name of the legendary bird of the Arabian desert, the phoenix, supposed to immolate itself every six hundred years and then rise, rejuvenated, from its funeral pyre. Derived from *phoinos*, meaning 'red with blood', *phoinix* also meant 'crimson'. Scholars have speculated, without conclusion, on whether the palm gave its name to the land of Phoenicia, or Phoenicia its name to the palm, once grown extensively in Palestine. ("Jericho, the city of palm trees". Old Testament; Deuteronomy 34:3) Since the tree reproduces itself by means of off-shoots which sprout (like verdant flames) at the base of the trunk, and bears fruit that (in many varieties) becomes a purplish-red as it ripens, it is easy to grasp how the tree might have become associated with the myth of fiery rebirth and the mysteries of blood.

Dactylifera simply means 'finger-bearing', a reference the long, oval date fruits.

Literary References to Dates

Like the olive, the date has long been eulogized for its economic value as well as its stately beauty. Solomon chose both trees for the embellishment of the house of the Lord: "The two doors were of olive tree; and he carved upon them carvings of cherubims, and palm trees and open flowers,... and spread gold upon the cherubims and palm trees." (1 Kings 6:32)

For the harvest celebration of Succoth, the Jews were told to "fetch olive branches... and palm branches...and make booths." (Nehemiah 8:15)

Hebrew for 'date', the feminine name Tamar evoked the elegance of the tree and perhaps had associations with fertility. "...Absalom the son of David had a fair sister whose name was Tamar". (2 Samuel 13:1)

"O prince's daughter... This thy stature is like to a palm tree, and thy breasts to clusters [of dates]...I will go up to the palm, I will take hold of the boughs thereof..." (Song of Solomon, 7:1, 7, 8).

"The righteous shall flourish like the palm tree... They shall still bring forth fruit in old age; they shall be fat and flourishing." Psalms 92:12, 14)

Palm fronds, members of a tree with regal associations going back to ancient Mesopotamia, were waved by the throng hailing Jesus as King of Israel (St. John 12:12, 13). As symbols of submission, rebirth, and thus, spiritual salvation, they were also held by the multitudes standing before the throne of God. (Revelation 7:9, 10).

***

OTTOMAN PRAYER HANGING

17th century (?)

Silk embroidery on moiré. In addition to the border of fruit-laden date-palms, note the palms within the 'footprints of the Prophet' at the bottom of the prayer-niche and the frond-like ornaments atop the columns at either side of the arch. (Collection of Topkapi Saray Museum, Istanbul.)

"For them there is a known provision, fruits. And they will be honored in the gardens of delight." (The KORAN, Sûrah XXXVII:41, 42, 43) It is in the Koran, which repeatedly promises the pious an afterlife in "gardens underneath which rivers flow," that we are given ideas of the culture and uses of the palm in Arabia in the seventh century A.D. Waves of Islamic expansionism took dates, along with Arab culture and agrarian technology, far beyond the Arabian Peninsula, across North Africa. There, and in areas like Mesopotamia and the Nile Valley, where the date had been cultivated for centuries prior to Islam, it gained in significance by virtue of its numerous associations with Arabia as the birthplace of Islam, and in particular, with the customs of the Prophet Muhammad, who is said to have satisfied himself in lean times with dates and barley bread. Various sources relate his fondness for fresh-ripe (rutab) dates, dates with cucumbers, and a confection of dates, butter, and milk.[4] Among the customary foods taken to break each day's fast during the month of Ramadhan are dates. When assessing their place in Middle Eastern diets, one must hold dates' acquired sanctity no less significant than their high glucose and fructose content. The palm itself, which today is posed over crossed sabres as the official blazon of Saudi Arabia, remains a symbol of Islamic culture.

***

The planting of palms not only for their own products but also to shade secondary crops unable to withstand the desert sun is an old and persisting practice: "We had assigned them two gardens of grapes, and we had surrounded both with date-palms and had put between them tillage." (Sûrah XVIII:33)

The Koran is replete with references to Allah's gift of water. Around the palm oases, settled Arabs developed complex subterranean irrigation systems that minimized evaporation and recycled water. "And in the Earth are neighboring tracts, vineyards and ploughed lands, and date-palms, like and unlike, which are watered with one water. And we have made some of them to excel others in fruit. Lo! Herein verily are portents for people who have sense." (Sûrah XIII:4)

"And we have placed therein gardens of date-palms and grapes, and we have caused springs of water to gush forth therein that they may eat of the fruit thereof..." (Sûrah XXVI: 34)

Long ago farmers understood that both male and female palms are necessary to produce a crop of dates. "Glory be to him who created all the sexual pairs of that which the earth groweth, and of themselves, and of that which they know not!" (Sûrah XXXVI:35).

Not only an essential food for Arabian peasants and nomads, dates, especially at a certain stage of ripeness, have been considered medicine. Extreme sweetness, which has made them, by association, requisite food at weddings and other joyous, 'sweet' occasions, also makes them an appropriate palliative for acute pain. The Koran relates the story of the Virgin Mary: "And the pangs of childbirth drove her to the trunk of the palm tree." A spirit directs her to "shake the trunk of the palm-tree toward thee, thou wilt cause ripe dates to fall upon thee. So eat and drink and be consoled." (Sûrah XIX: 23, 25, 26) Throughout the Islamicized world, expectant and nursing mothers are encouraged to maintain their strength by eating dates.

DATE CULTIVATION

To bear abundant fruit, date-palms must be grown in hot, sunny, arid regions in soil capable of absorbing water to a depth of at least two metres. Prolonged summers of high heat and low humidity are necessary for optimum ripening, as rainfall or night-time humidity after fruit-set can create conditions that make dates vulnerable to attack by fungi, bacteria, and insects. Ironically, the tree is dependent on irrigation and requires vast amounts of water. "An Arab proverb claims that the date-palm has to have its feet in the water and its head in the fire in order to flourish." [5]

The date is dioecious, with male and female flowers blooming on separate trees. Far more effective than natural pollination by wind or insects, manual pollination, the insertion of a few male flowering stalks amongst the flowers of a female palm, has been practiced for millenia. Assyrian reliefs (9th century B.C.) depict winged genii holding male palm spathes as they stand before The Sacred Tree. One male palm produces enough pollen to fertilize about fifty female floral clusters. Consequently, growers thin out male saplings, leaving only enough for pollen production.

Irrigation and protection from full sun are crucial when the palm is young. After the palm is well established and has sent down its tap root, it can withstand some drought. Constant irrigation usually leads to salinization of upper soil layers, but the date is quite tolerant of salt. Some date groves along the Persian Gulf are watered by tides.

In Iran and the Gulf countries, the ancient *qanat* or *aflaj* systems have long taken advantage of gravity to bring water from higher elevations to lower settlements by means of deeply dug underground channels. Requiring constant maintenance to keep them clear of obstructing rocks and mud, many of these systems are still watering date gardens. Once water reaches cultivated land, its use is strictly regulated by the community. (The gnomon, a primitve sundial, continues to be used to determine water-time in Oman.) Under the shade of the palms, evaporation is minimized and water flows in narrow, open ditches which are opened or closed off (often by pieces of palm-trunk) depending on the water allotments of individual farmers. The palms drink deeply and crops with shallower roots are planted as ground cover beneath their canopy. Citrus, guava, and stone-fruit trees, tomatoes, onions, beans, eggplants, and peppers thrive in the filtered light of the date gardens.

The growing of a nitrogen-fixing cover crop like alfalfa exemplifies the tightness of the date-centered ecosystem: If plowed under, alfalfa is a 'green manure' that fortifies the soil. More often, it is cut and replanted several times a year to provide fodder for donkeys and camels that are used to pull plows, turn water-wheels, and transport food and other necessities for the peasantry. The animals' manure is either spread upon the soil as fertilizer or gathered for use as fuel, to supplement the dry prunings of palms.

In addition to pollinating, watering, and fertilizing the palms, farmers must periodically remove older leaves because they are needed as raw material or because they are keeping too much sunlight from ripening dates. Unlike northern fruit trees, which can be pruned to keep their fruit within arm's or ladder's reach, palms' vertical growth makes harvesting and pruning increasingly more difficult. Since it has but one growing bud, a palm may not be 'trimmed back'. Another task requiring tree-climbing is cluster-thinning, which ensures better air circulation amongst the ripening dates and promotes growth of larger fruits. (The procedure is similar to grape-growers' handling of fruits on the vine.) In the Sudan, the date-climber achieves a lofty social status commesurate with the risks of ascending unbranched trees that may reach 30 metres in height. [6]

Depending on genetic and environmental factors, an adult palm bears from 50 to 160 kilograms of fruit each year. Dates take approximately six months to ripen to the dried fruit best known in North America and Europe. However, the Arabs recognize at least four different degrees of ripeness. At Aqba, while assisting the Arab revolt during World War I, T.E. Lawrence complained of "green dates [which] loaded the palms overhead. Their taste, raw, was nearly as nasty as the want they were to allay. Cooking left them still deplorable, so we and our prisoners sadly faced a dilemma of constant hunger, or violent diurnal pains more proper to gluttony than to our expedient eating." [7]

Date-palms at Indio, Calif. ready to harvest. The paper covers are put on to protect the fruit from rain during the ripening season. (Photo: California Date Growers Association)

From the green or *kimri* stage, the fruit swells to its greatest size and turns yellow, red, or purplish in the *khalal* phase of medium sweetness and slight astringency; Arabs enjoy several date varieties at this crisp stage. In the *rutab*, or 'moist' state, the date softens and wrinkles, turns brownish, loses some of its moisture, and develops a higher proportion of sugar to total weight. For most varieties, this is the point of greatest perishability, when the fruits must be quickly consumed or allowed to ripen and dry further to become *tamr*. At this point most dates can be stored in a dry place for months, if not years.

Though date pits germinate readily, the palm rarely bears true to seed. Trees grown from seed may yield good or indifferent fruit, but the date farmer would prefer to expend his energy and precious water on a tree with known potential. The only way to ensure propagation of a desirable tree is to cut away and plant the off-shoots which spring from the base of the trunk. The commercial date groves of California were all started from off-shoots collected in the Middle East in the early 1900's. Many of those palms, selected in the oases of southern Algeria and Tunisia, were packed in water-soaked palm fiber and strapped to camels for the long trek to coastal towns from which they were then shipped to California. The survival rate was remarkably high and suggests that while the majority of date groves in the ancient world probably originated from seedlings, cargos of live palms might have been transported long ago.

Though the Moors are credited with working the largest stand of dates in Europe, which graces the Spanish town of Elche near Alicante, dates had been introduced to Iberia centuries earlier by the Phoenicians.[8] In their ships laden with amphorae of olive oil and sacks of dry dates, were there possibly live palms? The Romans enjoyed dates, and they were an important crop in the Roman colony of Egypt, where many varieties were grown, and a type known as 'Syrian' was especially prized.[9]

USES OF THE DATE PALM

Their are literally hundreds of date cultivars providing a wide range of textures, flavors, and degrees of sweetness. The driest and least sweet, so-called 'bread dates' are consumed in quantity by peasants and Bedu, those for whom dates provide a large part of their daily nourishment. Choice, sugary dates are marketed in the cities. Left whole and unpitted or pitted and stuffed with nuts or clotted cream, they occupy an exalted place among Middle Eastern sweets. Softer or less perfect fruits are frequently pressed (by foot-stomping in the Sudan[10]) into huge, compact blocks that keep well and can be carried by train, ship, or camel. More refined date pastes made of fruits from which pits and surrounding fibers have been removed, are sometimes served with the accompaniment of butter (goat ghee in Oman). The Sudanese add hard, dry fruits of the local Barakawi date (whose apt name means 'blessing') to their water containers when they journey. The result is a sweet and refreshing beverage.[11]

Pressed from ripe dates, syrup, or *dibs*, is employed like honey, as a sweetener and as a preservative for other foods. In the Saudi oasis of al-Hasa, where even individual homes have special rooms called *jussah* for the storage of different kinds of dates, *dibs* is made by an ingenious method: A "storage building, called a *fida*, is a tower-like structure with a wooden trap door at the bottom, into which dates are dumped from the top. Due to the action of the heat, progressive maturing of the dates, and particularly their weight, the dates are pressed against the wooden floor, and a thick syrup, like molasses, is squeezed out." [12]

Date pits are milled and used in place of, or to extend grain flours. This may be done for flavor preference, as in bread baked at Hofuf[13] in Saudi Arabia, or out of necessity, during famine among the Baluchis in southwestern Iran.[14] Date pits may also be cracked, soaked, and fed to livestock. Dying palms or those with inferior fruit can be tapped and their sap drunk fresh (like sugar-cane juice) or fermented. In pre-Islamic Arabia, palm wine was widely consumed.[15] Nomadic tribes in western India today depend on a nourishing sort of beer which they brew from the sap of *Phoenix sylvatica*, a species closely related to *P. dactylifera*.[16] When reduced by boiling, the sap of either palm becomes a brown syrup from which jaggery can be made. The terminal growing bud of the leafy crown, the 'heart of palm', is a delicacy rarely eaten, unless an older palm has been felled for timber or knocked down by wind. To remove the heart is to kill the palm. Arabs have long eaten the sweet, soft pith of the palm.[17]

Palm trunks, though rather soft, are used as roof timbers and as framing for doors and windows in mud-brick houses. Hollowed out, trunks serve as beehives which are placed in the gardens. After harvest, the flower stalks on which the dates have grown can be softened by soaking, and their especially strong fibers plied for heavy-duty cordage. Coarse fibers *(lif)* at the basal sheaths of leafstalks where they join the trunk, resemble woven burlap and are used as packing material for young date off-shoots and other trees to be transplanted. *Lif* is also hand-plied into ropes used for the girdles worn by date-climbers; animal tack; and marine gear. When saturated with oil, *lif* cordage has been utilized, in traditional preference to nails, to literally sew together the teak hull-planks of Arabian dhows. E.W. Lane wrote that Egyptians would use small wads of *lif*, as strainers, in the spouts of sherbet vessels.[18]

Palm fronds are used in their natural state as awnings for markets and residences. Many of the Bedu tribes of Oman live not in in tents (their goats produce insufficient wool), but in square palm huts known as *burasti*. In al-Hasa, thorns growing at the base of the fronds are cut and used as "burglar protection. They are stuck vertically in the mud or plaster on the top of walls... as nails and pieces of glass are used elsewhere."[19] Wherever the palm is grown, the feathery parts of the fronds are woven into mats, fans, brooms, and baskets. In most instances the palms have supplied the very containers traditionally used to pack and ship their own fruit. The woody midrib of the frond is used to make household furniture and crates for transporting produce and live poultry. In Oman, the hulls of one-man fishing craft *(shasha)* have long been made entirely of palm-ribs lashed together with palm ropes. As the hull is not water-tight, heavy, but buoyant basal ends of palm-ribs are stowed, for increased flotation, beneath the palm-rib deck. (In 1979 along Oman's Batinah coast, it was still possible to have such a boat made, though there was an increasing tendency to use nylon ropes and unsinkable styrofoam packing from Japanese electronics.)

Forty years ago, in Baghdad along the Tigris, pieces of the basal palm-ribs would be strung, like large beads, around the waists of boys learning to swim. Their rough edges covered by linen casings, these ungainly but efficacious floats would be discarded, one by one, as a young swimmer gained in confidence and ability.[20]

Lastly, of course, any dry parts of the palm can always be burned. The date grows in regions where, until the discovery of petroleum and the extraordinary changes that discovery has wrought, there has never been surplus fuel.

***

Myriad representations of date-palms in Middle Eastern and Mediterranean art attest to the tree's long-standing importance as a symbol of vitality. Assyrian stone reliefs, wall-paintings and architectural details in Pharonic temples, Sassanian brocades, Roman coins, Byzantine mosaics, and Islamic miniatures and textiles feature both naturalistic and highly stylized palms. The cult of the the date-palm is an enticing conjunction of botanical, gastronomic, and art historical topics. Especially intrigued by the image of the fruiting palm, I realize that further investigation of its decorative role is outside the scope of this paper. It is however, a subject I hope to treat in greater depth within an expanded study of dates and date-palms.

*   Author's note: Much of the material on cultivation and usage of the palm is information gathered in 1979 during a six-week sojourn in the Sultanate of Oman. The American Museum of Natural History (New York) had commissioned me to acquire and document examples of Omani material culture.

THE DATE PALM: PILLAR OF SOCIETY

FOOTNOTES AND REFERENCES

1. Lunde, Paul; "A History of Dates;" *Aramco World Magazine;* Washington, D.C., March-April, 1978. p. 21
2. Kramer, Samuel Noah; *The Sumerians;* University of Chicago Press,; Chicago & London, 1963. p. 109
3. Simon, Hilda; *The Date-palm; Bread of the Desert;* Dodd, Mead, & Co.; New York, 1978. p. 33
4. "Ghidha; *Encyclopedia of Islam II,* Leiden & London, 1954- pp. 1058-9
5. Simon, op. cit.; p. 76
6. Mason, Silas C.; "Date Culture in Egypt and the Sudan;" *U.S. Dept. of Agriculture Bulletin* 1453; Washington, D.C., January 1927. pp. 64-5
7. Lawrence, T.E.; *Seven Pillars of Wisdom;* Doubleday, Doran, & Co., Inc.; New York, 1935. pp. 314-5
8. Simon, op. cit.; p. 79
9. Lewis, Naphtali; *Life in Egypt Under Roman Rule;* Clarendon Press; Oxford, 1983. p. 127
10. Mason, op. cit. p. 53
11. Ibid. p. 5
12. Vidal, F.S.; "Date Culture in the Oasis of Al-Hasa;" *Readings in Arab Middle Eastern Societies and Cultures,* ed. A.M. Lutfiyya & C.W. Churchill; Mouton & Co., The Hague, 1970. p. 211
13. Lunde, op. cit.; p. 22
14. Conversation with Jasleen Dhamija, Punjabi-born folklorist.
15. "Ghidha," op. cit.; p. 1061
16. ref. Jasleen Dhamija
17. "Ghidha," op cit.; p. 1060
18. Lane, E.W.; *The Manners and Customs of Modern Egyptians,* 1860 edition; J.M. Dent & Sons Ltd., London, 1963. p. 331
19. Vidal; op. cit.; p. 209
20. Childhood experience of Sami Zubaida.

Other references: *The Bible, Koran,* and *Oxford English Dictionary* (various editions).

** As they are from a variety of written and spoken sources, the Arabic words within this paper have not been transliterated in accordance with any single orthographic system. In quoting written sources, I have retained the authors' own diacritical marks.

Atolli: A Liquid Staple

By Sophie Coe

The last person to doubt the dominant role played by maize in the nutrition of Mexico and Central America was struck by lightning and killed on the top of the great pyramid at the Maya site of Chichen Itzá, in Yucatán, Mexico. It is said that the lightning came out of a clear sky, but that extra touch of drama seems superfluous. In that part of the world, maize figures in the origin myths as the substance forming the first humans, maize is the focus of endless gods and rituals, everything else is garnish for the maize.

Given the predominance of maize, the question becomes how this all-important staple was consumed. The standard answer is based more on the consumption of maize in modern Mexican restaurants than on the way it was eaten before the Spaniards landed on the coasts of Mexico. One would think that there are only two ways to eat maize, either as tortillas or as tamales. For a Transatlantic audience it would probably be well to describe these before going on to the alternatives. Many years ago I saw a book review in which the British reviewer lambasted the author for talking about some worthless American Indian maize cakes called tortillas, whereas everybody knows that a proper tortilla is a Spanish dish consisting of eggs, with additions of potatoes, sausages, and whatever else. The reviewer implied that the American dish was probably a fraud and a late introduction because it had a Spanish name.

Far be it for me to denigrate the Spanish tortilla, or its Italian relation the frittata for that matter. In fact, among the infinite edibles displayed for sale in the great Aztec market or *tianguis* in what is now Mexico City, the Conquistadors recorded none other than *tortillas de huevos*, so that it was a dish that existed on both sides of the Atlantic independently. The Spaniards gave the name to the flat maize cake because they found the method of cooking it, on a flat griddle, similar. The Spanish names for things they found in the New World should be taken as evidence for provenience with the greatest caution, if at all. They picked up the word *mahiz* in the Antilles, and proceeded to inform the Mexicans, in whose country maize had originated some 5500 years ago (if new and controversial dates are to be believed), that it was maize they were eating. They referred to New World pepper, *Capsicum*, species promiscuously as *ají*, another Antillean word; chilli, the Aztec term; *pimienta de las Indias* when they were not using that term for allspice; or *uchu* after the discovery of Peru. There is a perfectly good Aztec word for maize tortilla, which is *tlaxcalli* .

What is a maize tortilla as eaten in the New World? It begins with shelled maize, maize grains detached from the cob. The grains are then cooked with lime or wood ashes, which makes the transparent skin on each grain, technically known as the pericarp, detachable. After the decorticated grain is washed it is ground on the metate, a three legged stone, using a handstone. For extra delicacy the germ may be removed, presumably doing away with a good deal of the nutritional value. The dough is ground and reground until the desired smoothness is obtained. The lords had tortillas and small ball-like cakes so carefully prepared that they were transparent. The usual dough today, called *masa*, is either white or yellow depending on the maize used, and cakes made of it do not remotely approach transparency.

To make tortillas from *masa* you adopt one of two different strategies, depending on what part of Mexico you are from. In the central highlands a small ball of dough is patted into a flat disk by the tortilla maker, using her two hands. In the lowlands of the Gulf coast and Yucatán the tortilla maker uses a large leaf, nowadays a banana leaf, to shape the tortilla on. The banana was of course introduced by the Spaniards, but there are members of the same plant family that are native. The difference of technique is made puzzling by the fact that it is sometimes supposed that the dwellers in the lowlands did not use the tortilla until a period of heavy highland Mexican influence around the tenth century AD. Other evidence for the late introduction of tortillas into the Maya area is the fact that the Manche Chols were not making tortillas when the Spanish priests first contacted them in the 17th century, and that *comales*, the flat clay griddles for toasting tortillas, are rare to non-existent in archaeological sites in most of the area. But on the other hand, cacao beans are also toasted on *comales*, and we know the lowland people were heavy users of cacao, so we are left with the alternatives that they did not toast their cacao, or used some other method of toasting which they might have used for tortillas as well.

Once the tortilla has been shaped, by whatever method, it is toasted on the *comal*, which sits over the fire, supported by three stones. The toasting is brief, on both sides, and the tortilla is soft and flexible when it is removed. Everybody admits that they taste best hot off the fire, but the list of things that can be done with them afterwards is endless. Dried they become *totopoxtes*, a long-keeping item and military rations for the Aztec armies. Nowadays they are fried to form hard shells called *tacos*, which can be moulded into an infinite variety of shapes, and filled with an infinite variety of things. The tortillas themselves can be rolled, stacked, or shredded into soups and stews. As with anything in popular culture, all the variants have local names, and the names are often different in different regions. Not only are the names and methods of manufacture different according to their place of origin, but because the tortilla is not only a food but an implement, the method of utilizing it also differs.

Spoons existed in pre-Columbian Mexico. Inventories of the treasures sent back by Cortés to the court of Charles the Fifth include gold spoons and tortoise shell spoons, all of them, alas, long since vanished. We do not know if there were more utilitarian spoons, and if there were, how widely they were used. Even today the poorest families in the slums of Mexico City use no implements for eating, relying instead on the tortilla, and this is where the regional differences show up. In northern Mexico the tortilla is rolled up and dipped into the sauce or stew. In central Mexico the tortilla is torn up and the pieces used as spoons. Von Tempsky, who travelled in the 1850s, described the difficulties of this operation, and the amusement his awkwardness caused the young ladies (G.F.von Tempsky, 1858).

If maize is not consumed in any of the protean variations on tortillas, what are the alternatives? Tamales, of course, is the answer today. What is a tamale, a name taken over (and for once uncorrupted) from the Aztec *tamalli*? Today it is a construction of the same basic maize *masa*, this time beaten with with as much as half its weight in lard, and either baking powder or an indigenous substitute, *tequezquitl*. Josefina Velázquez de León, who published innumerable Mexican cookbooks during the 1950's, much prefers *tequezquitl*, which she calls *tequezquite*, which is a natural product of the dry lake beds of central Mexico, consisting of sodium chloride, or table salt, and sodium carbonate. A small chunk of *tequezquitl* is put in a cup of water with ten husks of husk tomatoes, and just brought to a boil. When it cools it is strained and added to the dough, where it acts as leavening. Given the lack of gluten in maize, it is a little mysterious how this works, but we are assured that with plenty of beating the *masa*, the lard, and the leavening will amalgamate into a dough that will float on water. Plain tamales are now ready to be wrapped in maize husks or banana leaves rendered more flexible by gentle heating, and stacked in a large pot and steamed. Tamales may also be filled with just about anything that is handy, such as whole beans, or beans cooked and mashed and incorporated into the dough, or squash flowers and native greens like chaya (*Jatropha aconitifolia*) or any available meat or fish. There are sweet tamales, which can be filled with pineapple, or peanuts, or strawberries. A very special treat is the tamale that is not made of masa but of tender young maize.

Unfortunately all these can have nothing in common with what was eaten as a tamale in pre-Columbian times. The unctuous, if not just plain greasy, dough could not have been made then, because fat in any large quantity simply did not exist. We have confirmation of this from the account of an anonymous Dominican, written in 1569, and describing the austerities of his life.

"...without eating bread or drinking wine, but eating native maize tortillas roasted on the fire, and as there were no fisheries, as there are now, the religious contented themselves with some boiled eggs ...with neither oil nor lard, because there were none in that country..." (Anon, 1866)

Not only did they not exist, but the Indians found them repugnant when they were introduced by the Spaniards.

"So we must live with the things the Spaniards brought us, because they are the masters; and now we have prisons, and incarceration, and beating, and basting with lard..." (Gómez de Orozco, 1940)

Can we reconstruct a pre-Columbian tamale? It would be quite solid, because the *tequezquitl* was probably not widely exported. There is a Maya pot showing a palace servant with a plate of objects, about the same size as his hand, oval, and some yellow and some white. They are not represented in the way the Maya would show flat round things like tortillas. It is curious that they are shown unwrapped, rather than being served in their leaf wrapping as is always done today. Perhaps the fact that they were made without fat and leavening made them firm enough to present unwrapped, which would be impractical with the rather sloppy modern version. The leaf wrappings would almost certainly be more varied than just maize husks and banana-like plants. There are many other leaves available, some with their own distinctive flavors. Avocado leaves are still used as ingredients today, and so are the leaves of *acuyo, Piper sanctum*. Reading through the sources one comes across mentions of other wrapping leaves, a tantalizing parade of lost tastes.

Modern tamales have ritual importance. We have had chicken tamales for Christmas Eve in Guatemala City, and the Quiché, who inhabited that part of the world before the Conquest, still have ceremonial tamales. We have accounts of tamales so large that they enclosed whole turkeys and had to be wrapped in a mat to steam. Tamales also figure in Yucatec Maya rain ceremonies, but sometimes it is not clear how they are constructed. Often they seem to be made in layers, illustrating the Maya idea of a many-layered heaven and a many-layered underworld. The *masa* layers are separated by layers of ground squash seed, but the anthropologists unfortunately show their deplorable blindness towards things culinary, and sometimes describe it as a stack of tortillas, and at other times as a tamale. But then, the author of a well-known book on the Indians of Guatemala thinks a tortilla wrapped around a piece of meat becomes a tamale! I trust readers of this paper know better by now.

What else may be done with maize? The very first expedition to explore the coast of Yucatán found out on their very first encounter with the natives. The ships of Francisco Hernández de Córdoba were in need of water in 1517 when they came across some native canoes near the island of Cozumel. They were given some calabashes full of water, and "...they also gave them ground maize in balls, and dough of which they make something like a *zahina* or *poleada*..." (Las Casas, p. 403). Even the English translation of the Spanish words needs translation, because what we are dealing with here is the culinary equivalent of a dinosaur, an extinct dish. A *zahina* is an unthickened gruel, and a *poleada* is gruel that has been thickened, presumably by cooking. No wonder modern authors don't think to suggest these foods for the ancient inhabitants of America - they probably have never heard of them. The last time gruels were seen in American cookbooks was around the turn of the century, when they expired in the sick-room section.

The ubiquity of these preparations among the Maya is illustrated by the *Relación de las Cosas de Yucatán*, by the second bishop of Yucatán, Diego de Landa. He is a man who will forever be regarded with the deepest ambivalence by scholars of things Maya, because on the one hand he left us with an excellent account, and a syllabary that was crucial to the decipherment of the hieroglyphs, but on the other hand he was responsible for burning at least 27 Maya codices, and inflicting unspeakable tortures on his newly converted flock.

"...[they soak the maize] one night before in lime and water, and in the morning it is soft and half-cooked ... and they grind it upon stones, and they give to the workmen and travellers and sailors large balls and loads of the half-ground maize, and this lasts for several months merely becoming sour. And of that they take a lump which they mix in a vase ... and they drink this nutriment ... From the maize which is the finest ground they extract a milk and they thicken it on the fire, and make a sort of porridge for the morning. And they drink it hot, and over that which remains from the morning's meal they throw water so as to drink it during the day, for they are not accustomed to drink water alone. They also parch the maize and grind it, and mix it with water, thus making a very refreshing drink, throwing in it a little Indian pepper or cacao. They make of ground maize and cacao a kind of foaming drink which is very savory, with which they celebrate their feasts. And they get from the cacao a grease which resembles butter, and from this and maize they make another beverage which is very savory and highly thought of; and they make another drink from the substance of the maize both ground and raw, which is very refreshing and savory ... In the morning they take their warm drink with pepper, as has been said, and in the daytime they drink the other cool drinks, and eat their stews at night." (Tozzer, p. 90-91).

A full commentary on this passage would take hours. I will just discuss a few points. The statement about cacao grease seems to contradict the dislike of grease I have mentioned before, but one must remember that the grease was extracted by crushing the cacao beans, and then putting them in a container of hot water and skimming the top of the water. Hardly a way to obtain vast quantities of cacao butter.

Landa's account suffers from the fact that he obviously never had prepared any food in his life. Francisco Hernández, the personal physician of Philip the Second, who was sent to the New World, gave a much more systematic account of the cooked maize beverages, called *atolli* in Aztec. He lists at least twelve variations, with additions ranging from bees' honey and maguey syrup to chilli peppers, beans, and squash. There was an especially delicious *atolli* of young maize and another which was prepared by making a special fat tortilla, and then extracting the interior crumb and cooking it in water. The drink made of parched maize that Landa mentions was *pinolli*. In this recipe the step of making *masa* is omitted, and the maize is toasted, a frequent flavor-enhancing technique in Mexican cooking. The *pinolli* could then be flavored any way one wished. Today, it is spelled *pinole,* the national drink of Nicaragua.

Special attention should be paid to the sour lump of *masa* which Landa said was given to workmen and travellers. This is now called *pozole* (Spanish spelling) in southern Mexico, although in northern Mexico *pozole* is a kind of stew. Ulloa and Herrera (1986) go into the nutritional benefits of this substance in some detail. Basically the dish begins with the standard preparation of maize *masa*, but the *masa* is set aside for periods lasting from a few days to a few weeks. Not only is it set aside, but it is set aside in a special container, or wrapped in leaves which have been used for this process before. In other words, there is a continuous culture, consisting of yeasts, bacteria, and fungi, and their action apparently considerably enhances the nutritional value of the *masa*. To make ones meal all that needs to be done is to add water and salt, sugar, honey, chilli, or whatever else appeals. Ulloa and Herrera describe this as comprising almost the complete intake of some people today, who may consume up to a kilo of *masa* a day in this fashion. An additional benefit is that it does not require cooking beyond the preparation of the *masa*. We forget the constant struggle for fuel in our houses where the gas or the electricity comes automatically, unless we forget to pay the bills. Given the possibility that there was considerable deforestation, a food that required cooking once, rather than twice as do tamales and tortillas, must have been very attractive. There were also variants even within this class of soured dough beverages. Hernández says *xocoatolli*, sour *atolli*, which is what he calls Ulloa and Herrera's *pozole*, could be made by taking a measure of soured *masa* made of black maize that had stood for four days, adding twice as much fresh *masa*, and seasoning it with salt and chilli pepper. *Xocoatl*, which some misguided people have suggested is the origin of the word chocolate, is sour water, *masa* mixed with water and allowed to ferment over night, after which the water is strained off and consumed.

There is evidence that some form of *atolli* was consumed in Classic Maya times, about a thousand years before the Spanish conquest. Two bowls have been found with inscriptions proclaiming them to be made for *sac-ul*, "*white atolli*" We do not know if this is the same white *atolli* that has ritual significance in some places today, nor do we know the composition of either the ancient or the modern white *atolli*, whether it was made of sweet young maize, specially selected mature white maize, or something undreamt of.

But *atolli* is not a thing of the past. Even though Ulloa and Herrera think, on what evidence is not clear, that *pozole* is of Maya origin, the consumption of these starchy drinks is not limited to people speaking Maya languages today. An excellent compilation called *Nuestro Maiz*(1982), a collection of essays on maize written by country people all over Mexico, gives five different recipes from just one group of people who speak a language related to Aztec. Incidentally this should quash the story one hears that "the Aztecs didn't eat soup". If this statement does go back to some Spanish conqueror, one would like to ask him how the following dishes differ from what we would call soup. *Tliolatoli* is made of maize that is soaked without any lime or woodashes, *kojnexatoli* is made of maize cooked with new wood ash, well washed and ground, and then drunk with a piece of *panela*, unrefined brown sugar, held between the teeth.

Chileatoli is a toasted maize *atolli*, which might perhaps be called a *pinolli* elsewhere, with young maize and boiled squash added, but no sugar, because the young maize and squash add their own sweetness. *Eskimoli* is toasted maize of a special variety made into a puree with turkey broth, with *hierba santa*, *Piper amalago*, which has a vaguely licorice flavor, added. Finally there is *kuetlaxkolmoli*, where cooked intestines are mixed with toasted maize, chillis and herbs. It is obvious that the range of liquid maize dishes is infinite. This whole discussion should point out the danger of interpreting the food of other people using our categories. We do not today have a group of foods that are liquid staples, unless one would care to include the Metrecal of a generation ago, and the Optifast of today. This does not mean that we should overlook the fact that other people have had them, and that in fact they may have been a vital part of nourishment for many generations.

Bibliography.

1. Anonymous Dominican, *Relación de la Fundación, Capítulos y Elecciones*...Coleccion de Documentos Ineditos, vol V, Frias, Madrid, 1866.
2. Casas, Bartolomé de las, *Historia de las Indias*, Biblioteca de Autores Espanoles, vol 96, Atlas, Madrid, 1961.
3. Gómez de Orozco, F.(ed.), *Cronicas de Michoacan*, Ediciones de UNAM, Mexico, 1940
4. Hernández, Francisco, *Obras Completas*. UNAM, Mexico, 1959.
5. Rivera Aragón, Rosaura et al., *El Maíz en Cuapa Pinopa: Zongólica Vera Cruz*. Nuestro Maíz, vol. 2, Museo Nacional de Culturas Populares, Mexico, 1982.
6. Tempsky, G.F. von, *Mitla. A Narrative of Incidents and Personal Adventures on a journey in Mexico, Guatamala and Salvador in the Years 1853-1855*. Longman, Brown, Green, Longmans & Roberts, London, 1858.
7. Tozzer, A.M.(ed.) *Landa's Relación de las Cosas de Yucatán*, Peabody Papers, Harvard University, Cambridge, Mass. 1941.
8. Ulloa, M. & T. Herrera, 'Fermented Corn Products of Mexico', in *Indigenous Fermented foods of Non-Western Origin*, Mycologia Memoir 11, Cramer, Berlin, 1986.
9. Velázquez de León, Josefina, *Tamales y Atoles*, Ediciones J. Velázquez de León, Mexico, 1956.

Polenta - An Italian Staple

By Anna Del Conte

The classic polenta is a mixture of corn meal and water, although in some areas of northern Lombardy buckwheat flour is also included. In antiquity, however, polenta - *puls* in Latin - was made from spelt, a kind of wheat, and other cereals, while in the Middle Ages and the Renaissance, millet, chestnuts, broad beans, buckwheat and even acorns provided the flour from which polenta was made. Corn came later; it was introduced to Europe by Columbus, who wrote in the diary he kept after landing in the New World: "Corn has a pleasant flavour and all the people of this country live on it."

Within Europe corn was first cultivated in Spain, and the Spanish began to enjoy it so much that Charles V, who did not approve since he thought wheat was the right food for Christians, promoted the cultivation of wheat rather than corn by means of various incentives to farmers. From Spain, corn reached Italy, arriving first in Venice at the Rialto market. The bridge was not yet built, but the market was there then as it is today, and it was there that goods from abroad were unloaded.

This yellow grain was soon christened *granoturco* - Turkish grain, but why it was given this name is still the subject of great dispute. Was it called granoturco because of the similarity between the ear of the cob and the beard of a Turk? Or was it, as Pianigiani wrote in his 18th century Etymological Dictionary, "because of a mistake in the translation of the name the English gave to the cereal - turkey wheat - that is to say food for the turkey, a bird so called because of a certain likeness between the bird's neck and the Turkish turban"? To my mind the only credible explanation of the name is that in 16th century Venice so many things of foreign origin came (or were thought to come) from Turkey that *turco* became synonymous with foreign.

In Italy, *granoturco* was first cultivated in the Polesine, an area around the Po delta hitherto bare of crops because of its hot, humid climate. The Polesine was part of the territory of La Serenissima, and by the 16th century Venice was no longer the rich and powerful republic of previous times. Its decline was in part due to the fall of the Eastern Empire, once a rich source of trade, and also to the discovery of America and the opening of the new route to the Indies round the Cape of Good Hope. By the middle of the 16th century the Venetians, and particularly the inhabitants of the *terraferma*, were very poor, and so it was that maize came as the miracle crop, for it was found easy to cultivate even in the Polesine. From this *granoturco* was made the first yellow polenta.

Although the better-off among the locals were soon to match this new polenta with their old foods, polenta *e osei* (with birds) or polenta *e baccalà* (with stockfish), it was mainly eaten by itself by the poorer people, for whom it soon became their staple food.

Polenta was made in the evening with all the ritual surrounding an important event. *La mare* (Venetian for *madre*) cut it carefully with a taught cotton thread into either a *feta*, a *mesa feta* or a *bocon* (a slice, a half slice or a mouthful). These she distributed to the family, the smallest children getting their *bocon*. Polenta was sometimes eaten with milk, and, in the rare days of abundance, even with butter. For breakfast *la mare* grilled slices of left-over polenta, and at lunch this was eaten with pieces of crackling, beans, pumpkin, slices of lard or whatever was the cheapest seasonal accompaniment.

Polenta took longer to establish itself elsewhere in the north. Bergamo and Brescia, then part of the Venetian republic, were the first outlying provinces to adopt polenta, and they did so with such enthusiasm and totality that many people thought it originated there. A new dish was invented there to alleviate the monotony of polenta on its own. A cotechino or a salame was hung over the table and the family rubbed their mouthfuls of polenta over it for flavouring. It was called polenta *e pica-sö* (polenta and take it from above).

The conquest of the rest of Lombardy and of Piedmont was slower. By 1650, however, polenta was popular enough to inspire the Marquis Edoardo del Carretto to offer a dinner based on polenta, baccalà and onions to a band of tinkers from Calabria who had arrived in his village near Acqui in Piedmont. The Calabresi enjoyed the polenta feast so much that they made the Marquis a huge *paiolo*, the round-bottomed unlined copper pan in which polenta is traditionally made. Still, to this day, in that village, on the last Sunday of Carnival, a huge dish of polenta is made with 600 kilos of maize flour; it is eaten with a huge *frittata* of 600 eggs and - as at the Marquis's feast - with *baccalà* and onions.

Polenta stayed in its stronghold in the North, never managing to cross the Po so as to conquer Emilia-Romagna and the staples of that region, tagliatelle, ravioli and lasagne. Northern Italians became known as the *Polentoni*, as opposed to the *Mangiamaccheroni* of the South.

It was during the first half of the 18th century that the popularity of polenta began to wane when a strange new disease was first noticed and written about. Characterised by red skin lesions, digestive disturbances, weakness and even mental and physical degeneration, the disease acquired the name of pellagra, from *pelle agra*, meaning rough skin. The symptoms occurred in the spring of each year, after a winter of eating little but polenta. In 1786, Goethe wrote in his Italian Journey: "Their features spoke of misery and their children looked just as pitiful... I believe the unhealthy condition is due to their constant diet of corn and buckwheat or, as they call them, yellow and black polenta." Not unnaturally, polenta was blamed, and before long its consumption was forbidden by law. There followed widespread suffering as a result of malnutrition and hunger: in a few years the food that had been the main nourishment of many thousands of people had been taken away and declared fit only for animals.

It was not until early this century, when vitamins were discovered, that the truth emerged. Pellagra was not due to any substance contained in the corn, but rather to a serious deficiency of protein and vitamins that occurred when corn was the principal, or indeed the only, source of nourishment. A fascinating backwater of nutritional history is to be found in the fact that the American Indians, and the tribes of Central and South America, had somehow learnt to avoid pellagra by cooking their maize with ashes or lime. This alkaline processing makes it possible for the body to absorb the niacin that is in fact present in the maize but that is otherwise not available to the human body.

In the past, polenta was acclaimed, written and sung about, and altogether celebrated, more than any other Italian food. Societies were formed in Italy and abroad, some of which concealed political movements behind the pretence of discussions on the merits of a dish of polenta. The most famous society was the Ordine dei Polentoni, formed in Paris in the second half of the last century. The order numbered many well know people among its members, including the painter De Nittis, Emile Zola, Edmond de Goncourt and the musician librettist Arrigo Boito, who wrote a long poem in Venetian dialect on the art of stirring polenta. Polenta also inspired artists such as Magnasco and Longhi who, when representing a country interior, both showed golden polenta being poured out of a copper *paiolo* on to a white tablecloth as the centrepiece of their paintings.

The actual making of polenta is a ritual, described by Giovanni Arpino, a 20th century writer, in nostalgic mood. "Between the stirring of the paiolo, the eating and the next unavoidable hunger, you had to devote a day to it. In the slower world, in the world that ignored frenzies, polenta acted as a clock. Can you seek, or find, these virtues in a lobster or a soufflé or even a raviolo? It is the unique, golden, refuge-food." And Alessandro Manzoni, in *I Promessi Sposi*, wrote that when polenta was turned out on to the board it "looked like a harvest moon in a large circle of mist."

The golden moon is a very versatile food; it is so ready to be matched with others that in the province of Cuneo in Piedmont they call it La Traviata! It goes with butter and cheeses, with rich stews, with game and with salt cod. Left-over polenta makes rich layered dishes - polenta pasticciata - or it is fried or grilled to accompany many meat and fish dishes. Polenta is to an Alpine Italian what pasta is to the southerner. Sadly it has lost ground to the all-conquering pasta, and this is partly because, for all its virtues, polenta is certainly not fast food.

In some specialist shops in Italy one can buy three or four different varieties of ground maize so as to be able to make the right polenta to go with the right food. There is now even a factory in the Bergamasco that produces the 'original' polenta, made with maize cultivated as it used to be before the present years of intensive farming. The small cob of a reddish colour is ground between mill stones without being heated. The resulting flour is very coarse and granular and it produces a superb polenta with a full, earthy taste. And in Turin there is a restaurant that offers a menu totally based on polenta. The polenta is cooked over a wood fire, so that it takes up the aroma of the burning wood.

Gastronomes in Italy today are moving away from the trend set by the Nuova Cucina, partly shaped by foreign influences, and are turning their attention to traditional dishes of the *cucina povera* and the *cucina casalinga*. It is for this reason that polenta, one of Italy's staple foods, has become once again one of its most highly regarded.

The above is an expanded version of the section on polenta in my new book 'Secrets from an Italian Kitchen' to be published by Bantam Press in July.

Pinza

This is a cake made with a polenta of maize and wheat flour cooked in milk. It is a speciality of Padua and Vicenza.

Ingredients: (serves 6)

30 g (1 oz) sultanas
15 g (½ oz) almonds, blanched and coarsely chopped
15 g (½ oz) walnuts, coarsely chopped
2 dried figs, cut into small pieces
2 prunes, stoned and cut into small pieces
4 tbsp Grappa
60 g (2 oz) plain flour
120 g (4 oz) corn-meal, finely ground
pinch of salt
600 ml (1 pint) milk
75 g (2½ oz) castor sugar
75 g (2½ oz) unsalted butter, melted
2 tsp fennel seeds

Put the dried fruits and the nuts in the Grappa and leave to soak.

Mix together the two flours and the salt.

Bring the milk to just below boiling point. Add the flour mixture in a thin stream while beating hard with wooden spoon to prevent lumps forming. Cook over low heat for at least 15 minutes after bubbles have began to break through the surface. Beat constantly during the cooking. Draw off the heat.

Reserve 1 tablespoon of the melted butter and add the rest, the sugar, the fennel seeds and the fruit and nut mixture with its liquid to the polenta. Mix thoroughly.

Heat the oven to 180°C (350°F Mark 4).

Butter a shallow 20 cm (8 inch) tin and line with parchment paper. Spoon the polenta mixture into the tin. Level it out and brush the surface with the remaining melted butter. Bake for 20 minutes, then cover the top with a sheet of foil and bake for a further 20 to 30 minutes. Allow to cool in the tin.

Staple Foods of the American West Coast

(A Semi-historical Perspective;
or,
Culture Change in Action)

By John Doerper

It's hot in the Los Angeles Basin at high noon; I am thirsty and my stomach sends out twinges of hunger. Along a road in suburban Monterey Park, I spot that ubiquitous American food emporium, a supermarket. I pull up to buy a sandwich and a cold drink. But as I approach the market, I notice a difference. Instead of the neon name sign of the familiar supermarket chains, this market proclaims itself as the: "Hong Kong Supermarket." I look around. This is obviously not a chain store. But everything else appears to be standard here - the parking lot, the glass front... everything except the name. Then I notice that the large window signs advertising goods for sale (a familiar feature of American markets) are in Chinese. This ought to be interesting. I've been in a great number of Chinese shops up and down the West Coast, but I haven't encountered a Chinese supermarket yet.

The Hong Kong Supermarket does not sell just Chinese foods, it also carries a few Japanese, Thai, and Vietnamese items, and it has three bakeries - Thai, Vietnamese and French. Why a French bakery? I ask the counter person. She tells me the bakery is here to satisfy Vietnamese customers who learned to appreciate Western-style bread and pastries during the French occupation of their homeland.

The market's customers don't throng to the bakeries, but they line up three deep at the seafood counter (fresh fish and shellfish in live tanks; fresh fish on ice) and in the meat department (not only pork and chicken, but big, hefty steaks as well).

The produce section, too, bustles. I note the standard greens of Asian cookery, as well as a good selection of American ones: bok choy, Chinese cabbage, fresh water chestnuts, jicama, potatoes, taro root, carrots, celery, onions, chile peppers, et al. There's a dairy section - dedicated to soy milk and tofu (bean curd). Off in a corner, shelves are loaded down with beer, but there's little wine. Though the store sells bread of the yard-long French kind, it offers no ready-made sandwiches. My luncheon plans change. I settle for a [Philippine] lumpia and a can of jackfruit juice.

As I wander back across the parking lot, I muse about the staples of modern American cookery. What surprised me the most in today's experience was the realization that many of the foods I found here are not at all that uncommon anymore in mainstream West Coast supermarkets.

Since I first visited the Hong Kong Supermarket several years ago, West Coast supermarkets have undergone even greater changes. Recognizing that the tastes of West Coast cooks were moving away from the standard American staples of beef, bacon, ham, butter, sugar, potatoes, soft wheat bread, et al., they have tried to anticipate changes in taste, and often met the consumers' needs before the customers quite knew what those needs were.

Supermarkets as an institution are of prime importance in any discussion of the staple foods Americans consume. A gourmet cook may shop at specialty markets, drive directly to an oyster farm for fresh oysters, buy beef from a rancher, fish from a fisherman, vegetables from a truck gardener, wine from a vintner - but the average American cook will shop at a local supermarket (the closer to home, the better). This new availability of different foods has been of great advantage to my own cookery. With the exception of fresh fish (which I will buy at very few supermarkets) I don't have to go far or wonder where I should go, when I need ingredients like fresh ginger root, bean sprouts, tofu, chiles (mild or hot), chorizo, jicama,

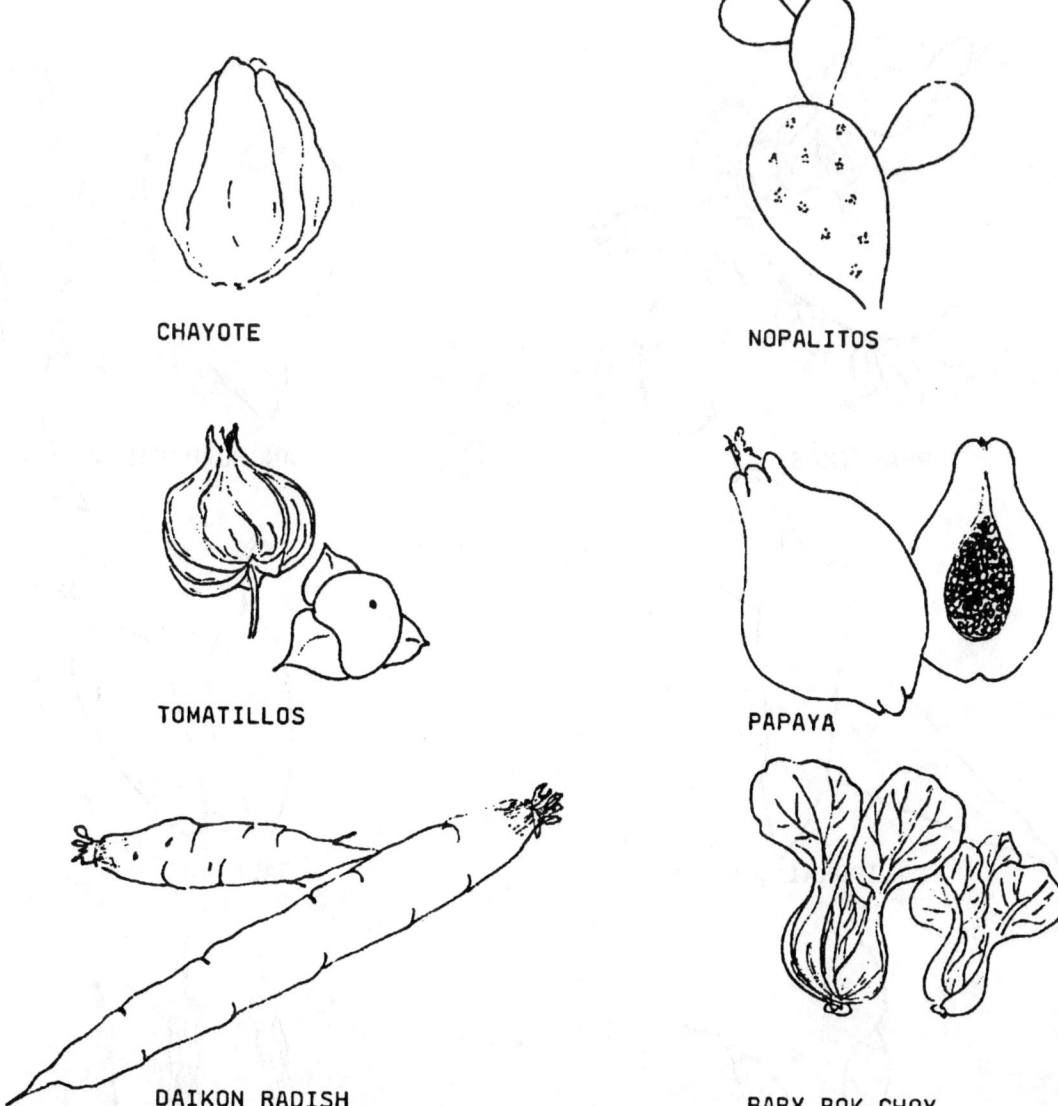

tomatillos, bok choy or oyster sauce - I just drive to the nearest supermarket, confident of finding what I need.

Such a culinary trend is of advantage to creatively minded cooks throughout the region. If such "exotica" as fresh ginger root, chiles, or oyster sauce were sold only at specialty shops, the average American eater would remain blithely ignorant of their availability; if they are sold in the local supermarket the home cook may wonder about their culinary properties yet buy them just to learn what they are like.

There's a further consideration to be taken into account when we follow the movement of staples across cultural barriers: specialty food shops rarely hand out recipes to their customers, rightly assuming that cooks who shop at such establishments know what they buy and why. Supermarkets, on the other hand, not only set up boxes with recipes next to unfamiliar foods, they play endless video tapes on food

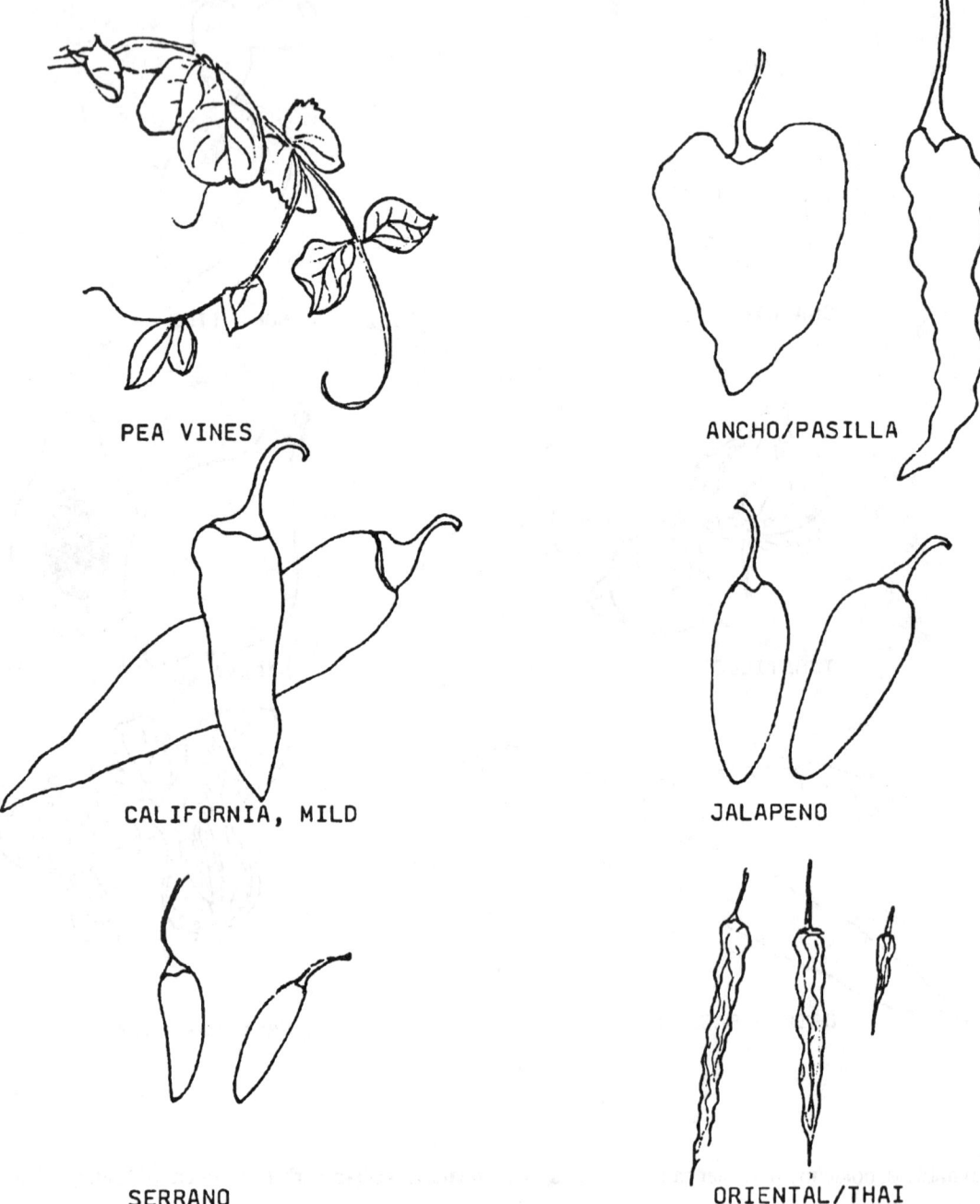

preparation and they periodically put on food demonstrations to show the home cook how to stir-fry or steam, say, Chinese greens, clams, or uncommon species of fish.

But this was not always so - even in the multi-cultural West. In the so-called Good Old Days, folks ate surprisingly modest fare. During the early period of colonization, West Coast foods were of the plainest kind. Early European settlers on the coast had little respect for native foods, and even the Spanish missionaries, who brought such Mexican Indian foods as maize, beans, and chiles to California, did their best to ignore the eating habits of the natives and teach them to like "civilized" foods instead. This often led to a rapid decline of the Indian neophyte population due to malnutrition and disease. For reasons not fully understood, the native Californian population did not thrive on wheat, maize and beans the way Mexican Indians did, nor could they adapt to the crowded living conditions at the mission compounds. At San Gabriel Mission alone 6,000 Indians died in an epidemic (out of a total Indian population of 100,000 for the entire area now included in the state of California)[1]. The cemeteries at other missions attest to similar losses. In Mexico's Baja California matters were even worse: here, where much settlement took place almost a hundred years before Alta California (the present U.S. state of California) was settled, many of the missions, including the once- flourishing establishments of San Juan Bautista Malibat, Santa Rosalia de Mulege, San Jose de Comondu, and San Ignacio Kadakaaman, had to be abandoned at or before the time the northern missions began to thrive, "due to a lack of [Indian] population."[2] Unfortunately, the Indians did not have time to adapt to European and Mexican mainland foods before they were decimated by nutrition-related diseases.

Yet not everything the mission padres did was detrimental. The long-range benefits were tremendous. The padres had an impact on the economic life of the coast which far exceeded their influence on the spiritual life: they were the first to introduce European and Mexican staples, foods which could support a large human population through agriculture, and they were also the first to bring cultivated fruits and vines to the region. In fact, the padres successfully raised many crops on the West Coast of North America from the late 1600s on, long before these foods became readily available in the more densely populated colonies and states east of the Appalachian Mountains.

Wheat, maize, beans, garbanzos, sugar cane, grapevines, figs and other fruits were cultivated at San Jose de Comondu, for example, as early as 1708. Barley, olives, onions, apricots, peaches, pears, pomegranates, and prickly pear cactus fruits came later at other missions. Fish, shellfish, and salt were obtained on the coast. Beef cattle, sheep, goats and hogs also thrived in the new territories. Beef did so well, it became not only a staple, its hides (sold to Boston merchants or traded for needed goods) were treated as ready cash. There was no charge for the meat of the animal because there was so much of it. As one historian writes,

> [The traveler] is fed the best of mission provender and is free to remain as long as he desires. There is no charge. If necessary, he is provided with fresh horses to continue his way. Should hunger overtake him before his next stop, he may kill a beef on the range; he is required only to hang the hide where the vaqueros can find it. The hides have value. There is a trade in hide and tallow[3].

By introducing staple foods which thrived in the semi-arid climate of the West Coast, the mission padres and the rancheros who followed them set many of the early food patterns of the region. Perhaps even more importantly, by introducing irrigation, they laid the foundation for one of the greatest agricultural empires the world has ever seen.

Climate is a major quirk in West Coast agriculture. Because of the peculiar nature of the off-shore currents, prevailing winds, and mountain ranges which intercept rain, much of the fertile lands from Baja California north to the Pacific Northwest are either too dry or too wet to satisfactorily grow crops on a large scale. The land either gets too much rain or too little. In the dry regions, rain often falls only in higher altitudes. But the high mountains which intercept the rain also channel the water into rivers or store it in slow-melting snow packs. The water can thus be captured in reservoirs, to be parceled out slowly throughout the year. In many of the West's growing regions, from California's Central Valley to Washington's Yakima Valley, a healthy snow pack assures a steady replenishment of these reservoirs, often into early autumn.

While the padres irrigation schemes were on a much more modest level, the missionaries were the first to recognize the value of capturing and directing the limited water supply into fields and pastures[4].

It can be argued that all of the West's current agricultural enterprises existed - if only in embryonic form - in the pastures, fields, gardens, and workshops of the California missions and ranchos.

Much nostalgia has been attached to rancho life, perhaps because it was based on a cattle economy - and Americans have ever romanticized the life on the range. But, in spite of the romantic myths, it was not an easy life. Yet food was plentiful, if only because the population was small enough to live on the margins of agricultural production. Here, too, patterns became established which still persist in the West today. As the Mexican settlers moved north, they brought with them the three traditional staples of Mexico, the foods which had served as the staffs of life of the ancient Aztecs and other Mesoamerican populations: maize, squash, and beans[5].

But this triad was quickly abandoned. Squash was relegated to the back burner. California burgeoned with edible greens - of both the wild and garden varieties - and greens could be eaten without cooking, in the form of salads, to accompany the ubiquitous barbecued or stewed beef, pork, goat, or lamb.

Maize required a lot of irrigation water in California, straining the resources of missions and small ranchos, while wheat grew well on dryland slopes, thriving on small amounts of natural rain water. Wheat, rather than maize, thus became the staple grain of early California, and the wheat tortilla, rather than the ubiquitous Mexican corn (maize) tortilla became the bread of the region[6].

Tortillas became scarce in California after settlers from the Eastern, Midwestern, and Southern United States flooded California. When tortillas reemerged as mainstream food in the nineteen-fifties, they were brought back by the new immigrants, Mexicans who had come to work on farms and in factories and stayed to start a new life in a region first colonized by their forebears. Since many of these new settlers came from the maize-eating regions of central and southern Mexico, they preferred the maize tortilla to the wheaten one[7].

Beans held their own, however. Unlike Texas, California never abandoned the bean in its chili. In California, as in Mexico, chili is based on chiles and beans, with perhaps some meat added to give depth to the dish; in Texas beans are frowned upon in chili. Beans survived even when Mexican-style cooking went into decline after the Yankees conquered the West[8].

Chile, fresh or dried, became an important staple. As Jacqueline Higuera McMahan, a direct descendant of a ranchero family relates,

> Chiles made the rangy beef cattle which fed on the wild fodder in the hills taste
> more palatable, and when spicy sausage, chorizo, was made from pork, chiles were
> the dominant seasoning[9].

Cooking oils, too, changed in California. After the Spanish conquered Mexico, they introduced the hog. This animal quickly replaced the Chihuahua-type dogs, which had been the native population's only domesticated quadruped, in the kitchens of Mexico. In fact, one might argue that few people have greeted the hog with greater culinary enthusiasm. Lard became the standard fat of the land. But in California, olive oil replaced lard, until American settlers reintroduced it. Today, lard again plays a major role in California/Mexican cooking. This is related to the culinary preferences of recent Mexican immigrants. Mexican markets always have coolers filled with large tubs or boxes of "manteca."[10] Yet olive oil, abandoned by mainstream West Coast cooks for lard for most of a century has been rediscovered and is making a strong comeback, and is of increasing significance in the regional cookery, due largely to recent concerns of too much cholesterol in the diet.

Cheese was also made during mission and rancho days. It was commonly of the simple kind, consumed before it had time to age. (Storage rooms cool enough to properly age cheese were a rarity.) In fact, Monterey Jack, a native American cheese which has become increasingly popular throughout the land, was first made during the mission period[11].

Yet while missions and ranchos tried to be as self-sufficient as possible, this goal was not easily achieved. Yankee traders, sailing around Cape Horn in small ships laden with trade goods, soon flocked to the California coast to exchange spices and luxury goods for cow skins. Henry Richard Dana, a Boston college student, signed up for such a voyage in the 1830s and has left us a vivid account of life in early California. He comments on the trade goods and disdains the lack of "Yankee enterprise" shown by the Californios:

> Our cargo was an assorted one, that is, it consisted of everything under the sun. We had spirits of all kinds (sold by the cask), teas, coffee, sugars, spices, raisins, molasses, hardware, crockeryware, tinware, cutlery,...
>
> The Californians are an idle, thriftless people, and can make nothing for themselves. The country abounds in grapes, yet they buy, at a great price, bad wine made in Boston and brought round by us, and retail it among themselves at a real (twelve and a half cents) by the small wineglass...Things sell, on an average, at an advance of nearly 300 percent upon the Boston prices...[12]

Dana's attitude was the attitude of the settlers from the United States who flooded into California during the 1849 gold rush and after. Mexican culture was soon a thing of the past, not to be resurrected until the 1950s and 1960s[13]. Like the native Indians, the descendants of the rancheros, the "paisanos," were shunted aside, but the foods the padres and rancheros had introduced stayed behind and influenced the way Americans eat to this day.

In the meantime, traders had colonized the Pacific Northwest. Following in the wake of James Cook, who came to the coast in 1778, British and American fur traders flocked to the Eastern Pacific, exchanging knives, axes and other metal trade goods for sea otter furs, which they in turn traded in China. But their most important contribution to the region related to food. The traders not only introduced flour, biscuits and rice to the natives, but they also introduced the potato, a starchy tuber which the natives could cultivate themselves.

> The potato was quickly and readily accepted by all the groups of the area, and soon formed such an important item of diet, that one suspects the Indians had previously felt a certain deficiency in starch foods[14].

In the Northwest as in California, the foods of the Indians found little respect among early settlers. Not even fresh Pacific salmon, recognized today for its excellence, could stir the hearts of immigrants accustomed to the foods of Europe and the eastern United States. The factors of the Northwest Company, a fur-trading predecessor of Britain's Hudson's Bay Company, even imported salmon from Scotland, since they deemed the local product unfit to eat[15].

Midwestern settlers reaching the coast disdained salmon as well. Oregon pioneer J. F. Galbraith comments on the surfeit of salmon her family ate after arriving in Oregon's Willamette Valley in the 1830's:

> They lived on salmon for several weeks while Grandfather was looking for a claim in the Tualatin Valley...When they got to the Ewing Young Place my grandmother found a quarter of beef hanging in the tree near the back door, with big crocks of cream and milk, lots of fresh eggs and a big panful of fritters that Sidney Smith had made. Grandmother could not keep the children out of the pan of fritters. They had been living on salmon straight, and they were famished for something good to eat. Pretty soon Sidney Smith came in, with Phil Thompson and Bill Doty. Grandmother said she would get supper for them, but Sidney Smith said, "Do not bother. I made up a whole pan of fritters. We can eat those."
>
> Grandmother said, "The children have already eaten those." So she got out the

Dutch oven and made a lot of biscuits and fried two or three pans of meat[16].

It is clear from Mrs. Galbraith's narrative that beef, cream, butter, and flour were already plentiful in the civilized settlements of Oregon in the 1830s. Much of the credit goes to the Hudson's Bay Company, or rather to Sir George Simpson, Governor-in-Chief of the Company's American operations. When Simpson learned, during an inspection tour of the Western trading forts during 1824/1825 that the Company's factors had run up huge bills importing foods from Britain, he decreed that henceforth the posts must be self sufficient[17]. Agriculture had come to the Northwest.

Parallel to the work carried on by padres and rancheros in California, the Hudson's Bay Company introduced beef, sheep, hogs, chickens, vegetables and fruits to the region. At first mission beef cattle and sheep were bought in California and hogs came from Hawaii[18]. But, under the leadership of Chief-factor John McLoughlin, the Company soon improved its breeding stock by importing prime breeds from Britain. These improved animals formed the nucleus of Northwest herds and were, in turn, traded to California and later helped stock the ranges of Wyoming and the grasslands of the northern Great Plains.

The vegetables introduced first by the Hudson's Bay Company and later by Protestant missionaries and Midwestern farmers were cool climate crops which not only thrived in the Northwest but in California as well. Among fruits, the most notable addition to West Coast eating was the introduction of the apple.

At the start of the California gold rush, the Northwest supplied California (whose resources could not deal with the sudden population influx) with food. Later, California made the Northwest's vegetables its own - so successfully that California has dominated the U.S. vegetable market for almost a century.

Salmon, too, was finally recognized as a valuable staple. But not in its fresh form, which was impossible to ship without proper refrigeration. There was a limited market for lox and other salted salmon, but the fish came into its own after the first canneries were built in the Northwest. Soon canned Northwest salmon flooded world markets. (Since the quality of canned salmon does not equal that of the fresh fish, it lowered the esteem in which Pacific salmon is held throughout the world. Modern shippers, who can ship fresh or frozen salmon almost anywhere without loss of quality, are fighting an uphill battle to overcome this bias.)

A curious thing happened on the way to the harbor. American settlers were reluctant to enter the fishing business, leaving the task to Scandinavian and Slovenian immigrants. (Even today, there are few "anglo" fishermen in the Northwest.) In California, fishing had not been high on the list of occupations during the Mexican period, perhaps because - like their U.S. brethren - most Mexican settlers came from inland regions. As a rule, the waterfront of coastal cities was little developed. Since California possesses few good harbors, boats had be launched through the surf, a dangerous task even in calm weather. Besides, the rough off-shore waters of California demanded uncommon skills in handling small boats. Fishing in California fell at first to Greek, Italian, and Portuguese immigrants, with the addition of Chinese families who fished mostly for the Chinese market and, more recently, to Japanese and Vietnamese fishermen.

Well into the twentieth century West Coast cooks shied away from exploring the full range of the regional largesse, preferring to stick to the basic staples of American cookery. An advertisement in an 1858 Bellingham Bay newspaper lists food staples which were as timelessly American back then, a few years after the Bay was settled by immigrants from Europe and the United States as they are now: flour, beef, pork, bacon, ham, lard, butter, rice, beans, cheese (from California), sugar, coffee, tea, Whisky, brandy, ...[19]. After settlement the diet of the Coast became dull and unimaginative. Alice B. Toklas, a native Californian, reports that she, upon joining Gertrude Stein in Paris, did her best to recreate the comfort foods of home:

> When in 1908 I went to live with Gertrude Stein at the rue de Fleurus she said we would have American food for Sunday-evening supper, she had enough French and Italian cooking; the servant would be out and I should have the kitchen to myself.

> So I commenced to cook the simple dishes I had eaten in the homes of the San Joaquin Valley in California - fricasseed chicken, corn bread, apple and lemon pie[20].

Perhaps West Coast cooks would never have left their protective culinary shell had it not been for the pervasive influence of Chinese and Japanese immigrants. While the Oriental settlers did not change the basic staples eaten on the West Coast, they changed the way they were grown, prepared and eaten and thus play as important a role in Western cookery as the missionaries, rancheros, and fur traders who introduced these staples.

The Chinese came early to the coast, even before the California gold rush, as workers for the Hudson's Bay Company. But the first great influx of Chinese immigrants came during the California gold rush and the succeeding period of railroad construction, when American business demanded cheap labor. Subsequent attempts to dislodge the Chinese living in the West failed, though San Francisco and Vancouver had anti-Chinese riots, and Northwest cities like Seattle and Tacoma actually forced their Chinese citizens out of town. Much of the anti-Chinese attitude was based on imagined economic competition (just as modern-day Americans are entering a period of "Japan-baiting") but a wide cultural gap also seemed impossible to bridge[21]. Northwest historian Murray Morgan describes what happened when the Chinese community invited several Tacoma business men to dinner on 17 February 1885:

> The businessmen...were bowed into a banquet room above a dry goods store and served a ten-course Chinese dinner. The comestibles had been carefully selected to give the least possible offense to an Occidental palate. Drinks were served during courses and between courses. Dinner over, the Chinese competed in expressions of humility and gratitude for the presence of their guests, in expression of wistful hope that the little dinner might strengthen the bonds of friendship and commerce that bound east and west.

Yet Joe Dieringer, a hay merchant who attended the dinner described his impressions of Chinese cuisine in unflattering terms:

> ...After regretting the absence from the feast of shoulder of fricasseed rat, he discoursed on the delights of "parboiled mouse, two months old, its teeth extracted, its tail pomaded with glue, its ears nicely set, the whole immersed in a sputtering, crackling lake of dragon's lard, dotted like an archipelago with ambrosial isles of waxed insects, tanned sealskins, swans' rudders, cormorant filets and jackass corns."

> Back to the rice brandy.

> It was in this tone that the Tacoma LEDGER reported one of the few efforts to create understanding between the white and Chinese communities on Puget Sound in the 1880s. The gap was deep and dangerous[22].

This is not the place to go into a detailed analysis of the causes and effects of sinophobia among West Coast residents, but we must note that the Chinese prospered (and continue to do so) in the West in spite of racial oppression of the worst kind. While many professions were originally closed to Chinese immigrants, restaurants were not. It was here that the natural Chinese disposition towards good food and hard work bore fruit.

> Before World War II, when job discrimination kept many a well-qualified young Chinese American from entering other occupations or pursuing a professional career, the enterprising ones managed restaurants or became waiters and bartenders. They were the ones who met the public. Their fluency in the English language and

their American upbringing enabled them to provide their patrons with gracious and hospitable service[23].

They also influenced the way the West eats. The Chinese "are probably among the peoples of the world most preoccupied with eating." They "have shown inventiveness in this area perhaps for the simple reason that food and eating are among things central to the Chinese way of life and part of the Chinese ethos."[24]

> ...the Chinese...are always very critical of their food and this habitual preoccupation is richly reflected in their art and literature. A Chinese gentleman is never apologetic about his having a good knowledge and skill regarding things gastronomic. In fact he is always proud about the fact.
>
> It is not an exaggeration to say that most Chinese rich and poor alike are gourmets, at least in the sense that they really enjoy good food to the full: by first anticipating it, then discussing about it and, having eaten it, commenting upon it[25].

It should be no wonder that, with a culinary philosophy like this, Chinese cooks were destined to eventually conquer the West. Even in the nineteenth century, Chinese cooks found jobs in mansions and on ranches alike. Their small restaurants might often be the only public eateries in the small towns of California's Central Valley, the Oregon coast, or the high plateaus of Eastern Washington. True, many might serve nothing more than steaks, hamburgers, chop suey or chow mein, but there would be fresh vegetables, stir-fried to crisp perfection, and fish and shellfish from local waters. Perhaps the impact of the Chinese cook on the small communities of the West had something to do with the fact that few organized Chinese communities existed far from the large cities. The Chinese might be "...truly accepted by the white community...because he lived in a white community away from most Chinese."[26]

The Chinese also gave the West Coast a new reverence for food, an esteem for its essential goodness. You can't walk through Chinatown and not note the exceptionally high quality of the food stuffs offered for sale, nor can you walk past a Chinese restaurant without being tempted by the pleasing aromas wafting from the kitchen. James Beard reminisced that

> I think, for example, of the mother of a Chinese friend of mine, a restaurateur of high rank in this country. Her husband was a Christian preacher in the wilds of eastern Oregon during the early part of the century, and for years they trekked to small towns and villages - wherever there might be a colony of Chinese workmen. She cooked Chinese food with almost none of the necessary ingredients, substituting what she found at hand, creating new dishes with a creative style of cooking. Some of her inventions, served in her son's restaurant, have also taken a place in the standard cuisine of this country[27].

The mother of James Beard's friend (both of whom he unfortunately leaves unnamed) did what many Chinese in America did: isolated from the foods and condiments of the homeland, they invented new dishes. This "Cantonese-American" cookery has often been frowned upon by culinary purists, yet it has also been recognized as a unique cuisine of its own.

It might be argued that Chinese inventiveness on the West Coast was limited to innovative cookery, since most staples had already been introduced by the time the Chinese immigrants arrived. But the Chinese were also the first to utilize species of tubers like the water potato or wapato which had been all but neglected by Americans, and by popularizing little-used species of seafood, starting with shrimp and squid in the nineteenth century (when few self-respecting Americans would eat either). Rockfish (commonly eaten whole), sea perch, octopus, sea cucumbers, and the giant geoduck clam were all popularized by Chinese cooks. The Chinese also introduced a number of vegetables: giant Oriental radish, soy beans, bamboo shoots, bok choy, Chinese cabbage, water chestnuts, and bean sprouts, to name just a few. And they gave us such condiments as soy sauce, ginger root, black bean sauce and oyster sauce. Last but not least, they gave us the fad health food dish of the late twentieth century, tofu (also known by the less appealing name of "bean curd").

The Japanese have also contributed to the cookery of the West Coast, but not by introducing new cooking styles. Though sushi is popular in some circles, Japanese cooking in general is still considered too "exotic" by mainstream diners. Tempura and sukiyaki, the Japanese dishes most popular among Western diners, are Japanese adaptations of European dishes or familiar ingredients. While the Japanese have contributed no staples to the cookery of the West Coast, however, they have turned the growing of vegetables and other staples into an art: vegetables and fruits have simply not been the same since Japanese gardeners captured the truck farming industry of the West in the years between the two World Wars. Unfortunately much good was undone by the internment of Japanese all along the West Coast during World War II. Seattle's Pike Place Market, for example, lost most of its farmers in 1942. "Old-timers say the market has never been the same."[28]

Much of the current excitement which pervades all of West Coast cooking, whether "California cuisine" or "Northwest cuisine," resulted from a merging of different cooking styles. This amalgamation has had a profound effect on the types and varieties of staple foods used in West Coast kitchens. Most significantly, perhaps, the three great culinary traditions of the West Coast - Mexican, European, and Asian - have come together in the modern American supermarket. While ethnic supermarkets - Chinese, Japanese, Mexican, and even Vietnamese - have joined the mainstream supermarket, this development may merely reflect an increased ethnic population, not a shift in taste. Of greater interest to the culinary development of West Coast cookery is the fact that many mainstream supermarkets now carry so many different "ethnic" foods that a housewife from the mid-1950s would feel quite bewildered if she were to enter such a market. This change is clear when you look at a sample of supermarkets.

I have tabulated the staple foods and condiments for sale in a number of West Coast supermarkets. Appendix A gives a short synopsis of some of my findings. Mexican food (including Mexican ways of preparing seafood) is the current "hot" trend on the Coast, from California north to British Columbia. Chinese food is not far behind.

A few short notes on the markets:

Uwajimaya is an upscale, mostly Japanese supermarket in Seattle, Washington.

Larry's Market and Queen Anne Thriftway in Seattle are innovative mainstream supermarkets.

Fred Meyer in Bellingham is a standard, mainstream market in a medium-sized Washington town.

Sun Wah is a brand new Chinese supermarket in the Chinatown of Vancouver, British Columbia. Like other supermarkets in its vicinity, it appears to suffer from the close proximity of Chinese produce, seafood, meat and specialty shops.

Saveon is a mainstream supermarket in Richmond, British Columbia (a Vancouver suburb), a community which has recently attracted many Hong Kong immigrants. Unlike the Safeway supermarket across the street, it carries quite a few Chinese vegetables and condiments. Mexican food is promoted at both markets, however - even though the tortillas are not locally made but imported from Oregon. Safeway in Roseburg, Oregon is just a small, standard chain supermarket in an average country town. Unlike its Richmond counterpart, it carries a great variety of ethnic foods.

The State Market in Davis, California (a Central Valley town) is where I did much of my shopping during my graduate student days. Owned by a Chinese merchant, this small supermarket carried an incredible selection of different foods - even Chinese "thousand-year-old" eggs (which were absent from Sun Wah during my last visit).

The Hong Kong supermarket in Monterey Park is a very Asian supermarket which seems to have found its niche in the burgeoning Chinese community of that Los Angeles suburb.

Tianguis in Montebello (and other locations throughout the Los Angeles area) is a supermarket conceived by the giant Von's chain as an answer to the buying power of the increasingly affluent Hispanic community. With its vast selection of foods it is a delightful place to visit, seeming at times more like a covered public market in Mexico than an American supermarket.

References:

1 Foster, Leo, *The Californian Missions*, Beaverton, Oregon: Beautiful America Publishing company: 1977, p. 26

2 Cesena, Margarita Henkel and Guillermo Villarino Henkel, *Las Misionas de Baja California*, La Paz, Baja California Sur: Editionas Aristos, 1977, p. 61ff.

3 Casey, Beatrice, *Padres and People of Old Mission San Antonio*, King City, California, The Rustler-Herald, 1957, p. 10

4 Vestiges of the padres' irrigation system are still visible at several missions, most notably San Luis Rey, Santa Barbara and San Antonio de Padua.

5 Coe, Michael D. Ancient Peoples and Places. *Mexico*. New York: Frederick A. Praeger, 1962;

   p.18 Naturally, the peoples of Mesoamerica followed a number of other customs which are rather widespread among New World Indians, but their typical method of food preparation as a unified complex appears to be unique. The basis of the diet was the triad of maize [corn], beans, and squash. Maize was, and still is, prepared by soaking it overnight with lime and grinding it with a hand stone (Spanish *mano*) on a trough- or saddle-shaped quern (*metate*, from the Nahuatl *metlatl*). The resulting dough is either toasted by the housewife as flat cakes known as *tortillas* or else steamed and boiled as *tamales*. Always and everywhere in Mesoamerica, the hearth is comprised of three stones and is semi-sacred.

   p. 45 For it was the cultivation of maize [corn], beans, and squash that made possible all of the higher cultures of Mexico, and, to a certain degree, those of Peru as well.

   This triad of foods was also encountered and adopted by the Pilgrims when they settled in Massachusetts and by the British immigrants who came to the Southern United States. But even Southerners moving West were unable to establish their "corn pones" and "grits" in the West. These foods are served, but as curiosities, or for variety; they are not staples.

6 McMahan, Jacqueline Higuera, *California Rancho Cooking*, Lake Hughes, California: The Olive Press, 1983.

   p. 3 Wheat was grown with much less effort than corn, which needed irrigation. Wheat flour tortillas became the early bread of California rather than the tortillas de maiz known to Mexico. To Grandmama, the subject was closed. Gentle folk ate the flour tortillas that would unfold like the petals of a flower from the basket lined with a white napkin.

7   Since U.S. Americans refer to "maize" as "corn," maize tortillas are known as "corn tortillas" in the United States. Curiously, wheaten tortillas are commonly referred to as "flour tortillas."

8   Steinbeck, John, *The Wayward Bus*, New York: The Viking Press, 1947,

    p. 136   Besides Alice was the only woman he had ever found outside of Mexico who could cook beans. A funny thing. Every little Indian in Mexico could cook beans properly and no one up here except Alice - just enough juice, just the right flavor of the bean without another flavor mixed up with it. Here they put tomatoes and chili and garlic and such things in the beans, and a bean should be cooked for itself, with itself, alone. Juan chuckled. "Because she can cook beans," he said to himself.

9   McMahan, p. 95

10 The Spanish word "manteca" can mean both butter or lard. In Mexico it generally refers to lard, while the diminutive "mantequilla" is used for [creamery] butter.

11 Jones, Evan, *The World of Cheese*, New York: Alfred A. Knopf, 1979

    p. 255   ...Monterey Jack, the cheese that originated in Monterey County, California, is considered indigenous to the U.S. It was known as queso del pais when first made in the seventeenth-century Spanish missions, but its manufacture was developed by other immigrants, among them a Scot named David Jacks whose surname was coupled with that of the nearest shipping point...

12 Dana, Henry Richard, Jr. *Two Years Before the Mast*, A Personal Narrative. Boston: Houghton, Mifflin and company, 1869 (1840). Chapter XIII

13 Though a somewhat spurious renaissance of matters Spanish and Mexican took place earlier, in the late nineteenth century, a romanticized sort of thing, deemed safe, because so few Mexicans were left in California at the time. Marinacci, Barbara and Rudy, *California's Spanish Place Names: What They Are And How They Got There*. San Rafael, California: Presidio Press, 1980

    p. 11f   The brash Americans displaced the Spanish-speaking Californios as fast as they could. Forty-niners called all Hispanic peoples "greasers" and treated them with the same contempt shown to Indians ("diggers") and Chinese. Some Mexicans who fought back became bandidos and earned place names on new maps. But the old order was passing. The land grants, disputed in the courts, were sometimes overruled. Ranchos were illegally divided by trespassing squatters. The rancheros mortgaged their only large assets, their lands - and often lost them to Yankee land grabbers. Some were developed into new cattle kingdoms, like those of Miller & Lux, Irvine, Bixby, and Hope. Many were subdivided into small farming tracts and hundreds of new settlements made possible by radical changes brought about by railroads and water control. Tourism and "boosting" became large California enterprises.

            At first it looked as if Spanish names would fade from view. But a new surge of interest in the Spanish past began in the 1880s, inspired by the enormous popularity of the novel Ramona. Land promoters saw commercial benefits in giving Spanish - or, more often, Spanish-looking - titles to localities old and new. California architecture incorporated elements of the mission and ranch adobes. Spanish "revival" became fashionable.

            In a very different channel, the state began to receive a new wave of Mexican settlers: contract laborers or braceros to work in fields and factories. The Chicano culture of today exerts its own energy, ideas, and needs regarding both language and land use - already affecting the place-naming process in California by producing new words and new places of Hispanic origins. Spanish names will remain with us forevermore.

14 Smith, Marian W., *Indians of the Urban Northwest*, New York: Columbia University Press, 1949, p. 21

But compare, Suttles, Wayne, *Coast Salish Essays*, Seattle: University of Washington Press, 1987, pp. 137-151, "The Early Diffusion of the Potato among the Coast Salish."

    p. 145   The truth may be that potatoes were accepted quickly and readily because in part they had a cash value at trading posts, and this in turn gave them a potential value and thus a superior status among roots even some distance from the posts.

15 Morgan, Murray *Puget's Sound*: A Narrative of Early Tacoma and the Southern Sound. Seattle and London: University of Washington Press, 1979 First paperback edition., 1981 p. 24

16 Lockley, Fred, *The Lockley Files: Conversations with Pioneer Women*. Compiled and edited by Mike Helm. Eugene, OR: Rainy Day Press, 1981, p. 238f

17 Morgan, ibid,

> Galled at the cost of shipping food from the British Isles to the fur posts - North West Company officers had even imported salmon from Scotland to the Columbia - Simpson decreed that the area be self-sufficient. Fur traders must henceforth farm as well...'
> [1824/25]

18 A triangle trade with Hawaii was established soon after the Northwest Coast was first explored by British traders, since it took less time, in sailing ship days, to sail from California and points south to Hawaii on the prevailing trade winds and then to run down the Pacific coast by way of the Gulf of Alaska, instead of laboriously beating up the coast in the face of adverse winds and currents. Modern sailing yachts still follow this route.

19 Reproduced on the back cover of Edson, Lelah Jackson, *The Fourth Corner*, Highlights from the Early Northwest, Bellingham, Washington: Whatcom Museum of History and Art, 1968.

20 Toklas, Alice B., *The Alice B. Toklas Cookbook*, New York: Harper & Row, 1984 (1954), p. 29

21 Norris, John, *Strangers Entertained*, A History of the Ethnic Groups of British Columbia, Vancouver, British Columbia: Evergreen Press Limited, 1971

    p. 213   The Chinese had come suddenly and in large numbers, they were economically disinherited by the slump following the building of the railway, most of them were single men in an alien society, subject to all the dark suspicions and social ostracism which such sexual imbalance inspires...Their strangeness was seen as menace; their passivity as "inscrutability"; their poverty as a threat to economic standards; their single state as a threat to morals; their numbers as a threat to racial purity.

22 Morgan, ibid p. 213 For the white Northwesterners' attitude towards the Chinese compare also

    p. 214f   Most Chinese lived on the waterfront on the outboard side of the Northern Pacific tracks on land leased from the railroad. (They were not permitted under law to buy real estate.) A few kept garden plots on the hillside. There was a handsome vegetable garden on the site later occupied by Rhodes Department Store at Eleventh and Broadway. They supplemented the truck from their plots and the catch from the bay - whites ridiculed their preference of bottomfish to salmon - with rice, steamed bread, and skunk cabbage.

              They kept apart. Most dressed in blue blouses, blue cotton pants, and sandals. They wore their hair long and queued. Few spoke good English; whites dismissed their language as bird chatter. Sometimes they held parties at which it was rumored that much gin was drunk and opium smoked. There was also talk of the kidnapping of white women, but on the one recorded occasion that police investigated the report that a Caucasian ingenue was suffering a fate worse than death in an opium den it was found that the lady in question had suffered the same fate professionally many times before. The police put her on a boat to Seattle.

23 Sung, Betty Lee, *The Story of the Chinese in America*, New York: Macmillan, 1974 (1971) p. 207

24 Kwang-chih Chang, in the introduction to Chang, K.C., ed., *Food in Chinese Culture*, Anthropological and Historical Perspectives, New Haven, Connecticut: Yale University Press, 1977, p. 13f

25 Peter C. Wong, in Lai, T.C. *Chinese Food For Thought*, Hong Kong: Hong Kong Book Centre, 1978. p. 9

26 Barlow, Jeffrey and Christine Richardson, *China Doctor of John Day*, Portland, Oregon: Binford & Mort, Publisher, 1979, p. 74

27 Beard, James, *American Cookery*, Boston: Little, Brown and Company, 1972. p. 5

28 Shorett, Alice and Murray Morgan, *The Pike Place Market*, People, Politics, and Produce, Seattle: Pacific Search Press, 1982, p. 111

Appendix A: Comparative Staples Chart, West Coast Supermarkets

GRAINS AND BREADS

	Uwajimaya Seattle	Larry's Mkt. Seattle	Queen Ann Seattle	Fred Meyer Bellingham	Sun Wah Vancouver	Saveon Vancouver	Safeway Roseburg, Or.	State 1 Davis, Cal.	Hong Kong Monterey Park	Tianguis Montebello
Grains, Breads										
Masa/Maize flour		★	★	★			★	★		★
Wheat flour	★2	★	★	★	★	★	★	★	★	★
Cookies	★	★	★	★	★	★	★	★	★	★
Noodles, Chinese	★	★	★	★	★	★	★	★	★	★
Noodles, Japanese	★	★	★	★	★	★	★	★	★	★
Pan Dulce (Mexican) 3										★
Pasta, Italian-style		★	★	★		★	★	★		★
Pastries, American 4	★	★	★	★	★	★	★	★	★	★

Footnotes:

1 - The data from this market come from the late 1960 to mid-1970, when this small, Chinese-owned, Western-style supermarket was in the vanguard of selling foods not considered profitable for supermarkets.

2 - Only in the form of pre-mixed flour with yeast added, the kind popular in Japan right now for instant bread bakers.

3 - Pan dulce, a yeasty, very Mexican pastry, shows up increasingly, though spottily, at supermarkets throughout the region. B

4 - These run the gamut, from very fancy European style constructions to plain, factory-made doughnuts and rolls, and include such American culinary curiosities as "twinkies" and "ding-dongs."

TORTILLAS

	Uwajimaya Seattle	Larry's Mkt. Seattle	Queen Ann Seattle	Fred Meyer Bellingham	Sun Wah Vancouver	Saveon Vancouver	Safeway Roseburg, Or.	State Davis, Cal.	Hong Kong Monterey Park	Tianguis Montebello
Tortillas										
made on premises										★
fresh, local 5	★	★	★	★			★			★
frozen		★	★	★			★		★	★

Footnotes::

5 - Perhaps the absence of wheat tortillas in Oriental markets can be explained by the fact that similar flat, thin sheets of wheat flour are sold under a variety of names: egg roll skins, wonton skins, et al. I have found no "corn" tortillas nor any other type of Mexican food products in British Columbia's Chinese supermarkets (though "Mexican" is much the vogue in mainstream markets), except for a few odd cans of generic "chili con carne" imported from the United States.

BEANS

	Uwajimaya Seattle	Larry's Mkt. Seattle	Queen Ann Seattle	Fred Meyer Bellingham	Sun Wah Vancouver	Saveon Vancouver	Safeway Roseburg, Or.	State Davis, Cal.	Hong Kong Monterey Park	Tianguis Montebello
Beans										
bulk, loose 6				★						★
prepackaged 7	★	★	★	★	★	★	★	★	★	★

Footnotes::

6 - Bulk foods are the coming thing in American supermarkets. Quite a few of the stores not listed here do have bulk legumes - dried peas, lentils, and several kinds of dried beans. However, in British Columbia, bulk foods tend to run towards candies, tea bags, noodles, crackers, and dog food.

7 - Generally the beans sold in Oriental stores differ from those sold in supermarkets with a mainstream orientation. While pinto beans, for example, are sold in both Mexican and mainstream supermarkets, they are rarely found in Oriental ones; mung and adzuki beans are uncommon in Mexican and mainstream supermarkets; black beans and black-eyed peas overlap, beansprouts are almost universal, though only Oriental markets usually carry both mung and soy bean sprouts.

MEAT, SAUSAGE, FOWL, AND DAIRY PRODUCTS

	Uwajimaya Seattle	Larry's Mkt. Seattle	Queen Ann Seattle	Fred Meyer Bellingham	Sun Wah Vancouver	Saveon Vancouver	Safeway Roseburg, Or.	State Davis, Cal.	Hong Kong Monterey Park	Tianguis Montebello
Meat										
service counter	★	★	★		★	★	★	★	★	★
fresh, cooler	★	★	★	★	★	★	★	★	★	★
Chinese sausage 8	★	★						★	★	
Chorizo (Mexican) 9	★	★	★	★				★	★	
Fowl										
fresh chicken 10	★	★	★	★	★	★	★	★	★	★
Cheese										
American types 11		★	★	★	★	★	★	★	★	★
Imported 12		★	★	★		★	★	★		★
Mexican types 13		★	★	★			★			★

Footnotes:

8 - Much of the West Coast's Chinese sausage is made in Vancouver, B.C. and distributed from here. Yet it has not found its way into local supermarkets, probably because Chinese cooks prefer to buy their meat at the butcher's.

9 - Mexican chorizo is in many ways quite unlike its Spanish/Basque ancestor. It comes in a variety of styles and is often quite spicy. The Tianguis supermarkets in the Los Angeles area have a counter devoted solely to the selling of different kinds of chorizo.

10 - Many West Coast shoppers prefer locally raised chickens - which are more expensive, as a rule - to the cheaper chicken raised in the American South. Local chicken is commonly sold fresh; southern chicken is frozen and thawed for sale.

11 - "American cheese" here refers to standard types produced in the United States and Canada: cheddar, longhorn, monterey jack, blue, cream, et al.

12 - Imported cheese in mainstream supermarkets tend to be of European origin; in Mexican supermarkets the bulk of the imported cheese is of the Latin American type and comes from Mexico, Guatemala, Costa Rica, et al.

13 - Mexican types, usually quite mild, are made both in the United States (as far north as the state of Washington), and imported from Mexico and Latin America.

VEGETABLES

	Uwajimaya Seattle	Larry's Mkt. Seattle	Queen Ann Seattle	Fred Meyer Bellingham	Sun Wah Vancouver	Saveon Vancouver	Safeway Roseburg, Or.	State Davis, Cal.	Hong Kong Monterey Park	Tianguis Montebello
Vegetables										
Standard American [14]	☆	☆	☆	☆	☆	☆	☆	☆	☆	☆
Mexican, general	☆	☆	☆	☆			☆	☆		☆
chayote	☆	☆	☆	☆			☆	☆		☆
nopalitos		☆	☆	☆						☆
tomatillos		☆	☆	☆		☆				☆
jicama [15]	☆	☆	☆	☆	☆		☆	☆	☆	☆
vegt. papaya [16]					☆				☆	☆
daikon radish	☆	☆	☆	☆	☆		☆	☆		☆
baby bok choy [17]	☆	☆			☆				☆	
pea vines [18]	☆	☆			☆				☆	

Footnotes:

14 - "Standard American" vegetables include all of the common kind we've seen in our markets for as long as we can remember: artichokes, broccoli, cabbage, cauliflower, et al.

15 - Jicama is a Latin American root vegetable which is not only popular in Mexico, where it is commonly eaten raw in salads but it has also entered Chinese kitchens where it is used in conjunction with or in lieu of water chestnuts.

16 - The small, bright yellow "fruit" papaya is found in just about every supermarket these days, but the large, football-sized (or larger) green vegetable papaya is very popular in Cantonese stir-fried dishes. These large green papayas (which, incidentally are merely a larger version of the "fruit") are a good indicator of how truly ethnic Oriental or Mexican a supermarket is.

17 - Bok Choy is widely sold in mainstream supermarkets, but the vegetable's stalks are huge - up to a foot long, and the leaves are equally large. Chinese cooks prefer small, "baby" bok choy in their cookery. They are a good indicator of how Oriental a supermarket is.

18 - Pea vines are an East Asian vegetable which is making VERY tentative moves into mainstream supermarkets.

CHILES

	Uwajimaya Seattle	Larry's Mkt. Seattle	Queen Ann Seattle	Fred Meyer Bellingham	Sun Wah Vancouver	Saveon Vancouver	Safeway Roseburg, Or.	State Davis, Cal.	Hong Kong Monterey Park	Tianguis Montebello
Chiles	★	★	★	★	★	★	★	★	★	★
ancho/pastilla [19]										★
California, mild [20]	★	★	★	★			★	★	★	★
jalapeno	★	★	★	★	★	★	★	★	★	★
serrano	★	★	★	★	★	★	★	★	★	★
Oriental/Thai [21]	★	★			★	★		★	★	★
dried packaged	★	★	★	★	★	★	★	★	★	★
dried bulk [22]										★

Footnotes:

19 - While most American supermarkets sell mild chiles, the richly flavored, meaty ancho/pasilla is hardly seen. The less appealing California/Anaheim chile seems to have become the mainstream supermarket chile of the West Coast. Small Mexican neighborhood produce markets throughout the region stock the ancho/pasilla.

20 - Strangely the mild California and other mild chiles are not found in British Columbia markets, Chinese or mainstream, while the hot jalapeno and serrano are common.

21 - The very hot pepper known as Oriental, Sichuan, or Thai is commonly sold in Oriental markets only. It is very hot.

22 - Bulk is a matter of definition when it comes to dried chile peppers: almost all markets sell large bags of dried (red) chiles; only Mexican supermarkets mound them into tall heaps in their produce sections.

TOFU

	Uwajimaya Seattle	Larry's Mkt. Seattle	Queen Ann Seattle	Fred Meyer Bellingham	Sun Wah Vancouver	Saveon Vancouver	Safeway Roseburg, Or.	State Davis, Cal.	Hong Kong Monterey Park	Tianguis Montebello
Tofu (bean curd) [23]	★	★	★	★	★	★	★	★	★	★
less than 3 var.						★				
four or more var.	★	★	★	★	★		★	★	★	

Footnotes:

23 - TOFU (also known as bean curd) is ubiquitous in American mainstream supermarkets; it is almost absent in British Columbia markets.

FISH

	Uwajimaya Seattle	Larry's Mkt. Seattle	Queen Ann Seattle	Fred Meyer Bellingham	Sun Wah Vancouver	Saveon Vancouver	Safeway Roseburg, Or.	State Davis, Cal.	Hong Kong Monterey Park	Tianguis Montebello
Fish	☆									
service counter	☆	☆	☆	☆				☆	☆	☆
fresh, live tank [24]	☆								☆	☆
halibut, fresh	☆	☆	☆	☆		☆	☆		☆	☆
salmon, fresh	☆	☆	☆	☆		☆	☆		☆	☆
swordfish, fresh	☆	☆	☆	☆			☆		☆	☆
tuna, fresh	☆	☆	☆	☆			☆		☆	☆

Footnotes:

24 - Live tanks for fish have not yet caught on in mainstream supermarkets. In British Columbia, the Chinese still prefer buying seafood at their local fish store. That explains the almost total absence of fresh (as opposed to dried or frozen) fish in Vancouver's Chinese supermarkets.

SHELLFISH

	Uwajimaya Seattle	Larry's Mkt. Seattle	Queen Ann Seattle	Fred Meyer Bellingham	Sun Wah Vancouver	Saveon Vancouver	Safeway Roseburg, Or.	State Davis, Cal.	Hong Kong Monterey Park	Tianguis Montebello
Shellfish	☆									
fresh, cooler	☆	☆	☆	☆		☆	☆		☆	☆
fresh, live tank	☆	☆	☆						☆	☆
clams, fresh	☆	☆	☆	☆		☆	☆		☆	☆
crab, fresh	☆	☆	☆	☆		☆	☆		☆	☆
crab, live	☆	☆							☆	
mussels, fresh [25]	☆	☆	☆							
shrimp, fresh	☆	☆	☆	☆		☆			☆	☆
oysters, fresh [26]	☆	☆	☆	☆		☆	☆		☆	☆

STAPLE FOODS OF THE AMERICAN WEST COAST

Footnotes:

25 - California supermarkets rarely stock truly fresh live mussels (though these tasty mollusks are raised along the California coast), preferring frozen New Zealand green-lipped mussels. The Northwest sells local mussels almost exclusively.

26 - In the Pacific Northwest, oysters are sold both in the shell and shucked; in California, they are sold mostly shucked, either loose or in jars.

CONDIMENTS, HERBS, SPICES, AND PACKAGED SAUCES

	Uwajimaya Seattle	Larry's Mkt. Seattle	Queen Ann Seattle	Fred Meyer Bellingham	Sun Wah Vancouver	Saveon Vancouver	Safeway Roseburg, Or.	State Davis, Cal.	Hong Kong Monterey Park	Tianguis Montebello
Condiments [27]										
black bean sauce	★	★	★	★	★	★	★	★	★	
chile salsa, green		★	★	★		★	★	★		★
chile salsa, red	★	★	★	★	★	★	★	★	★	★
ginger root, fresh	★	★	★	★	★	★	★	★	★	★
oyster sauce	★	★	★	★	★	★	★	★	★	
rice vinegar	★	★	★	★	★	★	★	★	★	★
soy sauce	★	★	★	★	★	★	★	★	★	★
Worcestershire sauce	★	★	★	★	★	★	★	★	★	★

Footnotes:

27 - The types of condiments may be similar in mainstream, Chinese or Mexican supermarkets, but the brands - and thus the flavors - differ widely between the different types of markets. Generally, condiments and sauces developed for American mainstream cooks do not possess the full, complex flavors of the originals.

WINE & BEER

	Uwajimaya Seattle	Larry's Mkt. Seattle	Queen Ann Seattle	Fred Meyer Bellingham	Sun Wah Vancouver	Saveon Vancouver	Safeway Roseburg, Or.	State Davis, Cal.	Hong Kong Monterey Park	Tianguis Montebello
Wine	★	★	★	★	★	★	★	★	★	★
local wine 28	★	★	★	★		★	★	★	★	★
imported wine	★	★	★	★	★	★	★	★	★	★
California wine	★	★	★	★	★	★	★	★	★	★
Northwest wine 29	★	★	★	★			★			
Beer	★	★	★	★	★	★	★	★	★	★
national brands	★	★	★	★	★	★	★	★	★	★
local "micro-brews" 30		★	★	★			★			★
imported	★	★	★	★	★	★	★	★	★	★
Sake/Rice Wine	★	★	★	★	★	★	★	★	★	★
imported	★	★	★	★	★	★	★	★	★	★

Footnotes:

28 - Local here is seen as opposed to regional: In Los Angeles, for example, "local" wine means wine from the Temecula Valley or other southern growing districts, but not wine from the Napa Valley or other northern California regions.

In Portland, Oregon, local means wine from the adjacent Willamette Valley and not from the Rogue, Umpqua, or Hood River Valleys.

On the other hand, in San Francisco, a city which is a ways distant from nearby wine producing regions, the wines of Mendocino, Sonoma, and Napa Counties as well as the Livermore Valley may be considered "local." The same is true for the Seattle, where the wine-growing regions lie east of the Cascade Mountains. Wines from the Yakima and Columbia Valleys to the east and the Nooksack Valley to the north are considered "local." There's some smudging of boundaries here, since quantities of grapes grown east of the Cascades are brought to Seattle and its suburbs for vinification.

In Vancouver, British Columbia, wine from the Okanagan Valley (also east of the Cascades) is considered to be local wine.

29 - The "Northwest" wine growing region is generally understood to include Washington, Oregon, and Idaho, but not British Columbia (which makes "southwestern" Canadian wine).

30 - Perhaps the best definition of "microbrewery" comes from Vince Cottone, *Good Beer Guide*, Breweries and Pubs of the Pacific Northwest. Seattle: Homestead Book Company, 1986, p 9.

Cottone, who prefers the term "Craft Brewery," describes this as "a small brewery using traditional methods and ingredients to produce a handcrafted, uncompromised beer that is marketed locally."

Curiously, despite the supposedly local distribution of these brews, supermarkets in the Northwest commonly stock many Californian "microbrews" while California markets carry virtually no Northwestern beers.

ETHNIC SUPERMARKET RESTAURANTS

	Uwajimaya Seattle	Larry's Mkt. Seattle	Queen Ann Seattle	Fred Meyer Bellingham	Sun Wah Vancouver	Saveon Vancouver	Safeway Roseburg, Or.	State Davis, Cal.	Hong Kong Monterey Park	Tianguis Montebello
In-House Preparation	★	★	★	★		★			★	★
sit-down	★	★							★	★
take-out	★	★	★	★		★			★	★
Takeout Sushi/Sashimi										
	★	★								

Different methods of bread-making in private households — a comparison of working time, quality and costs

by Martina Ehnle, Cornelie Pfau, Johannes Piekarski

1. Introduction

In Germany, in the middle of the 19th century, as much as 439g of bread were consumed per head per day[1]. In the past 100 years consumption of cereal products and bread decreased strongly. In 1973/74, only 188g were consumed per head per day[2, p. 151].

Since then, bread consumption has been on an upward trend again; today, the average is 220g per person per day; this corresponds to 4-5 slices of bread and one roll[3, p. 23].

Consumption in Germany corresponds to the average consumption of western Europe. The Finns and the Irish are in the lead, they eat 228g per person per day. People in Luxembourg bring up the rear, they eat 151g bread per day only[4, p. 21].

International nutritional guidelines, for instance, the "Dietary Goals" of the USA, the "Nutrition Report" of the German Society for Nutrition, and the "Report on the Nutritional Aspects of Bread and Flour" of the Department of Health's Committee on Medical Aspects of Food Policy in the United Kingdom, recommend that the consumption of cereal products and of bread be increased. The German Society for Nutrition recommends 300g of bread daily, that is five to six slices of bread and two rolls[3, p. 24; 5, p. 487].

Bread is an important supplier of nutrients: high-molecular carbohydrates in the form of starch, supply of which should be further increased; dietary fibre, protein, vitamins, minerals, trace elements, and little fat[2].

Besides the possibility of buying bread, bread can also be made at home. However, few households so far have been practising bread-baking at home. The proportion of homemade bread, or rolls, is about 1% only of the total consumption in a year[6, p. 38].

37% of 136 housewives questioned claimed to have experience in bread-baking. Of these, 17% bake regularly and very often, 43% now and again, 22% rarely and 17% did so only once[6, p. 37]. The main reasons for baking bread at home, so the women said, are fun and a sense of achievement after a successful operation. The flavour of homemade bread, furthermore, is regarded as much better[6, p. 38].

Baking bread at home allows one to vary recipes individually and adapt them to ones taste, and to have freshly baked bread also on Sundays and holidays.

Appliance manufacturers have developed special equipment for bread-baking to facilitate the job in private households. So two different methods of bread-baking are available to housewives: the conventional method of baking bread in an oven and the new method of using an automat.

A housewife decides to bake bread herself mainly - as has been mentioned before - for reasons of fun and because she expects to derive a sense of achievement. Hence fun and sense of achievement will be decisive for the choice of method. But choice of the method will also depend on a comparison of working-time required, quality of the bread and costs of the operation.

Figure 1. Construction of a Bread-making Machine

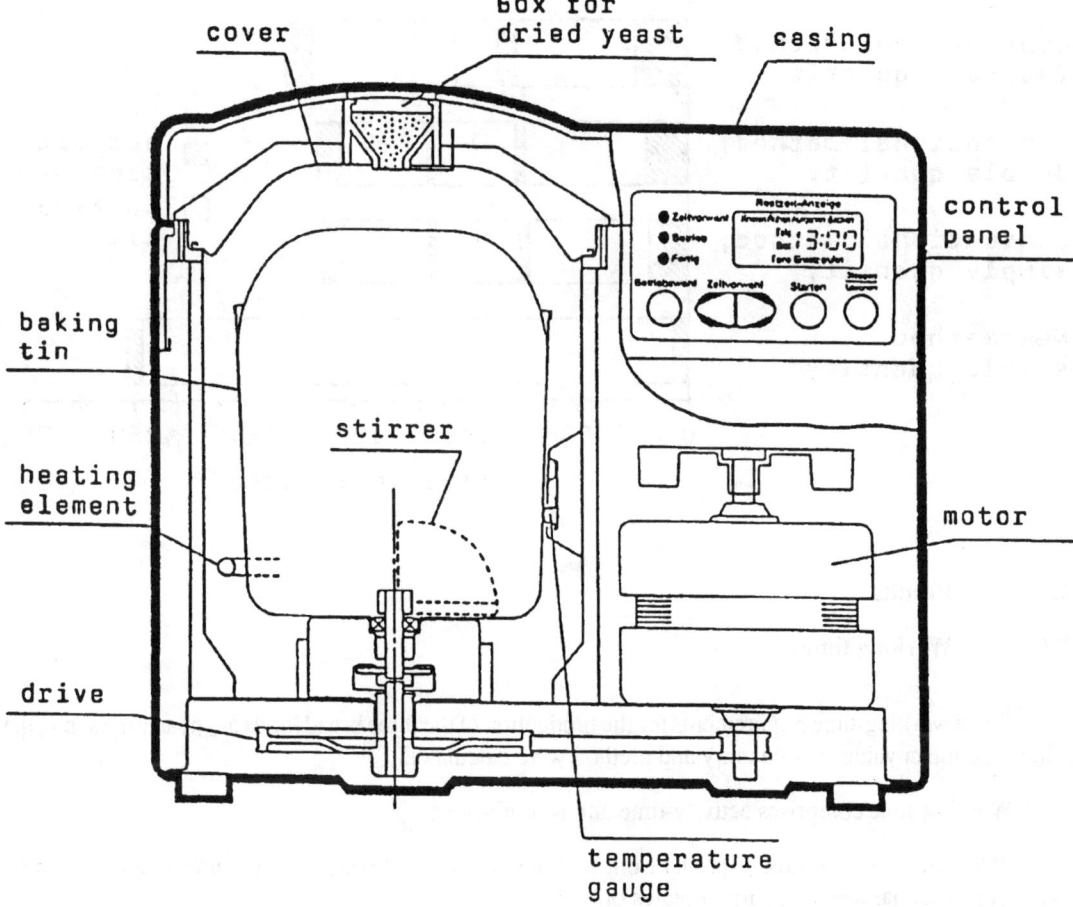

We conducted a study with the objective of analysing and comparing working-time, quality and costs of bread baking in an automat and in an oven, in order to help housewives in their decision-making for or against the acquisition of an automat.

To bake bread in an automat is simple. The ingredients are weighed and placed into the baking tin. The cover is closed and dry yeast is put into the yeast inlet at the upper side of the cover. Then the machine is switched on. Kneading, rising and baking processes are fully automatic. After an operation time of four hours, sound and light signals indicate that the baking process is completed. The bread is taken out. All that remains to be done is washing the dishes.

In the conventional way, the dough is kneaded manually and moulded; after "rising" in the mixing bowl first, and in the tin for a second time, the dough is baked in the oven. However, while only one loaf of 500g - 600g can be baked in the automat in one operation, several loaves may be baked in the oven.

Three different kinds of bread were produced with each method: white bread of light wheat flour, wheat and rye bread and wholemeal bread.

In the conventional way, two and four loaves, besides the original quantity of one loaf, were produced in one operation.

Figure 2. Working time for the production of bread by method and quantity

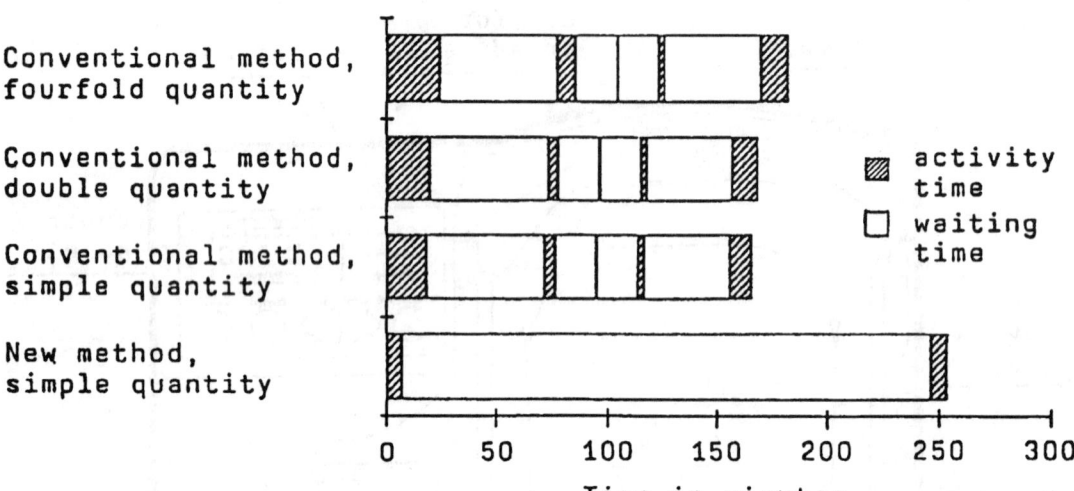

2. Results

2.1 Working time

Since working-time requirements for the production of the three bread kinds are nearly the same, the arithmetic mean values per quantity and method were calculated.

Working-time comprises activity-time and waiting-time.

255 minutes are required to produce one loaf in the automat; to produce the same quantity of bread in the conventional way takes 165 minutes only (Fig. 2).

With the automat, activity-time is shorter and waiting-time longer than with the conventional method. Waiting-time, furthermore, is uninterrupted (Fig. 2).

To produce one loaf, 14 minutes of activity-time are required with the new equipment, and 35 minutes using the conventional method. Activity-time includes the time of "cleaning and drying the dishes". Waiting-time is 240 minutes with the new, and 130 minutes with the conventional method (Fig. 2).

In the conventional method, waiting-time is three times interrupted; first, to "knead the dough" and to "place the dough into the tin"; secondly, to "switch on the oven"; it has been found that the result is better when the oven is preheated; and thirdly, to "place the tin into the oven" and to "pour water into the oven"; if exposed to water vapour, the bread crust becomes smooth and shining; otherwise it is unsmooth, dull and spotty.

Baking several loaves in the conventional way increases the working-time only slightly. Four loaves require only 20 minutes more than one loaf (Fig. 2).

Fig. 3 shows the difference in working-time between the two methods.

If several loaves are produced in the automat, the working-time requirements multiply. Four loaves in the automat take 17 hours, while in the conventional way only three hours are required.

Figure 3. Working time for the production of bread by method and quantity

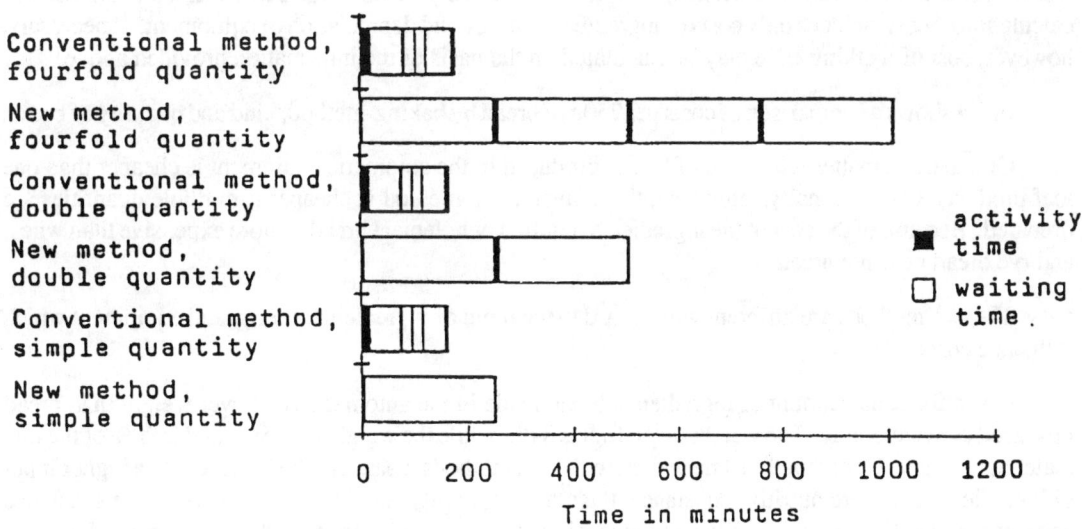

Figure 4. Costs per 250g bread by baking method, kind and quantity of bread

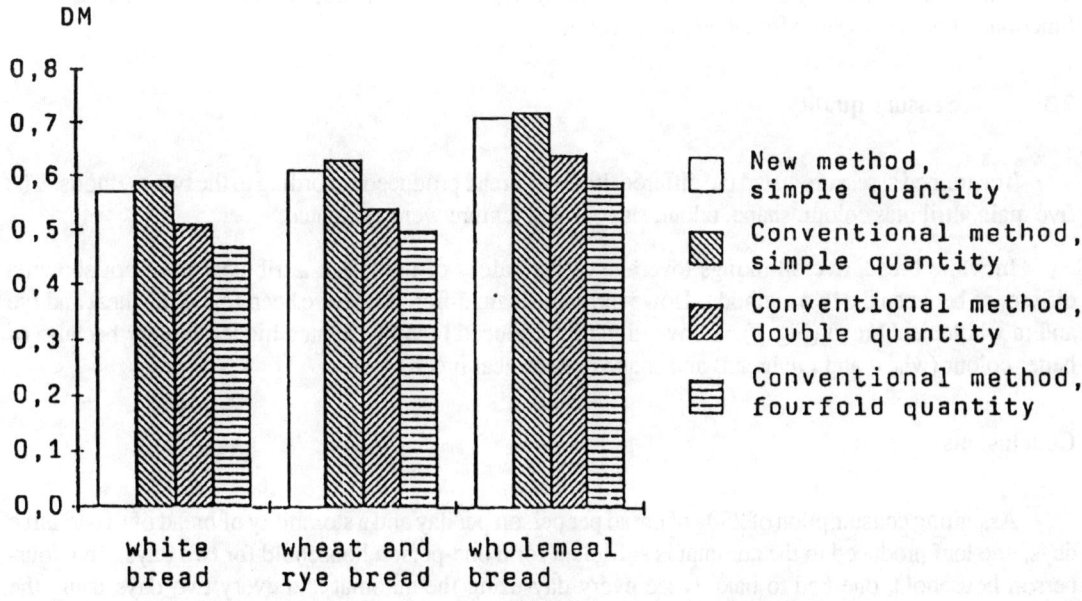

2.2 Cost

The cost of baking bread depends on the necessary input of ingredients, energy, equipment and working-time. In private households, working-time usually is not evaluated, therefore the present calculation of cost includes only costs of ingredients, energy and depreciation of equipment. If necessary, however, cost of working-time may be calculated on the basis of the information provided above.

Fig. 4 shows a comparison of costs per 250g of bread by baking-method, kind and quantity of bread.

One loaf, no matter what kind of bread, produced in the automatic equipment is cheaper than one loaf produced conventionally. However, the conventional method is cheaper if multiple quantities are prepared. Because of the cost of the ingredients required, wholemeal bread is more expensive than wheat and rye bread or white bread.

The two methods are different with regard to the input of ingredients, energy and equipment which influence costs:

Given the same amount of ingredients, bread made in the automat has a lower weight than bread produced conventionally. The weight of the first is 84%, while the weight of the second is 90% of the raw materials' weight. To obtain the same weight with both methods means for the first method a higher input of ingredients and more nutritive substance; the other possibility is to have the same nutritive substance and less weight. Consumers usually compare weights; so we decided to show the results of a calculation based on higher input of ingredients required for bread produced automatically. In general, the costs of the ingredients make up the largest part of total costs - between 78% and 90% (all kinds of bread).

Bread baked conventionally requires more energy, because milk must be heated and because baking in the oven requires more energy than baking in the automatic equipment.

In the Federal Republic of Germany, bread-baking automats are available at about the same price as an oven. But there are differences in economic life and frequency of use. The automatic device has a motor and several movable parts; information about its service life is not available. We estimated service life at half the economic life of an oven (10 - 15 years). By determining the amount for depreciation multifunctions of an oven were also taken into account.

2.3 Sensory quality

Ten trained assessors tested the different kinds of bread produced according to the two methods. The five main attributes colour, shape, odour, flavour and texture were evaluated.

In white bread, overall ratings (average of the values of the tested attributes) have not shown a difference between the two methods. However, significant differences have been found in wheat and rye and in wholemeal bread (Fig. 5). Conventionally produced bread was rated higher, mainly because of better colour (wheat and rye bread) and shape (wholemeal bread).

Conclusions

Assuming consumption of 250g of bread per person per day and a storability of bread of two to three days, one loaf produced in the automat is sufficient for a one-person household for two days. In a four-person household, one had to bake twice every day using the automat and every two days using the conventional method.

Figure 5. Sensory quality by baking method, kind and quantity of bread (weighted average)

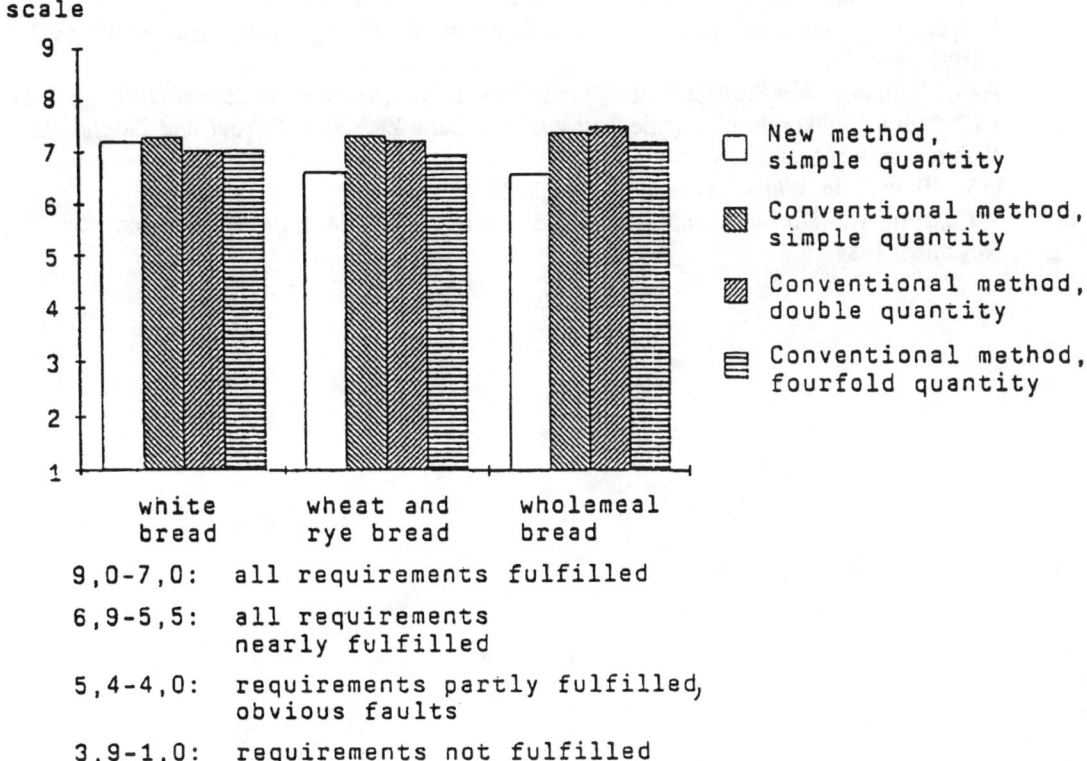

Our experimental results have shown that the conventional method is more favourable regarding working time, costs and frequency of the operation if larger quantities of bread are required.

If only little bread is consumed, or in one-person households, the automatic process has some advantages: activity-times are shorter, waiting-time is uninterrupted, and costs are lower.

As far as sensory quality is concerned, there is no difference between the two methods in white bread. Wheat and rye bread, and wholemeal bread produced conventionally are rated a little higher. These results apply to these three types of bread only. They may be different for other types of bread or when the recipe is changed.

However, the decision to buy an automat for bread baking should be based also on other factors, for instance, available kitchen space, frequency of use, or the multi-function purpose of an oven.

If the main reasons for baking bread at home are fun and, hopefully, a sense of achievement, the question arises whether a baking automat is the proper choice. All one can do is to "weigh the ingredients", "place them into the automat", "switch on the automat" and "remove the bread".

If, however, bread is made at home to make sure that freshly baked bread is always available, and further, because the bread can be adapted to one's taste or because one finds pleasure in techniques, acquisition of an automat may be worthwhile for small households or households in which not much bread is consumed.

References:

1. Auswertungs - und Informationsdienst für Ernährung, Landwirtschaft und Forsten (AID) e.V. (pub.): Broschüre *Mehl und Brot*, AID - Verbraucherdienst. Bonn 1987.
2. Becker, H.G.: 'Stellenwort von Getreide und Brot in der Ernährung.' In: *Getreide, Mehl und Brot* 5/1985, pp.151-4.
3. AgV e.V. (pub.): 'Was Sie über Brot wissen sollten.' In: *Verbraucher-Rundschau* 9/1988, pp.1-27.
4. AID (pub.): 'Vollkornbrot'. Begleitheft zur Film-Serie *Zwischen Zwiebel und Zweifel*. AID-Verbraucherdienst informiert. Bonn 1987.
5. O.V.: 'Bread'. In: *Which?* November 1985 p.487.
6. v. Boxberg, W.: 'Documentation Brot und Brödchen'. In: *CMA Mafo Briefe*. Kennziffer 423, November 1984.

The Metamorphic Potato:
A Revolutionary Root*

By Beatrice Fink

In a zealously patriotic speech given on 21 February 1794 (3 ventôse an II), Barère, a member of the famed Comité de salut public, eloquently proclaimed "La liberté et des pommes de terre! La liberté et des patates!", then proceeded to label the potato a "blessing of our saintly love for humanity."[1] This holy alliance between *Solanum tuberosum esculentum* and mankind was the rhetorical component of the Revolutionary government's protracted efforts to encourage the cultivation of the potato and promote its consumption by sending about town criers to spread the word. In a country plagued by chronic and drastic shortages of wheat, therefore of bread, it was cheap and easy to cultivate *ersatz*. The tuber was also glorified in the annals of cuisine. The *Cuisinière républicaine*, authored by Mme. Mérigot in 1795, was remarkable not only because it was the first cookbook to be published by a woman in France, but also because it dealt solely with the potato[2]. Does this backing by officialdom and the printed word then indicate that the potato and its derivatives were "in" at the time? In fact, rather the opposite was still the case, for the resistance to this food was particularly stubborn in what was to become the land of *pommes frites*. Without the turmoil of revolution and the ensuing changes in domestic and public eating patterns, the mainstream of the tuber would doubtless have proceeded far more slowly.

My main concern shall therefore be retracing the high-roads and by-roads of the potato's slow advance from countryside and tuber eyed with suspicion to the versatile food Parmentier did so much to promote, stressing the experimentation that went into producing pure potato bread, and focusing on contemporary publications treating the subject. Panification of the potato, however, fits into the more general framework of bread supply in eighteenth-century France. My final remarks will therefore centre on the broader question of how the Revolutionary government - specifically that of the Jacobins - addressed the problem of bread in political and ideological terms by legislating a "pain de l'égalité".

The potato plays a very special rôle in eighteenth-century French cookery[3]. Its social climb may be likened to that of a novelistic hero wending his way from provincial obscurity to the urban upper crust. At a time of population increase, when the scarcity and high price of wheat brought on food riots and a "guerre des farines", the tuber and its panification were regarded as a means of confronting the problem of sustenance[4]. An agricultural evolution, brought on by its relative ease of cultivation, ran parallel with a cultural one, a progressive change in the way it was perceived. Considered at first a forbidden food (a poison, a source of disease) it gradually turned into a heaven-sent one (nutritional, economical, antiscorbutic). When textually encoded, it generated converging discourses: social, economic, dietetic, scientific, and, last but by no means least, gustatory. These discourses are primarily though not exclusively inscribed in culinary and paraculinary texts.

Until the middle of the Eighteenth century there is no mention of the tuber in French cookery books or treatises even though its presence in France is recorded at the start of the Seventeenth century and it is cultivated in the Eastern part of the country[5]. In mid-century, de Combles treats it with contempt in his *Ecole du jardin potager* (1749), and the *Encyclopédie* refers to it as flatulent, then adds these oft quoted lines: "but what is a bit of wind for the vigorous organs of peasants and labourers?" ("pomme de terre", 1765). Yet in 1810 the gastronome Grimod de la Reynière states that the potato "can most singularly heighten the pleasures of the palate in savoury cusine."[6]

In spite of these flattering words, "savoury cuisine" potato recipes are few and far between prior to 1810. The very first are doubtless those provided by Parmentier, for whom potatoes constitute "a food that can be prepared in an infinite number of ways", one that may be "disguised in a thousand different manners". In his *Examen chimique de la pomme de terre* (1773), he lists dishes running the gamut from boiled potatoes to "tarts so like almond tarts that even the greatest connoisseurs are impressed". The cook, concludes Parmentier, "will find in potatoes the wherewithal to exercise his inventive genius"[7]. He lists recipes "doubtless little known" that include potatoes "en matelote", an "obviously inexpensive cake", and "a sort of preserve" (pp. 222 - 225). Many of them reappear in the *Cuisinière républicaine*. He even details an entirely potato-based meal where the tuber is metamorphosed into pâté, fritters, a brioche, and even into cheese-like and coffee-like substances. Mme. Mérigot extends the list by diversifying (à la Nanette, à la polonaise, à la barigoule, etc...). Grimod proposes an orange-blossom-flavoured potato cake with a *fines herbes* variant, but his primary aim is to promote certain derivatives such as potato rice or semolina. The *Cuisinière républicaine* likewise relegates gastronomy to the sidelines by prioritising the simplicity and thriftiness of potato dishes. While gourmand concoctions were indeed thought up, their primary purpose seems to have been to propagandise a cheap and healthy food by improving its gastronomic image.

Throughout the latter part of the century, the tuber as palatal delight in effect boosts a comestible linked with social conscience and technological efforts. This double thrust mirrors two differing representations of food preparation within cookery books. The first and better known of the two - best illustrated by Menon's *Les soupers de la cour* (1775) - situates cuisine within the confines of the *beaux arts*, is authored by chefs, and meant for the *cognoscenti*. The other, while conscious of quality cuisine, is more broad-based in terms of social class. It is manifest in Menon's *Cuisinière bourgeoise* (1746), a best seller in its time. Its declared aim is to "provide the public with a bourgeois cuisine", its targeted readership female. Its preface lauds "simple, savoury, and novel dishes" while stressing the desire to get across to the many. "My explanations", says the author, "are intelligible and understood even by those who know nothing of cooking." The *Cuisine et office de santé* (1758, attributed to Menon) is "specifically for those who are thrift and diet conscious". In other words, non-gastronomic criteria serve to heighten another set of priorities, one of which points to treatises on the potato, highlights the nutritional aspects of food, and raises civic consciousness (such words as "patriotic" and "citizen" appear in food texts well before the Revolution). A different kind of author emerges as well: instead of being a chef or "officier de bouche" he is a man of letters, a book seller, or even a scientist. In so far as the potato is concerned, its preparation comes under the category of ingredient (in a broth or gruel) or of panification.

Parmentier explains how to prepare a gruel with the "precious roots". This type of food ties in with a culinary tradition meant to satisfy the needs of those who must economise and cannot afford meat, or those in need of food both easily digestible and nutritious (the newborn, the sick). Recipes are generally given for sizable quantities of people, as in the case of kitchen soups. Vincent La Chapelle's *Modern Cook* (London, 1735), for instance, contains a gruel recipe for fifty people made of oats, barley, and rice, as well as directions for distributing the gruel to the poor. The 1755 edition of Louis Lemery's *Traité des alimens* includes directions for preparing huge quantities of a "dauphinoise" soup based on wheat, flour or rice. The potato adapted itself to this type of cooking all the more easily as wheat became a rarer and higher-priced commodity.

The tuber's versatility, its relatively simple cultivation, and its low price gave it an advantage over wheat. In 1772, that is to say just a year before Parmentier's prize-winning *Examen chimique de la pomme de terre*, a book entitled *La Cuisine des pauvres* and attributed to Varenne de Beost was published in Dijon. This is actually a compilation of contemporary writings meant to "compensate for the unforeseeable predicaments caused by grain shortages". It argues strongly in favour of potato consumption, and details various methods for preparing the tuber. These include a "means for preparing an inexpensive rice" with a lengthy supplement, and a prefatory note on "the manner of nourishing oneself cheaply and well" that refers to the inexpensive rice. In spite of its name, this broth's principle ingredient is potatoes. A recipe is given for preparing 444 pounds of said rice, and another where the cost of the dish is specified in both domestic and foreign currencies. It is described as being "basically preferable to the ordinary nourishment of common folk" given its salubrity and low cost. In addition, it is said to flatter the palate, lordly ones as well as those of ordinary people. In a letter dated 26 February 1769 Mirabeau says it is "comparable to the very best soups" (p. 34).

The most famous of these soups for the masses, in fact the first to have an international reputation and to be produced on a quasi industrial scale from the end of the Eighteenth century on, was soup "à la Rumford". It owes its name to the celebrated inventor of kitchen equipment and early methods of energy conservation[8]. Its ingredients are basically those of the inexpensive rice. The *Almanach perpetuel des pauvres diables* (1803) notes under the rubric "soups" that "the best known is 'soupe à la Rumfort' [sic]. It is a perfect blend, with seasonings that do not overly heat up the chest, and gives the belly a respectable degree of freedom." The Almanach specifies that there are several establishments in Paris where this soup is distributed, and that those in charge "are pleased to deliver a coupon with which poor devils may obtain a bowlful for ten centimes."[9]

It is not, however, until its metamorphosis into bread that the potato, already a symbolic food of the disinherited, becomes an ideological as well as an alimentary sign. Once it takes on the form of bread it also assumes the latter's metonymic function, that is to say it becomes representative of food as a whole. Beginning with 1767, which marks the publication of a *Mémoire sur les pommes de terre et sur le pain économique*,[10] there is hardly a cookery treatise mentioning the potato that does not also have something to say about potato bread. Sébastien Mercier labels potato bread "a priceless gift for the many needy [...] having the greatest influence on man, his freedom, and his happiness". This manna from heaven, he states, will purify bread by putting "the multitude's subsistence beyond the avid monopolist's reach", whereas wheat bread is suspect, since wheat serves the commercial purpose "of the deadliest greed"[11]. Speculation on wheat and wheat flour continued well into and after the Revolutionary period, in spite of elaborate anti-hoarding legislation.

If I insist on this non-culinary detail it is in order to stress the sociopolitical dimension of the potato, which Turgot among others was fully aware of. From the Seventeenth century on, and thanks in particular to the experiments and writings of Parmentier, potato-related technical know-how began to circulate throughout France. From 1761, when Faiguet presented a bread prepared with a blend of wheat, rye, and potato flours to the Academy of Sciences, until 1778, when Parmentier finally succeeded in preparing bread made exclusively from potato flour, experimentation with potato bread was going on throughout France. Mustel's 1767 *Mémoire* indicates that bread made with one part wheat and two parts potato flour is "quite edible". Parmentier brings up the subject in his *Examen chimique* and devotes his *Manière de faire le pain de pomme de terre sans mélange de farine* (1779) to describing the manner of preparing pure potato bread. A Swiss naturalist by the name of Samuel Engel, however, apparently solved the panification problem before Parmentier, as evidenced by two of his treatises[12]. It was Engel who authored the twenty-page "potato" article in the *Encyclopédie*'s *Supplément* (1777, vol. 4, pp. 473 - 93). This remarkable overview treats all aspects of the tuber (history, botany, cultivation, uses and preparation...) and dwells on its transformation into bread. Nothing better illustrates the potato's progress than the gap separating the original article of 1765 (by a doctor named Venel), in which the potato is dismissed in a few lines, and the lengthy erudite article of 1777. The latter article also surveys and evaluates potato processing devices invented at the time - thus a potato slicer labelled "coupe choux" (cabbage cutter) is deemed superior to Mustel's so-called "varlope". It may well have been that the former machine's inventor was Swiss... Promoting products by means of experts' recommendations is nothing new. There are likewise descriptions of potato-drying machines and of potato mills, complete with illustrations. The invention of mechanical contraptions mentioned by Parmentier, Mme. Mérigot and Grimod, *inter alia*, are too lengthy to be listed and often picturesque, clear evidence that panification of the potato stimulated what was then called the mechanical arts. Other mutations of the tuber go well beyond its panification, among them alcohol, glue, and wax[13].

In spite of its many appeals and the dire shortages of wheat and bread during the Revolution years, in spite of promotional campaigns and continuing technological thrusts, in spite of Mme. Mérigot's varied potato dish recipes and the upbeat introduction to her cookbook, the French population at large no more took to potato bread than it ever did to, say, corn on the cob. The era of *pommes frites* was not to emerge until the Nineteenth century. If potato technology and cultivation did make definite forward strides, the atavistic attraction and status of wheat were hard to overcome, and while Barère and the Jacobins touted the tuber, they were hard put to fight the desirability of a rare commodity desirable not only because of its rarity, but because white bread made of wheat flour both represented and concretised an age-old peasant's dream: eating the same bread as the privileged class. Wheat bread, a distinctive luxury in the war-torn Revolutionary era, was featured in festive meals - the so-called *banquets civiques* - and was always incorporated in allegories of food paraded during the great feasts of the Revolution, in particular that of the Supreme Being. It was also, and most interestingly, legislated into a "pain de l'égalité", a bread meant to institute equality where it counted the most - at the mouth. In line with its rigid policies regarding bread, the Convention decreed on 15 November 1793 (23 Brumaire an II) that a uniform bread would be baked throughout France. Here are articles five through nine of this singularly utopian directive:

- Bakers shall be authorised to produce and sell a single type of bread.

- In order to accelerate the provisioning of the armies in bread, and distribute it so as to maximise its preservation, three quarters of its flour shall be wheat-derived and the remaining quarter rye, except for those regions where rye is unavailable, in which case barley flour may be substituted.

- It is recommended that the war commissioners and all agents employed by the armies supervise all bakeries and all preparation of bread.

- Wealth and poverty having no place in a régime of equality, there shall no longer be produced a bread of the finest flour for the rich and a bran bread for the poor.

- All bakers are held responsible, upon penalty of incarceration, for producing a single and good type of bread, the bread of equality.[14]

Where bread is concerned, the potato was never able to make any serious inroads in the domain of wheat. In France, the Eighteenth century marked not so much an invasion, but a victory of the potato, or perhaps more modestly, a progress of Enlightenment in its fight against the scourges of superstition, prejudices, and alimentary disasters. A victory nonetheless, of the spirit of enterprise and of ingenuity, of a new social conscience with regard to the table. After the Revolution, and doubtless because of its many tribulations, the potato came into its culinary own: the great Carême includes it on menus meant for princely boards in his *Maître d'hôtel français*. And, quite naturally, in the form of French Fries.

Footnotes

* Revised, updated, English-language version of an article first published in *Dix-huitième siècle*, vol. 15 (1983), pp. 19 - 27, under the title "L'avènement de la pomme de terre".
1. *Archives parlementaires*. Première série, vol. 85 (Paris: Centre national de la recherche scientifique, 1964), pp. 312 - 13.
2. Mme. Mérigot, *La Cuisinière républicaine*. Paris: Mérigot jeune, an III.
3. See Redcliffe Salaman, *The History and Social Influence of the Potato,* Cambridge: Cambridge University Press, 1985; also M. Morineau, "La pomme de terre au 18 siècle", *Annales E.C.S.* (nov-dec. 1970), pp. 1767 - 85, and C. Vandenbroeke, "Cultivation and Consumption of the potato in the 17th and 18th century". *Acta historiæ Neerlandica* (1971), Vol. V, pp. 15 - 40.

4. See Steven L. Kaplan, *Bread, Politics, and Political Economy in the Reign of Louis XV*, The Hague: Nijhoff, 1976, and, by the same, *Provisioning Paris*, Ithaca: Cornell University Press, 1976; also V. Ljublinski, *La guerre des farines*, Grenoble: Presses universitaires de Grenoble, 1979.
5. See Olivier de Serres, *Théâtre d'agriculture et mesnage des champs*, Paris, 1600.
6. Grimod de la Reynière, *Almanach des gourmands* (Paris, Maradan, 1803 - 12), vol VII, p. 41.
7. I refer to a later edition of the *Examen chimique* that was published under the title *Les pommes de terre considérées relativement à la santé et à l'économie* (Paris, 1781), pp. 198 - 200 and 222 - 26.
8. This native American was a loyalist in the service of the King of England before passing on to that of the Elector of Bavaria. Born under the name of Thompson, he was later ennobled and became Count Rumford.
9. *Almanach perpetuel des pauvres diables, pour servir de correctif à l'Almanach des gourmands*, (Paris, an XI), pp. 30 - 31.
10. N. Mustel's *Mémoire*, which was published in Rouen along with a supplement, is one of the documents included in the *Cuisine des pauvres*. Mustel was instrumental in spreading potato cultivation in Normandy.
11. *Tableau de Paris* (Amsterdam, 1782 - 88), vol. IV, pp. 129 - 31.
12. Samuel Engel, *Traité de la culture et de l'utilité des pommes de terre*, Lausanne, 1771; *Instructions sur la pomme de terre*, Berne, 1772 - 4.
13. A list of these "mutations" full of surprises can be found in the table of contents of A. Payen's *Traité de la pomme de terre* (Paris, 1863), pp. 158 - 60.
14. *Archives parlementaires*, première série, vol. 79 (Paris, Dupont, 1911), p. 286, and Bénigo Cacérès, *Si le pain m'était conté* (Paris: La Découverte, 1986), pp. 96 - 97. Translation mine.

Rice And Traditional Ceremony In Japan

by Yoshiko Hirasawa
(Translated by Mitsue Tsuji)

Japan is sometimes known as 'the Land of Fresh Rice-ears'. Rice has been an important food for the Japanese nation for the last two thousand years.

In modern Japan, however, because many other consumer products are available, rice is not given due appreciation. On this occasion, it would be extremely interesting to have a fresh look at rice, a food inseparable from Japanese culture.

We take rice for granted. It is remarkable that a Westerner has noticed that, by exploring rice, a staple food for the Japanese, certain deeper aspects of Japanese culture can be revealed.

Dealing with rice as an inseparable part of Japanese life would inevitably entail considerable work, so I would simply like to give a brief account of rice with particular emphasis on the part it plays in festivities and celebrations.

In olden times, rice was revered as possessing power, spirit and energy.

When rice was harvested, first and foremost it was offered to the gods in thanksgiving. Rice dishes as offerings are closely linked with the customs of festivals and celebrations.

In the past, we paid land tax in rice. Even now, the rice price is a basic indicator determining commodity prices. There are many varieties of rice which are used for main dishes and for desserts. For main dishes, rice is simply cooked with water. This is 'Gohan' and accompanies every meal.

'Sake', the equivalent of Western wine, is made from rice. It is essential in our daily life, as well as for every festivity and celebration.

There are a great variety of foods made of or from rice: rice in its original form, chocolate and sweets, cooked with water, powdered, pounded, kneaded, vermicelli, brewed or fermented as with sake, 'Miso' and 'Mirin'. I would like to leave the subject of rice and day to day dishes for another occasion and concentrate on rice food associated with the gods and Buddha.

First of all, the New Year celebration. For us, the first three days in the new year are an important national festivity. It is customary for every household to have a decoration of round mirror-shaped rice cakes hence the name 'Kagamimochi' (Fig. 1). These are a thanksgiving to the gods and Buddha, as well as homage to our ancestors. Specially glutinous rice is selected, cooked and pounded in a mortar. The pounded rice is shaped into round, flat pieces. A small piece is placed on a larger piece, on top of which is placed a tangerine, a typical winter fruit.

Fig. 1

On the seventh day of the new year, one of five major ceremonial feasts, kagamimochi is cut into small pieces and cooked in a soup with rice and seven different spring herbs, 'Nanakusagayu'. The dish is to thank the gods for our having seen in the spring herbs, to pray for well-being in the new year. It is said that nanakusagayu can protect against all illnesses.

Incidentally, the five major ceremonial feast, 'Sekku', are 'Jinjitsu' on the seventh of January, 'Jôshi' on the third of March, 'Tango' on the fifth of May, "Tanabata' on the seventh of July and 'Chôyo' on the ninth of September.

For the New Year, kagamimochi is offered to the gods and Buddha and we have dishes using 'Mochi', rice cake such as 'Ozôni' and 'Oshiruko'.

Ozôni is a celebratory dish for which the only constant element is mochi and there are many varieties of flavour and ingredients developed throughout the islands. Grilled and softened mochi are cooked with a variety of ingredients in a hot soup which is either salt- or miso- based. Mochi is as glutinous and sticky as certain types of cheeses. One's fortune is told, and good fortune is told, and good fortune hoped for, depending on how mochi stretches.

Fig. 2 : Oseki han

Oshiruko is a sweet soup made from azuki beans served with toasted mochi. While ozôni tends to be enjoyed mainly around New Year, oshiruko is enjoyed throughout the year as a dessert, particularly by women and children.

On the 15th of January, 'Azukigayu' is served in farmers' households with wishes for a good harvest for the year. Equal amounts of rice and azuki are cooked with plenty of water until tender, and mochi are then added. This is a slightly sweet dish.

The feast of Jôshi on the 3rd of March is also known as Hina-matsuri or the feast of the peach, because it coincides with peach blossom season.

This is a festival for girls. A family with a girl has a decoration of a stand arranged in tiers with dolls clad in ancient court costumes. At the top of the tier sit an emperor and empress. The festival is to hope for a good marriage and a happy life as a woman.

On this occasion, 'Hishimochi' and 'Shirozake' are placed on the stand and 'Osekihan' is served. Hishimochi, shirozake and osekihan are all made of or from rice.

Hishimochi is an expression of prayer to the spirits and gods, as with kagamimochi. It has a diamond shape and three pieces, red, white and green, are placed on top of one another. These are offerings and are not normally for eating. However, a sweet of similar shape can be enjoyed.

Shirozake is a type of sake brewed using rice. It is cloudy

Fig. 3 : Chimaki

and milky, since the lees are not filtered out. This is cooked with rice to have a sweet flavour. Shirozake is a special beverage for the festival.

Oseki han (Fig. 2) is also known as 'Okawa'. Glutinous rice, mochi-gome, is cooked with azuki beans which give a red tint to the rice. Oseki han is popular not only for Hina-matsuri, but also for every festive occasion, whether public or private.

Hina-matsuri is a festival for girls, while the feast of Tango on the 5th of May is a festival for boys. A sweet-flag is displayed on the eaves of the house to fend off evil, and the occasion is also called the feast of the sweet-flag. Most families with boys display a doll of an ancient warrior, named after an ancient hero, and fly a carp streamer. Confections such as 'Kashiwamochi' and 'Chimaki' are served on this occasion.

Fig. 4 : Oha gi

Kashiwamochi is a ball of mochi with a sweet azuki-bean paste centre, wrapped in an oak (kashiwa) leaf. Jôshinko, rice flour, is mixed with water to a paste to form the mochi.

Chimaki (Fig. 3) is steamed glutinous rice, wrapped in a bamboo leaf and boiled. The lingering fragrance of the bamboo leaf joined with the natural flavour of the rice is enjoyed.

Both kashiwamochi and chimaki are regarded as lucky mochi: they represent a wish to protect children from evil and illness with the help of the fragrance of kashiwa and bamboo leaves.

Let us now turn to 'Higan', an occasion to pay tribute to one's ancestors. There are two equinoxes a year, in spring and in autumn. The Vernal Equinox falls around 21st March and the Autumnal Equinox around 23rd September. The periods are called 'Ohigan' and various Buddhist ceremonies are performed.

Fig.5 : Sakura Mochi

Higan is an occasion to devote oneself to Buddhist teachings aspiring to free oneself from worldly desires, as well as to honour one's ancestors. A confection called 'Oha gi' (Fig. 4) is placed before the tablet of the deceased and at the grave, and it is served at home.

Ohagi is a rice confection. Glutinous rice and ordinary rice are combined in equal quantities and boiled. The cooked rice is lightly mashed to form into balls. The the balls are covered with a thin outer layer of sweet azuki-bean paste, sweetened soybean flour (kanako), sesame seeds or green nori seaweed (ao-nori). Ohagi used to be served only during Ohigan. Nowadays it is enjoyed throughout the year partly due to the fact that it can be prepared relatively easily compared with other Japanese confections.

Apart from these festivals which fall on set dates, there are others associated with the seasons. Flower viewing when the cherry trees are in bloom and moon-viewing in autumn, for example. These two festivals also have special rice confections. For flower viewing, 'Hanami dango' and 'Sakura mochi' wrapped in a cherry leaf are customary.

Hanami dango is rice flour formed into little balls with a touch of spring colours and some herb fragrance.

Sakura mochi (Fig. 5) is a cherry pink-bean paste centre, wrapped in edible, brined cherry leaf. This is a typical Japanese confection full of the flavour of the season. Hanami dango and sakura mochi are enjoyed, accompanied by sake, under the blossoming trees.

Fig. 6 : Tsukimi Dango

Then, there is a moon viewing in autumn. In contrast to flower viewing, this is a tranquil custom for contemplation under the light of the full moon. In the lunar calendar, it was on the 15th August. On this day, 'Tsukimi dango'(Fig. 6) is prepared and offered to the moon. Using rice flour, Jôshinko, fifteen balls, which are made to look like the full moon, are prepared. The number fifteen is derived from the date.

An addition to such customs presents itself. Bitter green tea is inseparable from the sweet confections.

We can say that a great variety of beautiful confections were created thanks to the tea ceremony.

Most Japanese sweets, 'Wagashi', are made of rice flour. They also incorporate a variety of flavours and colours, hinting at the flowers and plants of the season, reflecting people's sensitivity to the different seasons. Wagashi is a silent and delicate means of appreciating the essence of the tea ceremony which is related to Zen.

Rice as an everyday staple food can be served in more than two thousand different ways. 'Sushi', a nationally favoured dish, now increasingly appreciated by many people abroad, is prepared using vinegared rice. Sushi has endless varieties in terms of shape (Nigirizushi, Gomokuzushi, Makizushi, Oshizushi, etc), as well as flavour and contents.

As I mentioned at the beginning of the paper, I have concentrated on rice related to festivities and ceremonial occasions, introducing a limited number of major items.

Tarhana - from Steppe to Empire

by Maria Johnson

Recently, while searching through my papers, I was struck by the fact that the 'T' file was much thicker than the others, overflowing with information (and misinformation) gathered over the years about *tarhana* - the saddle-bag staple of pastoral nomadic herdsmen who, it is thought, had no need of agriculture except perhaps to supply grain as supplementary feed for their livestock.

This paper is a modest attempt to establish the origin of *tarhana*, based not on archaeological discoveries - since no actual *tarhana* granules have survived to attest their high antiquity or actual composition - but on the few historical facts, theories and linguistic evidence available. The following points may prove useful in establishing the 'when' and 'where' of *tarhana*'s early history.

1. What is tarhana?

Tarhana could be described as a form of pasta, an ancient, still current, kind of convenience food quickly and easily turned into a nourishing meal or a substitute for bread. In its very simplest form *tarhana* is prepared solely from flour - based on the local breadcorn - and water; on a more sophisticated level it is made with durum or strong wheaten flour mixed with milk or eggs or yoghurt, and occasionally soured by fermentation with yeast or natural leaven. The dough is shaped into tiny shreds, rolls or pellets by grating, chopping up or rubbing it through a sieve or between the palms of the hands, which are then dried in the sun or semi-shade - the extent of drying depending on whether the product is needed for immediate consumption or for storage. The dried material is cooked either in a soup, or it is first fried and then allowed to simmer in as much milk or water as it will absorb - so no nutrients are washed out as in the case of conventional pasta which is usually boiled in fairly liberal quantities of water. The cooked dish is thick and gruel-like, the swollen pellets achieving almost the size of salmon ova.

Tarhana was a staple food, easily carried on long trips mainly by Turkic nomadic tribes. In the Balkans it is known, according to language, as *trahana* or *trahan* (in Albanian), *tarhana* (in the Bosnian dialect and in Turkish), *trahana* or *tarhana* (in Bulgarian), *trahanás* (in Greek) and *tarana* (in Macedonian and Serbo-Croat). The term is also widespread in most Turkic languages.

2. Who are the Turkic peoples?

The steppes and arid zones of Central Asia have been used as grazing lands for thousands of years. Most of the ethnic groups who inhabited these areas were expert horsemen and stockbreeders, traditionally nomadic or semi-nomadic, wandering to and fro between the northern summer pastures and those in the south which permitted winter grazing. However, even before the beginning of our era, a few groups of the steppe-dwellers maintained permanent settlements.

In the 11th and 12th centuries after Christ some of the Central Asiatic nomads were known as Mongols or Tartars, some as Turks, and each of these names was often applied to them collectively. Modern historians call them Turkic peoples because they spoke agglutinative languages, of the Altaic family, akin to Turkish. In 1924, in the Soviet Republic of Azerbaijan, the Latin script was officially adopted and the alphabet used became known as the Unified Turkic Latin Alphabet or UTLA, though in 1939 it was replaced by the Russian Cyrillic script with many additional symbols.

In Turkey, the Young Turk revolution of 1908 proclaimed the idea of 'pan-Turanism' (Turan being the ancient name for the region east of the Aral Sea and the river Oxus, syn. Amu Darya), which looked to the unity of all Turkic-speaking peoples whose origins could be traced to Central Asia, Siberia and the northern parts of the Ural mountains.

This great family of peoples includes the Huns, Khazars, Avars and Bulgar-Turks of former times, the present-day Turks of the republic of Turkey, the Uzbeks, Kazakhs, Azerbaijanis, Kirgizs, Turkmenis, Tartars, Bashkirs and Chuvash of the USSR, and a few others.

3. Did they practice agriculture?

Central Asia east of the Aral Sea was no land for agriculturalists; the sandy soils and the uncertainty of the rainfall rendered farming there a hazardous undertaking.

Early in our era, successive waves of Turkic and other peoples surged westwards from these unpromising lands to settle in the milder climate of the Russian platform (which comprises most of Eastern Europe from the Urals to the Carpathians), where the shores of the seas and the great rivers, like the Volga, made sedentism possible and offered amenable environments for agricultural experimentation. Large populations and multitribal state formations or empires based on mixed farming, hunting and fishing were established, which rose swiftly to power and just as swiftly disintegrated under the impact of other nomads who thrust out of Central Asia in the east.

It is certain that the Onogur (Unogundur) Bulgar-Turks, as well as the ancient Hungarians (of the Finno-Ugrian race) - who in the 6th and 7th centuries after Christ were part of the short-lived Onogur tribal federation - though predominantly herdsmen and horsebreeders were also engaged in marginal land tilling (Bourmov, p 29).

Not long after that, in the 7th and 8th centuries - according to László Makkai of the Hungarian Academy of Sciences - 'the Hungarians' neighbours, the Khazars and the Bulgars of the Volga region, gradually developed agriculture,' a fact which 'did not fail to make an impact on the Hungarian community. The last group of words of Bulgar-Turkish origin incorporated into the Hungarian language [before the 9th century] are all agricultural terms, including some related to wine growing' (Pemlényi, p 19).

Prof. Robert Smith in his 'Kazan Tatar Diet and Russia' also mentions that the food of the Volga Tartars had included 'many elements which appear to derive from Bulgar times' [8th to 13th century], and that both hard and soft wheat have been found dating from the same time (First International Food Congress, p 277).

It is obvious that in Eastern Europe some of the Turkic peoples grew not only the common hexaploid wheat, *Triticum aestivum* (*vulgare*), but also the high-gluten variety which could have made possible the preparation of top quality *tarhana*.

4. Is Tarhana a Turkic invention?

Not much is known about the early history of *tarhana*. Its origin seems to go so far back and to such remote places that it is difficult to trace. Conflicting information from the few available sources, none of which can be substantiated in any detail, has inspired several theories. One supposition is that the *tarhana*'s archetype was the Chinese steamed or boiled paste which, in an altered shape, has travelled west along the trade routes across Central Asia all the way to Constantinople; or that the priciple of making pasta was aquired directly from the Chinese during the centuries-long contact, when Turkic peoples occupied their ancient homelands to the north and west of China (AD 1st to 10th century and earlier).

Since there is no evidence to the contrary, another theory favours an independent Turkic origin, possibly in the 6th or 7th century of our era, when cereal crops began to be grown by sedentary Turkic tribes.

There is no single hypothesis, however, that will explain why Sir Gerard Clauson's *Etymological Dictionary of pre-Thirteenth Century Turkish* does not contain the word *tarhana*, though there are numerous entries of agricultural terms and paste products such as *ügre*, noodles, *süt ügre*, milk and noodles, and *tutma:ç*, some kind of farinaceous food, 'noodles, macaroni, vermicelli' and the like, 'a food well known to the Turks' (Clauson pp 112 and 457, his quotes). It appears, according to Charles Perry, that the oldest literary reference to *tarkhāna* (Persian spelling) might be in the works of the Persian poet Abu Ishaq Hallaj of Shiraz, better known as Bushaq (Perry, p 59), who lived at the end of the 14th and the beginning of the 15th century. In Bushaq's Divan (collection of verse), though, *tarkhāna* is barley bread, not the dried pellets of dough we know (Algar, p 19).

However this may be, it has been established that *tarhana* was first introduced into the Balkans by the Onogur Bulgar-Turks, who 'in their migration from Eastern Europe into the present Bulgarian lands [AD 7th century] had consumed, like all nomads, meat grilled over a wood fire, milk, dairy products and *tarhana*. They had also used various plants gathered from the wild' (Petrov *et al*, p 16). The Hungarians seem to have acquired *tarhana* from the Turkic peoples, probably when they were part of the Onogur federation (AD 6th to 7th century), and which staple food '...our ancestors brought with them over the Carpathian passes more than a thousand years ago...' wrote Mr Károly Gundel in the eighth edition of his *Hungarian Cookery Book*. 'Before and during the period of the great migration of the peoples, the ancient Magyars lived as nomadic tribes. Although we have no written records of those times, it is more than likely that they used to prepare some sort of food that could easily and quickly be cooked and could be kept during their long wanderings. One of these dishes still exists in our modern cuisine, under the name of *tarhonya*. It can either be made at home or bought in shops. The herdsmen of the Great Hungarian Plain still make it themselves... It is also known in the Caucasus, Mongolia and the Balkans under very similar names...' (Gundel, p 13).

Tarhana was reintroduced into the Balkans by the Ottoman Turks after the conquest of Constantinople in 1453, and into Hungary in 1526 when they overran the country. It is also consumed in Iran and nearly all the countries which once constituted the Ottoman Empire at the height of its greatness.

References

Bourmov, Prof Al, academician Kosev, D and academician Hristov, Hr, *Istoriya na Bulgariya* (History of Bulgaria), Narodna Prosveta, Sofia 1978
Pemlényi, E (editor), *A History of Hungary* - Corvina Press, Budapest 1973 and Collets 1975
First International Food Congress, published by the Turkish Ministry of Culture and Tourism, Ankara 1986
Clauson, Sir Gerard, *An Etymological Dictionary of pre-Thirteenth Century Turkish*, Oxford University Press, Oxford 1972
Perry, Charles, 'Tracta, trahanas, kishk' in *Petits Propos Culinaires 14*, Prospect Books Ltd, London 1983
Petrov, Dr L, Dzhelepov, Dr N, Jordanov, Dr E, and Ouzounova, S, *Bulgarska Natsionalna Kouhnya* (Bulgarian National Cuisine), Zemizdat, Sofia 1978
Gundel, Károly, *Hungarian Cookery Book*, Corvina Press, Budapest 1974
Algar, Ayla, 'Bushaq of Shiraz: poet, parasite and gastronome' in *Petits Propos Culinaires 31*, Prospect Books, London 1989

Beans Of The Southwestern United States Indians

by Mary Wallace Kelsey

Staple foods of modern America include foods which the Native Americans - American Indians - were growing or gathering before the European invasion of the continents. Three of these foods, maize (now called corn in the United States and Canada), squashes and beans, have been called the "Holy Trinity", or "The Three Sisters", among other names. They were the main staples used by so many of the Indian groups.

The Southwestern United States Indians are credited with having been the most advanced in agricultural development. It has been suggested that as long ago as two millennia, the Hohokam Indians were growing beans in the arid southwestern desert land of what is now the United States. The earliest recorded time of culturation of the tepary bean by the Hohokam is between AD 90-1100[1]. For irrigation, the Hohokam dug canals from the rivers to their crop land. In what is now southeast Arizona and southwest New Mexico, archaeologists have found the first sites in the United States where plants from Mexico were cultivated. Among the findings were *Phaseolus vulgaris*, the common bean. This bean grew well in regions above 2000 meters where there was enough rain for dry farming. At lower elevations, where temperatures were higher, irrigation was necessary to grow *Phaseolus acutifolius* variety *lactifolius* (teparies), *Phaseolus lunatus* (limas), and perhaps *Phaseolus coccineus* (scarlet runner bean).

The Anasazi who resided at Mesa Verde are known to have dug pools and ditches for gathering rainwater to irrigate their bean crops.

It is said that the most valued beans to the ancient Southwestern Indians were colors which symbolized the 6 cardinal directions: yellow-north, blue-west, red-south, white-east, multi-colored-zenith, and black-nadir[16,29].

The groups of Indians known as Pueblo because of the types of dwellings in which they resided are the Acoma, Hopi, Taos, Tewa and Zuni tribes[24]. Perhaps the most studied of these as far as food habits are concerned is the Hopi tribe. The Hopi's uses of beans will be discussed following general information about several types of beans.

Other Southwestern tribes include the Apache, Pima, Papago, Havasupai, Navajo, Mojave, Santa Clara, San Domingo and San Ildifonso[37]. Some of these will be included later also.

Teparies

Phaseolus acutifolius, wild teparies, grow well in dry climates, are disease-resistant, and nutritionally as good as most economic legumes and more nutritional than any other beans[20]. They are also apt to be resistant to insect infestations. Teparies are the only southwestern beans of the Phaseolinae which have been domesticated[2]. Nabhan et al. note that domestication began in 1699 A.D. Tepary beans are unusual because they grow well in arid desert land. The yield in such growing conditions has been 2,020 kilogrammes per hectare, but when the climate is better, more than 4,630 kilogrammes per hectare have been harvested.[20,41] Teparies are difficult to harvest, since the pods dehisce explosively, scattering their seeds (beans) in all directions. Also, the wild seeds are small[22]. Each pod contains 4-10 seeds[2].

The name tepary may have come from the Papago's word for beans, *pawi*. Castetter and Underhill suggest that when the Spaniards asked the Papago for the name of the seed, they said 't'pawi', 'it is a bean.' The Spaniards named the bean tepary[5].

The tepary seed is about half the size of an ordinary kidney bean, but the Indians preferred it as a more satisfying food than other beans. Teparies analyzed by Calloway and co-workers[4] contained 15 percent more nitrogen than pinto beans, and 10-40 percent more of most essential minerals expect iron, nickel and silenium. Because of this great food value from the least size, teparies were the choice for travellers to take along.

Teparies come in several shapes and a variety of colors. They may be oval, flattened or round. Colors include flesh color, speckled brown, dark brown, reddish brown, speckled yellow, purplish black, clay color with lavender speckles, greenish yellow and white[25]. Kavena mentions black teparies with wrinkled, veined surfaces as being grown by the Hopi. When teparies were planted, it was traditional to grow two for the rabbits and three for the table[15].

Teparies probably originated in Central America, say Nabhan and Felger. These beans have not been among reported finds in the prehistory of the Sonora Desert, an area of Mexico from which many plants were brought north to what is now the southwestern United States[20]. Wild teparies are usually gray with light brown mottling[41].

In current usage, white teparies may be used by the Hopis as the meal to break a fast. The beans are parched in hot sand, then mixed with salt water. Beans prepared in this way may also be used for observation of spiritual events[23].

Teparies have been grown commercially several times in southwestern states during this century. The only farmer to grow them successfully was W. D. Hood in Arizona, who with L. Romero was in the tepary business for about 12 years. He found early markets on Indian reservations, since most of the users were Native Americans. In the late 70's and early 80's, the "health food" markets stocked teparies for their Anglo customers.

Hood and Romero found that tepary growing was a labor-intensive business, which probably accounts for fewer Indians growing the beans[39].

Cooking teparies takes longer than for most beans because they dry out more than others. Niethammer suggests soaking for at least twelve hours, and cooking in a slow cooker for eight to ten hours. After the beans become soft, they will need about two additional hours to change the raw starch flavor to a more acceptable cooked flavor. A good suggestion is to pre-cook a large batch of beans, then freeze some, or dry them as the Indians do, for later use[40].

"Tepary Bean of the Brush"

Phaseolus metcalfei, known also as cocolmeca, tepary bean of the brush (*tepari del monte*) or *frijolillo* (little bean) in Spanish, have been used as an important food for many Indian groups in past years. Their seeds, larger than any other wild *Phaseolus*, are retained longer by the plant than those of the true tepary bean, and are dropped next to the plant when they are released. This makes harvesting of the *P. Metcalfei* easier than for the *P. acutifolius*. However, the seeds that have been dropped are vulnerable to rodent and bruchid beetle damage[22].

There is some evidence that the metcalf beans were eaten by the Western Apache, and that they stored large quantities of them for use in winter. Some have said that the beans are best when green because they are tough when mature. All of the Southwest Indians have used parching and grinding as a preparation method for both cultivated and wild beans. Metcalf beans and roots have been prepared this way, and used to make a beverage, sometimes fermented. The pods of the *P. metcalfei* plant have been used as food, too. Cocolmeca roots are still sold for use as folk medicine in the Southwest [22].

Lima Beans

Another bean with a long history in the Americas is *Phaseolus lunatus*, lima bean. These are larger flat beans which grow in pods shaped like scimitars and are known as limas, butter beans, or sieva beans[31]. Ford tells us that the sieva bean, which he calls a small lima, was domesticated separately from the large lima bean, and that it was rare in the Southwest United States, although found often in Mexico[8].

Not much data is available for limas in early agriculture except for the Hopi. There is even a commercial variety of lima bean called the Hopi Lima.

Varieties of beans found in archaeological digs in the pueblos are yellow-brown, purple and cream-color mottled with purple, but no solid-color purple was reported for the Hopi[31]. According to Kaveena, modern Hopi use black, red, bright yellow, white, cream, and light brown limas[15].

Screwbeans

The screwbean, *Strombocarpa pubescens*, is also called tornillo, screws, or screw pod mesquite. It gets these names because the seedpods are twisted in a spiral. The plants grow best in moist, saline soil, so do well in river valleys. The plants are shrubs or small trees, relatives of mesquite. Curtin tells us the pods are "rich in sugar" so that children like to chew them. Grazing animals like both pods and leaves. The beans themselves may be ground, blended into water, and drunk as a kind of gruel, making what Curtin calls a sweet and nourishing beverage[6].

Roots from the screwbean plant may be boiled and the liquid resulting used as a medicine. Also, dried screwbean roots have been powdered and dusted on sores to heal them[6].

Jack Bean

Jack bean (*Canavalia ensiformis*) seeds dating 1,000 years B.C. have been found in Arizona and New Mexico. Jack bean plants are hardy, tolerant of shade, and resistant to drought. They have a tough, thick seed coat, which requires long cooking periods to soften. Jack beans, also called horsebeans, contain about 30% protein[26].

In a storage house from an Arizona Hohokum site A.D. 900-1000, Bohrer et al., report finding jack beans, smaller than modern ones, and mention other finds in Arizona with many size and shape variations among the beans. They cannot establish that these were a food source, however. The authors comment that jack beans have not been found often in the southwestern archaeological sites, nor have the beans been found frequently in contemporary gardens. The Pima grow some, perhaps as a holdover from their ancient relatives[1]. Steen and Jones[31] report a find of tan jack beans streaked with yellow in a pueblo dig.

Hopi Indians

Kuhnlein and Calloway point out that the Hopis have been thought to have kept more of their traditional cultural patterns than have other Native Americans. In a 1977 study of Hopi food intake, these authors report that Hopis living on Black Mesa may be in the oldest continuously inhabited villages in the United States, perhaps dating from 1150 A.D. Maize and beans are the most popular of the Hopi's native foods grown. They use lima beans, teparies, kidney beans, green, or string, beans, and occasionally, scarlet runner beans[18].

Lima beans may be white, brown, yellow, red, or black, with the latter being used for sprouts. The teparies grown are white, blue-black or mottled colors. Rather than the typical red kidneys, the Hopi grow white kidneys; small, large, or flat ones.

String beans are white, flat yellow, purple, or red. Blue ones have been used to dye materials for making baskets[15]. The string beans are used whole and dried whole, in the pod, or the pods may be used for "culinary ash".

When the Hopi eat string beans, they usually eat them as finger food, removing the strings as they eat by drawing the beans through their teeth[23]. Strings are put on the edge of the plate[15].

Hopi culinary ash is made by burning vines and pods of harvested beans, or juniper branches, sheep dung, corn cobs, or wing salt bush. The resultant ash is used to make piki bread or other foods[19]. Ash from four-winged salt bush (*chamisa*) is preferred because it adds a blue color, which has religious significance to the Hopi[15].

In dietary recall, Hopis in the Kuhnlein study mentioned beans or green beans at least 15 times, but specified lima beans only about 5 times. They eat boiled bean sprouts along with beans as a part of their contemporary breakfast. A dish called *badipsiki*, boiled beans and hominy is served[17].

Traditionally, the Hopis have had as many as 21 folk taxa of *Phaseolus* beans which they named and grew. Some of these were inherited through clans. Most of the beans grown now by the Hopi are saved for ceremonies, or for when the cook wants a different variety. Classification of the beans is made difficult because the colors and shapes may have been influenced by the environment. Nabhan, Weber and Berry[23] have cited findings of beans in the same field, or the same pod, which had been listed by other researchers as separate varieties.

Within the last decade, the Hopis were still growing about 14 varieties of beans on their reservation. They purchased or traded for other varieties. Most of the special, home grown types are saved for ceremonies. The beans used regularly are pintos[15].

Cooks usually soak the dry beans overnight, then simmer them with salt pork or other meats, and season them with chili powder and salt. The mixture may be dried, ground and re-cooked at a later time as a source or eaten with onions[15]. Women prefer using fresh beans when they have them because it takes too long to cook dried ones[15].

Both the Hopi and Zuni have masked kachinas (male "priests"), who are the center of ceremonies. When they put on the masks, they are impersonating gods. The Hopi ceremony for bean planting is *Powamu*, a ritual to reawaken the sleeping earth. Sixteen days before the ceremony, beans are planted in *kivas*, underground chambers, and nurtured so they will sprout, a symbol of the power of the kachinas. Sprouting also predicts a successful bean crop. At sunrise the day Powamu begins, the crow mother kachina will bring a basket of cornmeal to the shrine at the village edge. Other kachinas appear, giving sprouted beans and other gifts to each household. Everyone eats stews of bean sprouts and corn, called *hazrugive*. To avoid partaking of this dish may foster bad luck[35]. Men eating in the Chief's house are brought boiled beans in special trays. The Chief's wife and others bring the food, which breaks a fast for the men dancers[23,34,36]. Gravy (*wotaka*), symbolizing ponds and rivers, must be eaten first[35].

Later in the day kachina "ogres" will threaten to take away children who do not obey. Mothers appeal to the kachinas to spare the children by offering gifts of corn on the cob, cornmeal or piki bread. These offerings are distributed to the priests and others in the village. The ceremony ends with dancing at night[24].

For the last 50 years, the Hopi have used the following beans. Names given in parentheses are the commercial variety names: trader's beans (Mexican Pinto), red bean, black bean, orange bean (Sulphur), pink bean (Robust), chewed-color bean, cow bean, little dog bean (Jacob's Cattle), red string bean (Red Kidney), white string bean (White Kidney), grease bean, dye/spot bean, and black, dark, white, and trader's teparies. Also grown are grey, red, white, orange, black and spotted lima beans, and Scarlet Runner beans which the Hopi just call large beans[23].

The Hopis simmer pinto beans with meat or mix them with corn or watermelon seeds. Pink beans and little dog beans may be prepared in the same manner. Kavena's recipe requires toasting watermelon seeds, grinding them and putting in a sieve. Boiling water is poured over them and the resulting liquid is used for cooking the beans[15]. White teparies may be parched in sand and mixed with salted water. Dry gray and orange limas are often cooked with pork, tomatoes and chili peppers; these limas may also be sprouted in *kivas* and the sprouts cooked with white corn or simmered in beef broth and eaten as part of ceremonies[23].

A special Hopi dish of beans and hominy, eaten with onions, is called *patupsuki*. Sweet corn baked in pits, then dried, may be used with pintos and salt pork[15,35].

One of the Hopi beliefs is that if smoke from a fire drifts especially toward one man, it is a sign that he has eaten too many beans[35].

An interesting non-food use of beans is to make a paste from them and to anchor turquoise in jewelry with bean paste instead of pine pitch[35].

Mojave Indians

The Mojave Indians, one of the Yuma language group, lived along the lower Colorado in "aboriginal" times. They practiced flood water farming, as did other tribes in the region. It has been estimated that the Mojave may have grown as much as half of the food they needed. Supplementation of the diet with wild plant foods, fish and game was probably easy.

In years of drought, when the floodwaters did not irrigate the land, wild plant foods became very important to the existance of the Mojave. The two most important of the foods gathered were screwbeans and mesquite pods. As with most hunter-gatherers, women collected most of the plants. Stewart was told by the Mojave that screwbeans don't taste good when they are young, so they were allowed to decay in a pit about four feet deep, lined with arrow weeds, for about four weeks. The beans turn red in color and take on a sweet taste after storage. After being dug up, the decayed beans were air-dried for about about a week, then pounded in a mortar to make a flour. Water was added to the flour to make a beverage[32].

Among the crops grown in the rich silt left by the receding river was the tepary bean, second only to corn in importance. The Mojave grew white and yellow teparies. Later, after contact with the Spaniards, these Indians grew cowpeas (*Vigna sinensis*), a legume which originated in the Old World. The large cowpeas, known as black-eyed beans or black-eyed peas in the United States, were called quail beans (*Marik axma*) by the Mojave.

Unlike other Indian groups, the Mojave had no ceremonies or special observances prior to planting their crops. No tabus were observed and no prayers were recited for planting. The Mojave made fences of brush and arrowweeds around the plots where teparies were grown, to discourage rabbits who seemed particularly fond of the bean plants.

When the beans were harvested, the Mojave pulled the plants up, root and all, and let them dry in piles for a few days. Then the women trampled the beans and pounded them with large sticks in order to separate seeds from pods. Last, the beans were winnowed in baskets and stored in gourds in a separate place from where the corn was stored[33].

Havasupai Indians

The Havasupai, Plateau Yumans, grew a big red variety of bean as well as yellow beans. On the same day the corn was planted, the beans were planted in the cornfield[30]. More recent information labels the bean types as navy and pinto.

Beans and corn were harvested at the same time, with extra beans stored in stone receptacles for winter. Enough beans were saved from the store for spring planting.

One of the Havasupai superstitions was that new mothers must avoid eating bean mush, since its mealiness would prevent the milk from being free-flowing[30].

Papago and Pima Indians

The Papago and Pima Indians may be descendants of the ancient Hohokum tribe. Because they have a similar language, and both practice agriculture, they are often grouped together. The Pima Reservation is south of Phoenix, Arizona. There is little water for irrigation, so farming is difficult, and not much food has been raised for consumption[10,24].

The Papago are known as the "Bean People" because beans have been the basis for their diet throughout their history. In 1944 bean consumption was estimated to be 300 grams daily per person. Analysis done in 1980 showed that beans account for a higher protein, iron, thiamine, calcium and riboflavin intake than any other components of the diet[23].

Castetter and Bell[5] wrote extensively on the agriculture of the Pima and the Papago. The beans grown in largest amounts were brown teparies, white teparies and common pink beans. The Papago planted teparies once a year, in July, and harvested them in October. The Pima planted twice a year; once, when the mesquite leaves first began to show, then again after harvesting the saguaro. Beans from the first planting were harvested the end of June, while the second batch was gathered in October. The Northern Pima were specialists in growing teparies[20].

There were separate plots for the pink beans as well as for each of the brown and white teparies. Each variety was harvested separately.

The beans were harvested by first pulling up the whole plants, then letting them dry in piles while the threshing floor was prepared. This was made in the field or near the house by weeding the ground, leveling it, then tramping on the moist dirt.

The beans were piled on the threshing floor and beaten with a long, thick pole to separate pods from the vines. The vines were set aside for a few days for the immature pods to dry. Then the vines were beaten again to release the remaining pods.

All the pods were placed in piles on a mat, where they were beaten with a smaller stick to release the seeds from their hulls. Winnowing was done from a basket held about 4 feet in the air.

All of the beans were dried on mats for several days. Storage was in baskets or ollas sealed with salted dirt. The seal was to protect the beans from rodent or insect damage. Each variety of bean was stored separately[5].

Freeman learned that teparies grown by these tribes yielded 450-700 pounds per acre when dry farming was done, but would give 800-1500 pounds per acre when irrigated. No matter what the growing conditions, teparies outyielded the common bean by as much as 4:1[9].

The Pimas and Papagos thought the pink beans needed to be cooked with meat, especially fatty meat, in order to be good. Meat was scarce, so pink beans were not used as much as teparies. The teparies were most often cooked alone, although they may have been combined with corn. The Pimas seemed to prefer the taste of white teparies over the brown ones. Both varieties of teparies absorbed more water than the pink bean, perhaps because of the thicker coat on the pink bean[5].

The Pima and Papago tribes also grew cowpeas and lentils (*Lens esculenta*) for many years. These seeds were brought to Mexico from Europe by the Spaniards. Small and large cowpeas were grown, along with a hybrid that was reddish-brown with a big white spot at one end of the seed, instead of around the hilum[5].

The black-eyed beans were harvested as they ripened and the plants would continue growing and producing until damaged by frost. The Indians could pick the first pods from the base of the plant eight weeks after planting. The major harvest was done in the same manner as described for the teparies and pink beans.

The Pima planted lima beans in late March, the Pagago in July. As with the garbanzos, the limas were harvested throughout the growing season, until frost or dry growing conditions affected the plant's production. The Pima did irrigate when the plants stopped producing pods, and this stimulated growth. Also, the Pima had more success with growing limas since more water was available to Pimas than to Papagos.

Usually, limas were shelled by hand. If there was a large quantity to be prepared, the pods would be piled on a cloth on the ground and trampled. Smaller amounts of both limas and black-eyed beans were planted by the Indians, than of teparies and pink beans.

Garbanzos were planted in smaller amounts, and even fewer lentils. Both Pima and Papagos planted garbanzos as a winter crop, anywhere from November to February. Harvesting began in April or May, so not much irrigation was needed for them. Garbanzos were picked as they ripened. The first to ripen were rubbed by hand to remove the seeds from their pods. For the major harvest, the routine of pulling up the plants and beating the vines, as for previously-mentioned beans, was followed. A yield of 1000 pounds of garbanzos per acre was usual on good soil[5].

Lentils were also planted as a winter crop. The Pimas planted a second crop as well. Lentils are drought-resistant so were easier to grow than other beans.

In spite of all this agriculture, the Pimas needed to gather foods to supplement their food supply. Mesquite beans (*Prosopis chilensis*) were a favorite item[5].

A year later than the Castetter article, Pijoan et al. reported that the Papago also grew yellow teparies, pinto beans (*Prosopis vulgaris*), mesquite beans, and screw beans (*Prosopis pubescens*). The Papago diet was reported to consist mainly of beans and bread, supplemented by occasional meat (small amounts used as seasoning), cooked pumpkin or cactus fruit[27].

In their 1935 paper, Castetter and Underhill reported as many as 15 different varieties of teparies, all different colors, being grown by the Papagos. The one considered to be most valuable by them was the white bean, which grew as a climber as well as a bush plant[5].

The Papagos sometimes added dried tepary leaves to their tobacco. It was said that when a man smoked this mixture, he was wishing for a large crop of teparies next season.

In 1959, Hesse[11] reported that the Pima land south of Phoenix had little water, so not much land could be irrigated. Cotton was a major crop and not much food was being grown for consumption. Food was being purchased at trading posts, with little perishable food bought because not many Pimas had refrigeration. Beans were still the basic food, eaten at least twice daily by all members of the family. Limas, pintos, kidneys and black-eyed beans were all used. The common method of cooking was to add about 1 teaspoon of lard per serving. Lard was the only fat used at this time. Hesse found a very low incidence of coronary atherosclerosis among the Pima.

In 1971, Reid and a group of researchers[28] analyzed the nutrient intake of Pima women. They found that, although the Pimas were still using some brown and some white teparies, pink beans, limas, black-eyed beans, garbanzos and white beans, pintos were used most often, and were served with tortillas and chili peppers. Kidney beans were not used much because of a general dislike for them.

Reid and workers found that working Pima women did not use dry beans as much as had been traditional because they took too much preparation time. Also, the women were earning enough money to be able to purchase more meat. Government surplus foods available to the Pima included white beans, but these were not popular. The vegetable used most was canned green beans. Canned vegetables were used more often than fresh or frozen, perhaps due to ease of storage and preparation[28].

When surplus commodity foods were made available to the Papago, they, too, learned to use the surplus pinto beans and no longer grew many of the traditional beans. Also, external pressures are making floodwater farming more difficult for the Papago. Now their values are changing so that subsistence skills are not as prestigious as they used to be. More Papagos pick cotton for wages[21].

When teparies are available, they are valuable. In 1978, they sold for $1.39 a pound in Papago markets. This is a high price for beans[21].

Bean Preparation Methods

A variety of sources are available for bean preparation ideas. It is often impossible to determine whether the method described is a traditional or a modern method. In most cases, I suspect there's been at least a modern influence.

The one bean dish that is all-American, coming from Indian tribes all over the United States, is succotash. Root and de Rochemont suggest that it appeared everywhere in the country, spontaneously[27]. No matter what else it contains, the dish is basically a mixture of beans and corn. Having grown up with succotash made from canned green lima beans and corn (which changed to frozen vegetables when they became available), I was surprised to learn other forms of this ubiquitous bean dish.

The word succotash has come from these Indian names: *sukquttahash*, *msakwitash*, and *M'sick-quotash*, and may have meant whatever was in the pot at a particular time[27].

Johnson and King[13] tell us that the original succotash was made with kidney beans and corn cooked in bear grease. Cushing reports deer suet being used by the "more eastern tribes."[7]

Hughes' recipe for Zuni succotash uses green beans, corn and beef cubes with crushed sunflower seeds to thicken the mixture[12].

A succotosh recipe in Kavasch's *Native Harvest*[14] has onion, green pepper and nut butter in addition to lima beans and corn. Yet another version, in *Indian Cookin* by Whisler[38] calls for tomatoes and meat drippings to be added to the beans and corn.

A recipe for Pueblo succotash which Williamson and Railsback have labeled "traditional" combines cooked dried corn, lima beans, sunflower seeds, tomatoes, onions and chili powder[35].

Other bean preparation includes many soups and stews. A recipe for Pueblo chili stew has dumplings, containing both corn meal and whole kernel corn, cooked in a stew of pinto beans, other vegetables and ground lamb - still another version of the corn-bean combination[12].

The Maricopa Indians included cholla buds (from the spring Cholla cactus) with their bean and corn stew[24].

Other soups are black bean with leeks and garlic; lima bean-tomato soup with salt pork and seasoned with chili pequins, those tiny but powerful peppers; a pink bean "spoon soup" with lamb; and *pojoaque* cream soup, made with mashed pinto beans, evaporated milk, and spices. The gourmet variation of the latter is to be served cold with a topping of sour cream[12].

A marinated bean salad starts with pinto and green beans mixed together and seasoned with chili powder. Many of the Southwestern bean dishes use some form of chili peppers for flavoring.

All the Southwestern Indian groups seem to make the "refried" beans which we often associate with Mexican cooking. Soft, boiled beans are mashed to make a paste, and then fried in hot animal fat until the fat is absorbed. One way to serve this dish is spread on Indian fry bread, or on tortillas[25].

No matter how they are prepared, beans may be used at one-third to one-half of all meals for the Southwest Indians[3]. They have been an important staple throughout the history of these people.

Bibliography

1. Bohrer, V., H. C. Culler and Jonathan D. Sauer: 'Carbonized plant remains from two Hohokam sites, Arizona BB:13:41 and Arizona BB':13:50. *The Kiva* 35: October, 1969.
2. Buhrow, R: 'The wild beans of Southwestern North America'. *Desert Plants* 5:67, 1983.
3. Calloway, D. H. and J. C. Gibbs: 'Food patterns and food assistance programs in the Cocopah Indian community'. *Ecology of Food and Nutrition* 5:183, 1976.
4. Calloway, D. H., R. D. Giauque and F. M. Costa: 'The superior mineral content of some American Indian foods in comparison to federally donated counterpart commodities'. *Ecology of Food and Nutrition* 3:203, 1974.
5. Castetter, E. and W. H. Bell: *Pima and Papago Indian Agriculture*. Albuquerque, University of New Mexico Press. 1942.
6. Curtin, L. S. M.: *By the Prophet of the Earth. Ethnobotany of the Pima*. Tucson, University of Arizona Press, 1984.
7. Cushing, F. H.: 'The Nation of the Willows'. *Atlantic Monthly* 50:374, 1982.
8. Ford, R. I.: 'Gardening and farming before A.D. 1000: patterns of prehistoric cultivation north of Mexico'. *J. Ethnobiology* 1:6, 1981.
9. Freeman, G. F.: 'Southwestern beans and teparies'. *University of Arizona Agricultural Experiment Station Bulletin* 68:1918.
10. Greenhouse, R.: 'Preparation effects on iron and calcium in traditional Pima foods'. *Ecology of Food and Nutrition* 10:221, 1981.
11. Hesse, F. G.: 'A dietary study of the Pima Indians'. *American Journal of Clinical Nutrition* 7:532,, 1959.
12. Hughes, P.: *Pueblo Indian Cookbook. Recipes from the Pueblo of the American Southwest*. Sante Fe. Museum of New Mexico Press. 1972.
13. Johnson, J. and F. B. King: *Green Corn and Violets. Amerindian Recipes for Camp and Kitchen*. Springfield. Illinois State Museum Society. 1976.
14. Kavasch, Barrie. *Native Harvest. Recipes and Botanicals of the American Indian*. New York. Random House. 1977.
15. Kavena, J. T.: *Hopi Cookery*. Tucson. University of Arizona Press. 1980.
16. Kimball, Y. and J. Anderson: *The Art of American Indian Cooking*. New York. Avon. 1970
17. Kirkpatrick, D. T. and R. J. Ford: 'Basketmaker food plants from the Cimarron District, Northeastern New Mexico'. *The Kiva* 42:257, 1977.
18. Kuhnlein, H. V. and D. H. Calloway: 'Contemporary Hopi food intake patterns'. *Ecology of Food and Nutrition* 6:159, 1977.
19. Kuhnlein, H. V., D. H. Calloway and B. F. Harland: 'Composition of traditional Hopi foods'. *Journal of the American Dietetic Association* 75:37, 1979.
20. Nahban, G. P. and R. S. Felger: 'Teparies in Southwestern North America. A biogeographical and ethnohistorical study of Phaseolus acutifolius'. *Economic Botany* 32:2, 1978.
21. Nabhan, G. P., J. Berry, C. Anson and C. Weber: 'Papago Indian floodwater fields and tepary bean protein yields'. *Ecology of Foods and Nutrition* 10:71, 1980.
22. Nabhan, G. P., J. W. Berry and C. W. Weber: 'Wild beans of the Greater Southwest: Phaseolus metcalfei and P. ritensis'. *Economic Botany* 34:68, 1980.
23. Nabhan, G. P., C. W. Weber and J. W. Berry: 'Variation in composition of Hopi Indian beans'. *Ecology of Food and Nutrition* 16:135, 1985.

24. National Geographic Society: *The World of American Indians.* Washington, D.C. 1974.
25. Niethammer, C.: *American Indian Food and Lore.* New York. Collier Books. 1974.
26. Obizobon, I. C. and N. Obiano: 'The nutritive value of jackbean (Canavalia ensiformis)'. *Ecology of Food and Nutrition* 21:265, 1988.
27. Pijoan, M., C. A. Elkin and C. O. Eslinger: 'Ascorbic acid deficiency among Papago Indians'. *The Journal of Nutrition* 25:491, 1943.
28. Reid, J. M., S. D. Fullmer, K. D. Pettigrew, T. A. Burch, P. H. Bennett, M. Miller and G. D. Whedon: 'Nutrient intake of Pima Indian women: relationships to diabetes mellitus and gall bladder disease'. *American Journal of Clinical Nutrition* 24:1281, 1971.
29. Root, W. and R. de Rochemont: *Eating in America. A History.* New York. The Ecco Press. 1976.
30. Spier, L.: 'Havasupai ethnography'. *Anthropological Papers of the American Museum of Natural History* 24:81, 1928.
31. Steen, C. R. and V. H. Jones: 'Prehistoric lima beans in the Southwest'. *El Palacio* 48:197, 1941.
32. Stewart, K. M.: 'Mohave Indian gathering of wild plants'. *The Kiva* 31:46, 1964.
33 Stewart, K. M.: 'Mojave Indian agriculture'. *The Masterkey* 40:5, 1966.
34. Titiev, M.: *Old Oraibi; a study of the Hopi Indians of the third Mesa.* Harvard University. Peabody Museum of American Archaeology and Ethnology. Paper 22, No. 1. New York. Kraus Reprint Co. 1971.
35. Titiev, M.: *The Hopi Indians of Old Oraibi; Change and Continuity.* Ann Arbor. University of Michigan Press. 1972.
36. Voth, H. R.: *Oraibi natal customs and ceremonies.* Chicago Field Columbian Museum Publication 97, Anthropology Series 6, No. 2, 1905.
37. Williamson, D. and L. Railsback: *Cooking with Spirit. North American Indian Food and Fact.* Bend, Oregon. Maverick Publications. 1988.
38. Whisler, F. L.: *Indian Cookin.* Tennessee. Nowega Press. 1973.
39. Burgess, M. A.: 'The Tepary Connection: A visit with W. D. Hood'. *Desert Plants* 5:3, 1983.
40. Niethammer, C.: 'Tepary cuisine'. *Desert Plants* 5:8, 1983.
41. Teiwes, H. and G. P. Nabhan: 'Tepary beans, O'odham farmers and desert fields'. *Desert Plants* 5:15, 1983.

The Flavour of Japan.
One Western View Thereof.

By Max Lake

"To find and enjoy a new flavour adds seventy days to your life."

Japanese folklore.

The visit of a young Japanese honeymoon couple to Lake's Folly became the inspiration to delve more deeply into the flavours of Japan, to reassess the staples at the core of an important, and as yet not widely known, cuisine. As a would-be flavour professional, I was already aware that there are some obvious differences in the Japanese palate and approach to the pleasures of the table. *Umami* ("oomarmi") is but one of these, a fifth perception of taste, in addition to the sweet sour salty bitter ones we all recognise, and which is still something of an enigma outside of Japan. The simple reason is that there is no English word to identify it. A single sensation of 'deliciousness' seems to express it. This taste was first labelled early this century, (1908-9), and now has many descriptions. It is that character in foods which produces similar effects to MSG on flavour recognition, seen for example in reduced protein and vegetable stocks, tomato puree, certain mushrooms, soy sauces, grated mature Parmesan, etc.

There also seems to be a preference for a higher level of sweetness than is common in the West, perhaps a function of the fact that sugar was virtually unknown to ordinary Japanese until fairly recently. Another important racial difference is that some Japanese lack the enzyme, alcohol dehydrogenase, socially useful in Western cultures to speed the metabolism of alcohol.

Previous experiences in Japan had given me a clear idea of both Japanese foods and the entrancing ceramic and lacquerware in which they were served. When Joy and I were able to offer two attractive young visitors an unplanned lunch of a jellied codfish tail, a few snow peas, a flower of broccoli, some unpolished rice, and a dab of green dynamite *wasabi* arranged on handmade tenmoku glazed stoneware (from a previous visit to a Tokyo folk potter), the nostalgia of such unexpected hospitality so far from their homeland opened the floodgates. This started a friendship, one demonstration of which was to be the ultimate manifestation of Japanese hospitality at a dinner at their apartment in Tokyo, when one particular course displayed over sixty different components. The evening concluded with an ambrosial liqueur made from dried ripe persimmon fruit by a grandmother of my young friends. How can you adequately thank anyone so generous?

There can be few ethnic cuisines where so much respect is paid to the primeval forces of Nature. M. F. K. Fisher observes that the preparation and serving of fine as well as routine Japanese food is more obviously mixed, than is ours, with other things besides hunger, such as aesthetics, religion, tradition, history, seasonal changes, sea storms, a guest's birthplace, or one's childhood. Consider also the special wooden box to measure the 15 to 20 gallons of soybeans which will be used to make the next day's *tofu*. It is beautifully formed in the typical fashion of Japanese woodwork, with fine dovetailing or precisely mitred corners. Burnt into one side are images of the deity of craftsmen, tradesmen, and fishermen, symbolising hard workers who earn their living by honest toil; and the other, a happy god who is so well off that he does not mind rodents nibbling at his sacks of grain.

There is a north/south rivalry to which any nation can relate, but which is sometimes difficult to understand. The Kyoto locals are said by Dr. Umesao to believe that the flavours of their dishes are more subtle than anywhere else in Japan.

To get some idea of tradition as it is seen in today's Japan, let me quote from Soei Yoneda, abbess of the Sanko-in Temple, introduced by Robert Farrar Capon:

There is, for example, a small plate of assorted cooked vegetables, beautifully arranged: a slice of mashed yam roll in *nori* seaweed, an arrow feather made of burdock root, a slice of lotus root tempura, a piece of simmered *tofu* and three lightly cooked snow peas. There is *rōbai*, a delicious concoction of deep-fried wheat gluten, softened in a soy-flavoured stock and dressed with hot Japanese mustard. There is a mock "*tofu*" made of sesame "milk" and a *nori* roll filled with flavour-simmered spinach and served with a lemon-soy sauce. There is an altogether remarkable steamed savoury cup in which grated turnip, chopped rice, ginkgo nuts, *shiitake* mushrooms, carrot slices, snow peas, fried waterchestnut balls, and thickened *dashi* stock are turned into a delicate "custard" punctuated with an emphatic dot of *wasabi* horseradish. There is a starkly simple shredded ginger tempura, served crisp with mustard-soy for dipping. And there are clear soups and miso soups, and bamboo rice, vinegared turnips and pickled Chinese cabbage...

Every vegetable, however deftly or mysteriously sauced, tastes first of all like itself - even in dishes where several ingredients are combined. And every presentation, even the most artful (there isn't one that's not: even three bits of fried, shredded ginger on a plate are arranged so as to have a front and a back) is worlds away from mere artiness. Everything looks like food meant to be relished both by eye and palate.

The flavour preference of the ancient and formal culture of Nippon derive from the produce of the islands that arise out of bountiful seas. It is rather a small landmass of mountains and fertile farmland to support the needs of many millions of the most diligent and selective individuals on the planet.

Japanese food was traditionally based on grains, sea produce, and greens. It is worth noting that while ever rice was a medium for local barter, and payment of labour, the farmers had little or no idea of variety in regional flavours. This changed radically with the establishment of brand differences only a century ago. Rice tended to be eaten at home, and *soba*, buckwheat, was consumed in noodle restaurants. *Tempura* could be bought from street wagons.

Cattle and dairy produce were virtually unknown to the majority of the population because there was was not enough space to farm them, and there were religious objections. Another good reason for the scarcity of cattle, however indelicate the subject, is the simple fact of life that the body odour from a diet of cattle meat and dairy based products, so common in the West, has been offensive, at a cultural level at least, to Japanese sensibilities. No nation in the world spends more time and effort to ensure a natural and almost neutral body smell, and the rancid odour of the unwashed Westerner in confined metropolitan transport (a turn-off even to their own ethnic group) is the pits, and the subject of coarse humour. So, no cheese at all, nor milk for adults, until the adoption of Western foods.

Even as the rest of the world watches in fascination, the tastes of the Japanese people are now shifting from traditional to international, from local and home made to mass production. Those of us who think about food and flavour and suchlike, have much to learn, and perhaps even something to teach. I first visited Japan in 1968, and there have been major changes in a mere twenty years, but my personal experience is limited virtually to the seven towns which have now become modern Tokyo. Foreign food was uncommon, and as an example, an initial search of the big department stores turned up about six wines, all of which were less than promising. Prior to that period, I have been told that if any wine was enjoyed, it was a form of "port", a sweet fortified locally manufactured product, and the less said about that the better. At that time Western table wines were considered "too tart, too sour". A few years ago a similar exercise revealed forty local wines, some of which were quite good. Beef was now in to the extent of McDonald's hamburgers, and even Aussie meat pies. Cheese is still not widely found in any but the largest supermarkets, and then mainly where the wealthy, and foreigners, do their shopping.

The waves of Western influence on the austere culture that derived from the imperial court of ancient Kyoto, began around the turn of the seventeenth century. With the second, late last century, neither did much to Japanese food preferences, and certainly nothing like the third wave that took place after 1945. It is quite fascinating to learn from Shizuo Tsuji of a reverse thrust, that the secret seasoning of the ceremonial banquets of Louis XIV of France was soy sauce imported to Europe from Nagasaki by Dutch traders! To many Japanese today, Western food has been synonymous with French, with little inspiration from anywhere else. But this is changing rapidly.

Paradoxically, at the same time as English speaking peoples are turning to leaner meats, the Japanese import and market at centres like Kobe (I can never work out what *rubbing* the cattle with beer does, it is probably a bit of foreign leg-pulling) lot-fed beef heavily marbled with fat for extra flavour. With the astronomical prices that a fine steak fetches on the restaurant plate in Japan, there is not probably much risk to national health.

Geography is the natural arbiter of flavour. Broadly speaking, intense stimulation from curries, chilli, spices, and suchlike are Equatorial. The milder flavours of animal fats, fish, and herbage derive from cooler zones, with striking exceptions like Szechuan and Korea. Today's commercial transport has blurred the distinctions, and nowhere is this becoming more common and apparent than in the Japanese islands.

The traditional staples are still to be seen in the flood of Japanese restaurants which now exist worldwide. Rice, wheat, and buckwheat noodles: the bounty of the soybean in the form of miso paste, more than three thousand different brands of soy sauces (there were five thousand!), and wet (*tofu*) and dried curd, and soybean oil. Japanese food is happily fried in any light fresh oil, stopping short of olive, which is considered too heavy and flavoursome for the subtlety of traditional dishes. *Tofu* has developed its own circle of connoisseurs, as knowledgeable and devoted as any group of winelovers in the West. Fish leads the vast range of seafoods, either wriggling fresh, barbecued, flaked and smoked, fried, or dumplings; various kelps, and seaweeds, head the list of a vast collection of green and growing things.

Freshness of ingredients is a special feature of Japanese food. The only way to appreciate what I mean is to grow one's own vegetables. And as far as seafood is concerned, have the privilege of eating at places like Wheelers in London or the Grand Central Station in New York to clearly understand the Japanese pursuit of freshness. Their use of fish is classified by time out of the water. An early morning visit to the Tsukiji fish markets in Tokyo is a revelation, among the most striking of the market visits I have ever made. For example, fish to be served as *sashimi* must be absolutely fresh. Any older and it is consigned to various cooking methods that relate to the time since it was landed. The skill and sensitivities of a chef may be judged solely on the care in the selection of the primary ingredients at the market. And this assessment applied equally in the home. You will see the celebrated *fugu*, Pacific puffer fish, auctioned for service later in the day by those licensed to prepare it, to those adrenalin junkies who take the risk of dying fairly smartly if any quantity of the neurotoxins remain. The last death known was in 1975. Martin Franc reports that seven thousand tonnes of it is consumed yearly!

The large white *daikon* radish would be near the top of the vegetable list, along with yams and a fascinating collection of mushrooms, shiso leaves (which has tones of mint, aniseed and Dr. Pepper type sarsaparilla), the spice the Chinese call Szechuan pepper (whose pungency lies in the marjoram/sage/thyme area rather than black pepper), *wasabi* horseradish, garlic and other members of the onion clan, and ginger. Plenty of flavour interest, but more compact than the range available to equatorial cultures.

Some inquiring Westerners, used to stronger flavours in meat stocks, sauces, and so on, may find the natural foods and drinks of Japan tend to be rather low key. (The standard stock for noodles is made of shaved flakes of dried bonito fish and kelp.) After weeks of a Japanese diet, I find many of the flavours tend to be less assertive than one is used to. I started to see the difference in terms of colours of the spectrum. The flavours available to the Western kitchen run the gamut from red to purple. The traditional flavours of Japan lie in the more central yellow/green/blue range. For those who argue they had the best Indian curry of their life in Roppongi, or the best French food in the world is available on top of the Okura Hotel, look at it this way. Where else could you start a meal with five kinds of seaweed on a plate?

They all have a common marine flavour, so what is the attraction? They all have a different form, their colours are different, and when you chew them they all have a different mouthfeel from slimey to brisk crunch, that's what. This texture difference is another part of my thesis. Much of the interest and excitement of Japanese flavour lies in the variety of textures available, and it is here that the West has much to learn. This subtlety is one facet of Japanese aesthetics, like the possibilities of *ikebana* based on a single twig, or the curving strokes left by a rake in the white pebbles of an indoor courtyard. Chris Driver tells us, like the "al dente" borrowed from Italian, the Japanese have perhaps ten or more words for degrees of texture that do not exist elsewhere. And nowhere else will you see such exquisite presentation of food, the zenith of Japanese style. It verges on vandalism to disrupt some of the creations, for which I sometimes try to atone by a flash photo beforehand, for a permanent reminder of such inspiration. How many plates in good restaurants and homes around the world acknowledge their debt to the perfection that is so inspiring in much everyday Japanese food?

It has been well said, "show me what you eat and I'll tell you what you are." The same with drink? From Kyushu to Hokkaido, the Nipponese equivalent to Land's End to John O'Groats, the foreigner gets the impression that an awful lot of sake is being consumed. Anyone in a Japanese subway later in the evening, or having a midnight snack of *yakitori*, believes this. But the figures show that there is four times more beer drunk, no doubt because it is very good. Increased wine production from grapes has paralleled their appreciation of its nuances, and I believe this will continue because table wine has about two thirds of the alcohol of sake, and has a far more complex flavour profile. Quite a lot of domestic and imported whisky is taken, and a distilled spirit called *shochu* is a popular tipple because people tend to forget its revolting hangover.

Sake is likely to hold its position while ever tradition and culture retain their importance in Japan. The enjoyment of wine from fruit may have begun by the accidental fermentation of its sugars by yeasts in the air. Monkeys are reputed to ferment grain and drink the result, and elephants are known to make pilgrimages to get tanked on fermented berries, complete with hangover! Sake certainly was made by chewing rice and spitting it into a tub, just as kava is made today in the Pacific islands. In the case of sake, the enzyme ptyalin in saliva converts starch into fermentable sugars, after which yeasts take over to produce sake. In 300 A.D. *bijinshu* "beautiful women's sake" was made by young virgins who did the chewing. The making of sake became part of certain festivals, doubtless fostering a marvellous community spirit. The gods took a special interest in the conversion of rice to sake, and it became the drink of choice at weddings, dedications, for special gifts, and to be enjoyed with friends. Today, at table, it is more than polite for your friend to pour your first sake cup and to keep it topped up, and for you to do the same. This delightful bit of neighbourliness seems to have translated to cans of beer.

One special evening, Naoya and Yuriko took me to an inn where a wide range of sake of different regions and ages was kept. I went through the lot, including one that was fourteen years old and another that either the host or some ancestor had made and treasured. My notes are not clear. Anyhow it was a great honour. As near as I can recall, it seemed to be intended as a tribute to my international reputation as a wine judge. It did not seem fitting to remonstrate that any such reputation was based on the fact that wines being assessed are rarely swallowed, and the substantially higher alcohol of the sake was no help. What did strike me was the fact that the range of flavours of even the greatest sakes was far less than that of our good creature of the West, wine, and thereby less likely to evoke the discussion and general interest that it does. Again, as with the traditional food, we seemed to come to rest in a limited part of the flavour spectrum. As a final comment on sake, its aroma, and that of other rice residuals, typifies, to this traveller, the smell of Japan, in the way for example that the aroma of garlic and Gauloises is an intrinsic part of the ambience of rural France.

The range of teas, green and black, is phenomenal. One shop in the maze of independent food stalls under the Ginza has over two hundred. But it is in exploring the coffees of Japan that the keen Western palate experiences its greatest shock. The best coffee is imported and drunk in Japan, without a doubt. They take 98% of the seventy ton annual production of the rare mountain coffee of the island of Jamaica, and it is available in Japanese speciality shops at about three times the price of lesser coffees, both in an exquisite white shell porcelain cupful of filtered bliss, or by the kilo of beans.

Which brings us to the current flavour of Japan, in such a state of flux that the observant visitor becomes aware of the turbulent transition. The nation's traditional foods, which include the staple grains, greens, and fruits of the seas around the home islands, are meticulously selected for freshness and visual impact, to be presented in a unique style which individualises flavour and delights the eye. A marvellous mélange with their flavoured Western goodies is occurring. And perhaps the best French food in the world is to be found there. As they say in Japan:

"Let little seem like much, as long as it is fresh and beautiful."

My warmest thanks for the help of Tak Nishizawa who has absolutely no responsibility for any error or malapropism. Then perhaps neither have I. "I was out of the country at the time."

Maize as a Staple Food

By Elisabeth Lambert Ortiz

Corn (maize) *Zea mays* has been a staple food crop in the Americas for an immensely long time, having first been cultivated in the Valley of Mexico about 7000 years ago, in 5000 BC. It was not until Christopher Columbus arrived in what is now the Dominican Republic in the 15th century and found the grain growing there that Europe and the rest of the world learned of its existence. In the time between 5000 BC and the 15th century maize had spread from Mexico throughout the Americas and the Caribbean. The people we call the Incas in what is now modern Peru were brilliant agriculturalists and developed their own strains of corn from the Mexican original.

The name maize derives from the name given to the plant in the local language of the Indians of the Dominican Republic, *mahis*. This became *maíz* in Spanish and maize in English, though in the United States it is corn. It is just as well that this simple name was the one adopted by most of the world rather than the original name in *Nahuatl*, the language of the Aztecs, which was *tlaolli*. The ancient Mexicans developed many different types of maize. *Iztactlaolli* was white maize, *yauhtlaolli* was black maize, and so on. The Spanish were faced with a barrage of infinitely complicated Indian names for different types of tortilla. *Tatonqui tlaxcalli tlacuelpacholli* meant a large white corn tortilla, only one of a great variety of tortillas lovingly described by the Spanish priest Fr. Sahagun when he studied Mexican cuisine in the market place of Tenochtitlán, the name of the old Mexican city which was built on a lake, rather like Venice, with islets connected by canals. Fr. Sahagun had the good fortune, our good fortune too, to be in Mexico for some time before the city was destroyed and was able to sample dishes using tortillas and tamales, stuffed maize dough steamed in corn husks or plantain leaves. The Spanish, faced with the complexities of Nahuatl fell back on the simple name of little cake, tortilla, which, being an economically minded race when words are involved, they also use for omelettes. This has caused some confusion though not as much as there might have been if they had adopted the names the Aztecs used. They also made another contribution to kitchen simplicity by inventing the tortilla press. Originally tortillas were patted out by hand from the soft, moist dough before being briefly baked in a *comal*, a griddle, for about one minute on each side. The Spanish invented a simple wooden press consisting of two circles of wood hinged together and with a handle to allow the tortilla-maker to bring the right amount of pressure to bear on the round ball of dough between the wooden plates. I never became an adept at hand-patting a tortilla and I use an old cast-iron press dating from Mexico's first industrial era, and I also have a less efficient very modern aluminium model. The wooden ones are prettier and qualify as antiques, and there are still many Mexican women who pat out tortillas by hand.

Tortillas, which took the place of bread in pre-Columbian Mexico are unique in being the only bread that is made with cooked flour. Today few cooks bother to make their own *masa harina*, literally dough flour, by the old method where shucked dried maize was boiled with lime, then rinsed to get rid of the skin around each kernel, and laboriously ground on a 3-legged inclined volcanic grinding stone, a *metate* with a volcanic stone rolling pin, a *metlapil*. Nowadays the masa can be bought freshly made by machine in markets or dried and packaged in supermarkets though the maize is always boiled with lime.

Mexico has had wheat bread since the Spanish introduced wheat in the early days after the Conquest which was in 1521. A type of French bread was introduced during the brief, ill-fated reign of Maximilian and Carlotta and there are many of the Spanish sweet breads, but for all that tortillas have never lost their place and maize remains an important staple. Even though fewer types of maize are grown than in the past, what can be called the Corn Kitchen is still of great culinary importance and is all that many people outside Mexico know of the cuisine. When the Spanish found how many roles the tortilla can play in the kitchen they called them the *antojitos*, little whims or fancies. They are so flexible and so varied that I don't think a better name could have been invented. There are the simple *tacos*, fresh warm tortillas stuffed with beef, pork or chicken, chillies and various sauces, and cheese. These are folded over and eaten by hand. They may be rolled and secured with a toothpick and fried in lard or oil. The more elaborate *enchilada* is a tortilla dipped in sauce, briefly fried, stuffed, rolled and heated in a sauce. *Chilaquiles* are leftover tortillas, cut into strips, lightly fried and combined with a variety of ingredients usually including chillies and tomatoes and grated cheese, then baked. Yucatán has a quite elaborate tortilla dish, *papatzul* in which tortillas are stuffed with chopped, hard- boiled egg then masked with a pumpkin seed sauce, and a fresh tomato sauce and garnished with a few drops of green oil squeezed from grated pumpkin seeds *(pepitas)*. *Quesadillas* are unbaked tortillas stuffed and fried, or less usually baked on a *comal*. *Tostadas* are large tortillas fried and covered with several layers of filling. Other tortilla dishes include *panuchos* where black (haricot) beans and tortillas are combined with chicken breasts and hard-boiled eggs, and variously shaped tortillas such as *chalupas*, which are stuffed. Tamales, whether in dried corn husks or plantain leaves, have persisted from the days before the Conquest. A favourite dessert of Emperor Montezuma was a corn husk tamal stuffed with strawberries and probably sweetened with honey.

There is evidence that sweetcorn was developed in pre-Columbian Mexico and had already spread to the eastern seaboard of the United States when it was still a colony. A soldier travelling north from New York found it grown and eaten by Narragansett Indians. However it was not popular at the time as the so-called garden corn was the favourite. Corn on the cob however was popular enough to inspire Charles Hay in *Hints on Etiquette*, 1844 to deplore the habit of picking up a corn cob and gnawing at it. He wrote that the kernels should be scored with a knife, scraped off into the plate and eaten with a fork. Probably by then sweetcorn was popular and certainly did not merit this barbarous suggestion. The Shakers dried corn then soaked it and cooked it with sugar and cream. There were many maize recipes popular round about the time of the American Revolution, and still popular, including Corn Pudding, a savoury dish, Corn Custard with Tomatoes and Corn Oysters, a sort of fritter, all made with fresh corn. Succotash, made from fresh lima beans and fresh corn was probably one of the first dishes the Indians taught the Pilgrim Fathers and it is traditionally served at Thanksgiving dinner. It is clear that maize (corn) was an important staple in all of North America. Cornbread remains a firm favourite all over the United States.

Though everyone knows about tortillas which are sold tinned or frozen and about the *masa harina*, sold packaged ready to be made into tortillas, few outside Venezuela and Colombia know about *arepas*, the corn bread of Venezuela and Colombia in South America. The *arepa* flour is made from the large white rather starchy corn developed from the Mexican original by the Incas of Peru. I have been wrong in claiming that tortillas are unique as the only bread made with cooked flour. The flour for making *arepas* is made from corn cooked in exactly the same way as corn for tortillas, but the cooked result, the *arepa* is exceedingly different. The *arepa* can be eaten as bread split open with the doughy centre pulled out and buttered, or simply split and buttered. They can also be stuffed with meat or cheese and thinner ones can be fried when they puff up and are called *infladas*. Despite the popularity throughout Latin America of rice which the Spanish and Portuguese introduced, maize remains an important staple. Paraguay has a very robust corn bread curiously called *Sopa Paraguaya*, which translates as Paraguayan soup. It is made using two cheeses, cottage cheese and the local Munster, cornmeal (maize meal), fresh grated sweetcorn kernels, sautéed onions and eggs. There are pancakes made with sweetcorn, notably from Venezuela, and corn on the cob, sliced, appears in many stews and especially in Peru, as an accompaniment to potato dishes. Corn is used in soups, as a vegetable and, in Chile, in combination with fresh cranberry beans and pumpkin. In Brazil it is used in soups and seafood dishes and in cakes and biscuits. It is an extremely versatile staple and it would be impossible to imagine the cuisine of the Americas without it.

It was the Portuguese who took maize to Africa where it has become a staple food, mostly using ground white corn (maize) meal; *kenkey* and *banku* from Nigeria are examples. A number of corn (maize) dishes called mealies are popular in South Africa. Maize is also a staple food in South East Asia, especially in Indonesia. It is not clear who first took maize to China but it was probably the Portuguese.

In Europe it is increasingly popular as fresh sweetcorn on a seasonal basis, though it has generally been eaten as ground dried maize made into a maize (corn) meal porridge - *mamaliga* in Romania, *polenta* in Italy. Caesar's legions made *polenta* of mixed cereal grains, and, though the name *polenta* was retained, the grains were supplanted by maize once that cereal was introduced. In Portugal, especially in the north, ground maize (corn meal) is used in a popular bread - *broa*. Spain may have conquered Mexico, but the maize (corn) of the early agriculturists of ancient Mexico has conquered the world as a useful staple food.

Corn - A Staple From The New World

by Janet Laurence

Rich golden maize, easy to grow, easy to harvest, eaten for thousands of years as a staple for the Aztec, Maya and Inca civilisations; it seemed to offer as much nutritious bounty as the potato to the Spanish conquerors of Peru and Mexico. Unfortunately it was a staple that packed an unsuspected health hazard.

In most countries of the world, corn is a generic term which can refer to a number of different grains. In the US it refers solely to Maize or Indian Corn (Zea mays). The wild grain from which it was originally cultivated no longer exists, what we eat today is a result of hybridization. The name comes from the American Indian ma-his.

Grown in ancient times in both North and South America, carbon dating of corn remains found in caves takes cultivation back at least 3,000 years[1].

The Spanish Conquistadores brought maize to Europe in the late 15th century, where it found instant popularity wherever the weather was warm enough and the rain sufficient for successful cultivation. In eastern Europe it was known as "Turkey" corn, though the Turks apparently called it "Egyptian" corn and the Egyptians, "Syrian"[2]. The Portuguese took it to Asia where it was equally popular.

Today maize is still grown in most of these regions, much of it for animal fodder. But some 50% of the world crop is produced in the corn belt of the Mississippi basin; Illinois, Indiana, Missouri, Iowa, Kansas and Nebraska. We learnt of the immense size of the great American Corn Belt in school geography lessons. I imagined vast prairies of waving, golden wheat, oats and barley. Great was my disappointment many years later to drive through the corn fields of Iowa. Mile after dreary mile of thick, dense, green leaved plants, some "as high as an elephant's eye". Every so often the dead straight road would take two right angled turns that would send it on in the same direction but on a fractionally different latitude. The only reason we could work out for these deviations was that they were to prevent drivers from falling asleep.

Maize is extremely easy to cultivate, both plant and fruit are large and it is reckoned that the crops of the early cultivators required no more work than one day a week per person, so that the remainder of the time was available for building the monumental pre-Columbian temples and fortifications[3]. Certainly the Incas of Peru, the Mayas and Aztecs of Mexico, the Mississippi mound builders, the cliff dwellers of the American Southwest and many North and South American semi-nomadic tribes depended on corn as a dietary staple[4].

Its pedigree as a nutritious staple must have seemed firmly established when it was exported to the Old World and its ease of production made it highly popular in poor areas. It seems that as the Irish peasant grew potatoes, many southern European peasants grew maize. And, particularly after the Civil War, it provided the main diet in poverty stricken areas of the cotton growing southern states of America.

From the mid-eighteenth century into the early years of the twentieth, outbreaks of a disease that was called pellagra occurred in many areas where maize was the staple crop: northern Spain, southern France, Italy, the Balkans, Egypt, southern Africa. It was named by the Italians after the roughened skin the disease caused (pel agre), in Spain it was known as "mal de la rosa", and in 1776 the Patriotic Society of Milan offered a prize for a solution to the problem of pellagra[5]. The disease brought dermatitis, diarrhoea and dementia, causing it to be known also as the disease of the three "D's", and could end in death. Though as early as 1762 it was suggested by Gaspar Casal that diet could be a cause of the disease, his observations were made against a general belief that pellagra was a contagious infection. Though part of the Milan prize mentioned earlier was given to a Dr. Viemar who cited poverty and malnutrition as the cause, and there was a French doctor who observed that the disease appeared in areas where maize was the staple crop (by the end of the nineteenth century, few Frenchmen ate corn), the prevailing belief remained that pellagra was an infection.

Then in 1914 the US government invited Joseph Goldberger to join an expert commission that was studying pellagra, at that time causing 100,000 cases a year in the depressed areas of the south, 10,000 of which resulted in death[6]. It was, the commission reported before Goldberger began his investigations, "an intestinal infection, transmitted by contaminated food"[7]. The contamination was thought to be similar to ergot, a fungal disease to which various grasses and cereals can be susceptible.

After a long and bitter battle with accepted medical opinion, Goldberger finally proved that pellagra was due to a dietary deficiency and was suffered by those who were fed on a poor diet consisting almost entirely of maize. It appeared most commonly in the spring, after a winter diet unsupplemented by fresh foods. When milk and a little meat were eaten as well, the pellagra disappeared. The problem was that such items were too expensive for most of the sufferers to introduce into their diet. But Goldberger discovered that an inexpensive extract of liver and yeast contained a factor that would prevent the disease amongst maize eaters and named it the P-P (pellagra-preventing) factor. He died in 1929. In 1937 nicotinic acid was identified as the P-P factor by Elvejham[8].

The maize protein, zein, does not contain tryptophan, which enables the body to synthesise nicotinic acid and although maize contains a certain amount of the nicotinic acid itself, not much but more than white bread, oats or rye, the vitamin is present in a bound form as niacytin, which does not release the nicotinic acid into the alimentary canal[9].

The puzzle was how civilisations could exist for thousands of years on maize without suffering endemic pellagra. The answer was found to lie in their handling of the grain, for the nicotinic acid is released if maize is prepared with an alkali.

In Mexico tortillas are made from maize mixed with lime water. Hominy, probably brought to the southern and southwestern states of America from Mexico, is made from corn kernels that have been treated with wood lye, that is ashes and water. Hulled corn is first soaked, then boiled; after which the outer skin can be rubbed off. The swollen grains can then be used immediately as a starch food similar to porridge, eaten either on its own or as an accompaniment to meat, or dried then coarsely ground for future use as hominy grits[10]. It was when maize was treated as any other grain and preparation with an alkaline substance was omitted that pellagra caused such trouble amongst poor diets.

Pellagra notwithstanding, the popularity of maize has led to the development of hundreds of varieties that range in size from tiny popcorn cobs and the miniature sweetcorn currently so popular, to the long cobs of the Mexican Jala valley which reach 50-75cm (2 ft).

Maize is mainly golden in colour but can be so pale as to be almost white and also red, blue or brown. There are five main kinds of corn. I quote Harold McGee: "Each [is] characterized by a different endosperm composition. Pop and flint corn have a relatively high protein content and hard rather than waxy starch. Dent corn, the variety most commonly grown for animal feed, has a localized deposit of soft, waxy starch at the crown of the kernel, which produces a depression, or dent, in the dried kernel. Flour corn, with little protein and mostly waxy starch, is grown only by native Americans for their own use. What we call Indian corn today are flour and flint varieties with variegated kernels. Finally, sweet corn, very popular as a vegetable when immature, stores more sugar than starch, and therefore has translucent kernels and loose, wrinkled skins (starch grains refract light and plump out the kernels in the other types)."[11].

Much of today's maize production is used for animal fodder (I have no information on steps taken to prevent pellagra in cattle, Goldberger made use of observations of a similar disease in dogs during his investigation of the disease), some of it goes towards a variety of other uses, these are covered a little later in this paper, but a certain quantity is still used for various traditional dishes in many parts of the world although in recent decades it has become a less important part of the diet in the US and southern and eastern Europe:

Mexico:

I understand Elisabeth Lambert Ortiz has produced a paper on corn in Mexican and South American food so I will leave this area to her infinitely more expert knowledge.

United States Of America:

Three-quarters of the US production of corn is now used for animal fodder but maize is still a staple food of the southern and southwestern states of America.

Mexican tortillas, corn chips and tamales are prevalent in the Mexican influenced south western states and the southern states adopted corn to the almost total exclusion of wheat. Hominy was for years a great southern dish. But it may no longer be quite as popular as it was when I was travelling through in the 60's, when hominy or hominy grits were on every restaurant menu in the modest establishments we patronised. James Beard suggests that by the 70's the raw material was getting more difficult to obtain[12].

Corn bread is still very popular, sometimes made today with a mixture of wheat flour to add gluten, which does not exist in corn. Corn bread is best consumed immediately after baking and is sometimes eaten direct from the tin with a spoon and naturally called Spoon Bread. Pioneers cooked small cakes of a simple corn dough on hoes heated over an open fire, these became known, of course, as Hoecakes.

Cornmeal mush is the south's version of Italy's great national dish, polenta, and is often served with butter and/or grated cheese. The meal for making cornbread or mush needs to be much coarser than the modern cornflour, whose production was made possible by the invention of roller mills in the late nineteenth century, which reduced the grain, usually with the germ extracted, to a fine powder.

Never flagging in popularity, even perhaps gaining, is Popcorn, for which there are a number of varieties, all small, hard and flinty with the grains entirely surrounded by a very hard skin. Popcorn was apparently the first of the cultivated corns. Harold McGee makes an educated guess as to why popcorn actually pops. He suggests that as the protein-to-starch ratio in popcorn is higher than in other varieties, so the protein matrix in which the starch grains are embedded is probably stronger. "When the kernel is heated in hot oil, the small amount of moisture it contains partly gelatinizes the starch grains; this happens at around 150°F (66°C). Then, as the kernel temperature reaches the boiling point, the water vaporizes and expands rapidly in volume. The hard protein matrix holds until the pressure becomes too great, at which point the kernel bursts open and the endosperm expands in volume on account of the sudden pressure drop. At the same time, the already cooked starch granules are dried out as the water vapor escapes, and the endosperm texture becomes light and crisp. If, however, popcorn is cooked in a covered pan that offers no escape for the water vapor, the spongy endosperm will absorb it again, and chewy, tough popcorn is the result. Popcorn pops well only when it falls within the narrow range of 11 to 14% moisture content, to which the better brands are adjusted before being packed in air-tight containers."[13]. And I always thought popcorn had been treated in some clever way to make it swell up so dramatically! It is now possible to buy popcorn which can be popped (literally) in the micro-wave.

Sweetcorn is the corn used as vegetable and is becoming increasingly popular. It has a relatively high sugar content. Grown mainly in the cooler areas, it is picked and eaten unripe and fresh. As corn (and peas) lose up to 40% of the sugar content in room temperature within six hours of picking, either by conversion into starch or energy, it is no illusion that sweetcorn is best cooked straight after picking (as are peas). Sweetcorn for the commercial market is cut and instantly chilled to hold the sugar content stable, enabling transportation without deterioration[14].

There is apparently little record of early Americans eating corn on the cob although James Beard says it must assumed that it was done and occasionally the term "roasting ears" is found, maybe to indicate the corn had its silky floss removed, was rebound in the outer leaves and then roasted. An early recipe for boiling corn on the cob uses this method of preparation[15]. However the popularity of the vegetable was definitely established by the turn of the century. "I remember," says James Beard, "the times when yellow Bantam corn was the first of the season and people looked forward to fall with great appetite for the Shoe Peg and Country Gentleman corn, which were supposed to be so much better. In the case of Shoe Peg, few varieties have outdone it for sheer tenderness and sweetness. Nowadays the season for corn is year-long."[16] And there are over 60 varieties of sweetcorn. It is sold vacuum packed, tinned and frozen as well as fresh. Recipes are abundant for the golden niblets in salads, fritters, soups, souffles, gratins, buttered or creamed and in many other guises.

Corn assisted America's great gift to gastronomy, the invention of the breakfast cereal, first introduced in the 1850's by Dr. John Harvey Kellogg, a sanatorium director at Battle Creek, Michigan. Granola, which failed to catch the general imagination, was made of wheat and oatmeal in addition to maize. Cornflakes themselves made their entrance in 1899. Their main contribution to the diet is their requirement for milk to be added. On the negative side they both contain sugar and usually have more added at the table[17].

Italy:

The Italians are the most corn conscious of the European nations. Maize is grown extensively in northern Italy, where it is a staple food north of the River Po, in a strip south of the Alps and running through Piedmont, Lombardy, Trentino and the Veneto. To quote Anna del Conte: "In the Middle Ages and in the Renaissance, polenta was made with spelt, chestnut flour, millet and even ground acorns. It was also made from buckwheat, as it still is in Valtellina and other Alpine valleys, where it is known as polenta taragna. The Friulani in north-east Italy were the first to make polenta with the maize from the New World that was unloaded at Rialto in Venice. In Veneto, polenta is also made with very fine flour which is sometimes white. It is called polentina bianca, and being thinner it is spooned rather than cut. Although equally good, it is not as striking in appearance as the golden mountain that is proudly placed on a board in the middle of the table. Yellow polenta is made with coarse-ground maize, of which there are two or three different grades on the market. The right degree of coarseness should be chosen for each dish."[18]. Cold polenta can be recycled in a variety of ways, cut and fried or grilled or used to make gnocchi. The various combinations in which polenta is used and its sauces are many. Traditionally it is made over a slow heat in copper pans (known as a Paiolo) and stirred with wooden paddles.

Romania:

The Romanian dish of Mamaliga is their equivalent of polenta. Usually a simple peasant dish, often cut with a thread as in Italy, there are gourmet versions dripping with butter and melted cheese.

Spain:

Cornmeal can still be found in Spanish cooking. Elisabeth Luard has a recipe for sweet cornmeal pudding from Milhassou and also mentions a Basque dish, Broyo, a cornmeal mush served with Boeuf en daube[19].

Hungary:

The Turks imported maize into Hungary along with many other new plants, such as tomato, paprika and cherries, and for centuries it was known as "Turkish wheat". George Lang gives recipes for cornmeal dumplings and a corn cake from Transdanubia that is a sweetened corn bread[20].

Egypt:

I have no information on the use of corn in Egypt.

Africa:

"Mealies", the staple diet of the poorer Africans is their version of the cornmeal mush of America and the polenta of Italy. It has caused pellagra problems.

Other Uses For Maize:

Corn Oil:

Maize is a rich source of oil, some 50% of the germ is oil. 50% is polyunsaturated, 35% monounsaturated and 15% saturated fatty acids. Strong flavoured, like other corn products, it is not everyone's favourite oil and though it emulsifies easily, perhaps more easily than any other culinary oil, its flavour is not ideal for mayonnaise. But it is sweet and rich and good for frying.

Whisky:

Alcohol is made from most grains and various ingenious processes have been used to break down the starch from corn in order to produce fermentation. Pre-Conquest, ground corn was chewed, the human saliva being used to break down the starch into glucose and maltose (a double glucose sugar). Not surprisingly, such a time consuming production method did not catch on generally. Distillation is the method preferred in the U.S., where Kentucky corn whisky was established by 1780. The taxing of distillation in 1791 resulted in illegal moonshining, especially popular in the poor hilly regions of the south where distillation offered the best return on the small amount of corn that could be grown there. Today America's favourite whisky, Bourbon, is made from barley malt and ground corn and there is also corn whisky together with a number of other varieties (A liquor can legally be given the name of a particular grain if that grain constitutes 51% or more of the grist.).[21]

The Future:

It is possible that maize may, like other grains, now move into industrial fields as a renewable resource for the production of ethanol as a petrol additive to replace lead, and a starch base for the production of a wide range of products including: pharmaceuticals, glues, enzymes, artificial resins and plastic materials, paper and paperboard and miscellaneous chemical products. Development will depend on relative costs in the market place and financial incentives by governments.

It would be nice to think of an ancient American staple food becoming an industrial resource for the twentyfirst century.

Notes

1. STOBART, Tom, *The Cook's Encyclopaedia*, pub. B.T. Batsford Ltd., London, 1980, p. 243
2. Ibid
3. McGEE, Harold, *On Food and Cooking*, Charles Scribner's Sons, New York, 1984, p. 240
4. Ibid
5. GRIGGS, Barbara, *The Food Factor*, Viking, Middlesex, 1986, p. 40 (Ms Griggs has a lively account of the problem of pellagra and its investigation by Goldberger)
6. YUDKIN, John, *The Penguin Encyclopaedia of Nutrition*, Viking, Middlesex, 1985, p. 183
7. Ibid
8. Ibid
9. Ibid
10. BEARD, James, *Delights and Prejudices*, Victor Gollancz Ltd., London, p. 76. Also: STOBART, Tom, cf p. 209
11. McGEE, Harold, cf p. 241
12. BEARD, James, *American Cookery*, Little, Brown & Co., Boston, 1972, p. 580
13. McGEE, Harold, cf p. 242
14. BEARD, James, *American Cookery*, cf. p. 512
15. Ibid
16. Ibid
17. YUDKIN, John, cf p. 84
18. DEL CONTE, Anna, *Gastronomy of Italy*, Bantam Press, London, 1987, p. 257
19. LUARD, Elisabeth, *European Peasant Cookery*, Bantam Press, London, 1986, p. 275 & 273
20. LANG, George, *The Cuisine of Hungary*, Penguin Books, Middlesex, 1985, p. 303 & 391
21. McGEE, cf p. 490

In researching the use of corn a wide number of books on various cuisines has been consulted covering Indian, Far Eastern, Middle East and European cookery as well as that of the American continent. I feel listing them would be tedious and of little value. Unfortunately I have not had access to books on African food and cookery.

A Dictionary of Edible Aroids

By Jenny Macarthur

I've always found eddoe, cocoyam, baddoe, dasheen, malanga and yautia names particularly confusing. Before starting on this paper I had bought and cooked the unfamiliar roots for sale in my local London market and found them mostly rather dull. But undoubtedly they are staple foods in the Tropics and the aroids are probably the most unfamiliar of all. So I thought I would investigate them for myself and share my findings with other symposiasts. Clearly I am no expert but I hope these notes and names will be helpful. For a long time I have collected names of the basic raw materials, so that if a recipe says "add a pound of chopped elephant ear" I might have some idea what is required but most of the information included in this paper is either the result of reading or the kindness of the traders and shoppers of Shepherd's Bush Market, who were tremendously enthusiastic that I should learn about Caribbean food but a bit shaky on taxonomy. A list of the principal books studied is included at the end.

The *Araceae* are a large family of monocotyledons. There are about 110 genera and over 2000 species. Many are spectacular and grown as garden ornamentals or as house plants. These are not, though, the kinds cultivated for eating which seldom flower. The botany can be confusing but I have done (very) rough drawings based on our own common British arum, the cuckoo pint (Arum maculatum), which I hope shows the unique shape of the spike, spathe and spadix and how the tubers/corms/cormels grow. Cuckoo pint (it was originally cuckoo pintle, ie cuckoo's prick, before the puritans got to it) is, with our other English names such as Wake Robin, a good example of the many lewd names attached to the plant because of the spadix, which grows (in some varieties) large and evil smelling. (See the elephant yam). Although most parts of all arums are poisonous raw, even our own cuckoo pint tubers were once treated to produce a starch known as "Portland sago", used as a thickener and to make a soft drink, like barley water, known as salep.

As far as I know no-one has ever found cuckoo pint leaves palatable, unlike the young leaves of taro and malanga, eaten in South America and the West Indies and the principal ingredient of the famous West Indian soup, Callalou, and the South Pacific coconut cream bundles known as Palusami. The leaves, as bought in Shepherd's Bush in February, were an unexpected delight. They are neither tough nor truly tender, slightly glutinous and probably one of the next foodies' fad finds. Some of these, especially those with purple stems, may contain harmful saponins and need to be thoroughly cooked. Both leaves and tubers contain calcium oxalate crystals which can cause intense irritation to the skin, even more so to the mouth. Cultivated varieties may need less cooking but wild varieties are always suspect and should be given a prolonged boiling even after the recommended (for some varieties of tubers) soaking for several days in an unpolluted stream. Cook leaves as spinach, as a wrapping for fish, in earth pits, in chicken casserole. Cook small roots, eg eddoes qv, as potatoes but for longer. Dasheen is usually peeled and parboiled before being fried or baked.

As always I hope that symposiasts will help my researches both by providing more local names of the foods described and by pointing out errors.

Taro

Colocasia esculenta   *Araceae*

There are eight species and at least a thousand known cultivars of taro which fall into two main groups. The first has a relatively small corm surrounded by large cormels. These are known in most of the West Indies as eddo. The other, known as dasheen, has a large central corm and fewer, smaller, less important cormels. Many writers claim that these are separate species but it seems to be generally accepted

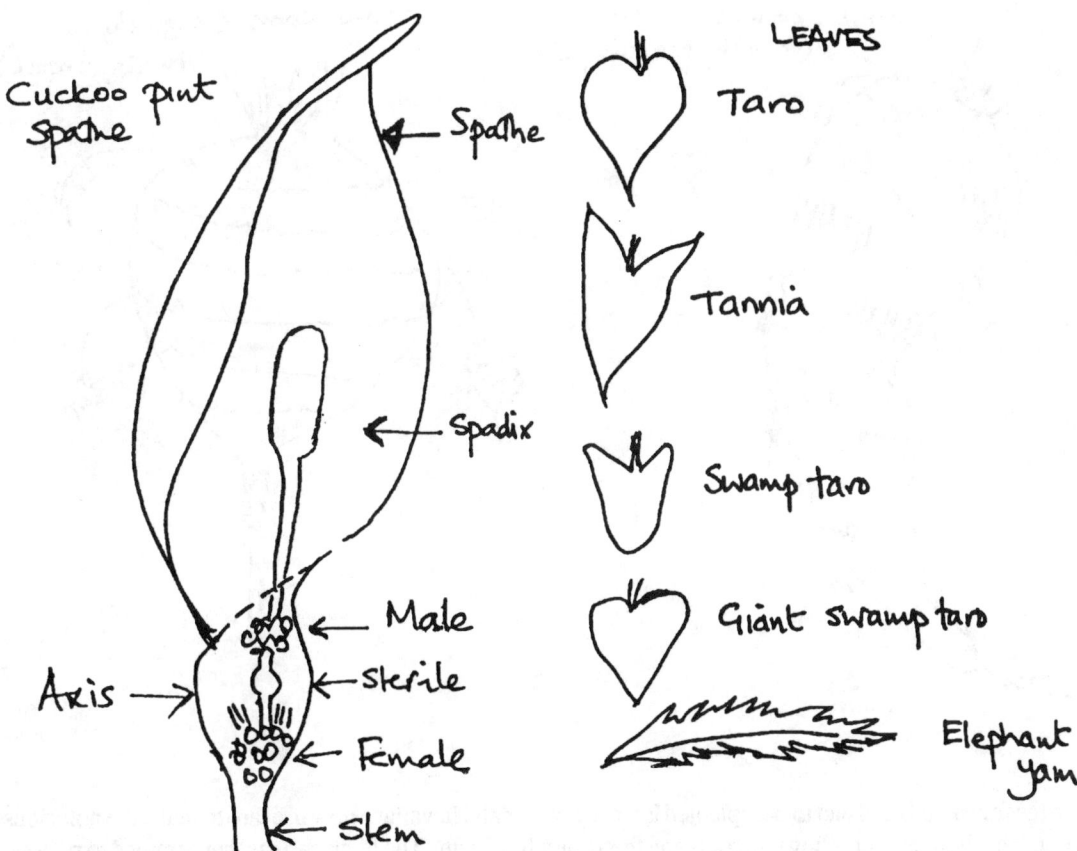

that they are both *Colocasia esculenta* with eddo being *C. esculenta var. antiquorum* or *C. esculenta var. globulosum* and dasheen being *C. esculenta esculenta*. It is not clear whether *C. macrorrhiza* which identifies most of the taro grown in the South Pacific is a synonym or a separate species. Equally the Kolocassi of Cyprus and the Kolkas of Egypt are simply given as *C. esculenta* but the cultivar is different, the main cormel being usually more carrot shaped, although the side shoots are also eaten locally. The plants have heart shaped, or sometimes spear shaped, leaves which are a popular vegetable when young, preferably barely unfurled. Taro tubers, which vary enormously in size from 50 grams to 5 kilograms have a brown hairy skin, sectioned a bit like an armadillo and may have flesh coloured white, pink, cream or lilac/purple. There is little difference to the uninitiated between the old (world) cocoyam and the new (world) cocoyam, commonly known as tannia, see below, except that tannia are more pear shaped and do not have such a ridged skin.

Alphonse de Candolle first suggested in 1886 that South Sea Island taro was a domesticated form of its Asian wild relatives and subsequent botanists and archaeologists have traced the migrations of the Polynesian people and their taro through Indonesia, Australasia and the South Pacific. Taro is most important of all in Hawaii, where, in their version of the creation myth, taro is the first born of Father Sky and Daughter Earth. Humans came second. The cultivation of taro was men's work. Women were allowed to cultivate other crops, but only men might touch (and presumably eat) the sacred taro. Maoli lehua, the most commonly produced red variety, was once the special prerogative of the chiefs and reserved exclusively for them. In times of scarcity wild taro from the forest swamps could be eaten but it belonged

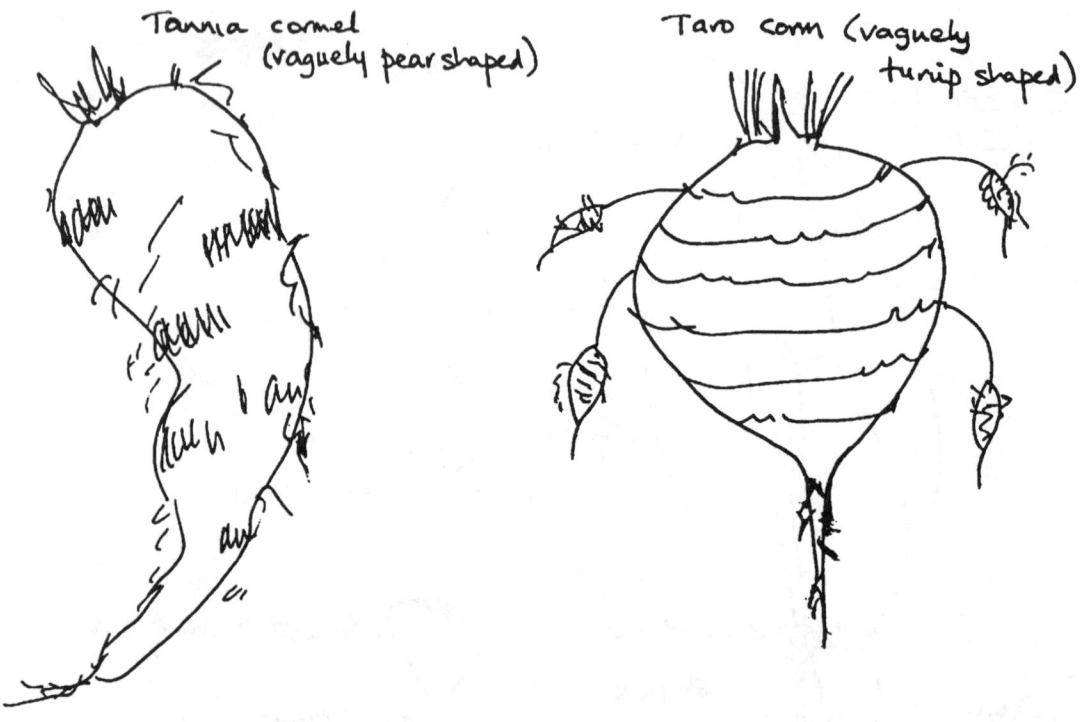

to the forest gods and had to be replanted later. However all Hawaiians now use taro to make the notorious poi. This is made by boiling taro roots and then pounding them. The thick paste is then strained to remove fibre and excess lumps. This may be eaten as it is or allowed to ferment and sour for two or three days, when it is usually eaten with salty or spicy foods and seaweed (which the Hawaiians seem to love almost as much as the Japanese). Each island claims that its own taro cultivar makes the best. Poi may be one, two or three finger. One finger is thick enough for one finger to collect a mouthful from the pot; three finger requires three fingers to get a mouthful. Most Europeans find it pretty unpalatable. It resembles grey wallpaper paste and has little taste when fresh. Fermented it is rather worse than 'an acquired taste'. No symposiast who tasted the sample I brought had anything good to say about it. However the early Hawaiians are said to have consumed over ten pounds (four kilos) a day without ill effect. Nowadays it is almost entirely produced commercially. Recipes say open a packet and add water to taste. Taro has as less romantic legend in the West Indies. It was imported as cheap food for slaves. Dasheen has become **the** soul food on many of the islands, eaten mashed with hot chili sauce. I have found no evidence so far that taro was similarly imported into the U.S.A., although it is grown in the South now.

Although taro is native to the more humid tropics, cultivars have been developed which will grow in most warm areas from Australian rain forests to Egypt. It is not of major commercial importance but valued in Polynesia, the Far East, Hawaii and the West Indies, and to a lesser extent in Egypt and Cyprus (the only European country to show any interest), in India and in parts of West Africa where, on the whole, it is considered inferior to other root starch vegetables, although West Africa is probably the world's largest producer. Taro is said to have been known in ancient Egypt as Kolkas, hence the scientific name but the earlier known vegetable of that name, mentioned as Kolokasia by Diascorides, is thought to be lotus root. It sounds like another case of "It's a bit like the cod we had at home so we'll call it cod" syndrome to me. Most taros in the West Indies and South America seem to be grown both for their roots and leaves, although some are more favoured for one rather than the other. With some varieties the central corm is not eaten at all, only the smaller "side" corms. These, when boiled and dried, are said to taste of mushrooms. The mature tubers may be forced in damp soil or moss when they produce delicious white shoots, eaten like asparagus. Taro starch replaces cornstarch/cornflour in many tropical cuisines. Indian cookery books seem mostly to translate taro (arvi) and elephant yam (zaminkand) as yam, whereas true yam is chupri aloo or khamealu. The leaves are translated as colocasia leaves. I can find few Indian recipes for the leaves, unlike the cormels, but there is one famous Gujarati speciality variously called patrel, patrail or pateer which has "black", not "red", stemmed leaves sandwiched in layers with a paste of gram flour, tamarind, palm sugar, spices, onion and sometimes fruit. These "sandwiches" are rolled up, tied and steamed for twenty minutes. When cold the rolls are sliced and fried and served with coconut. I have another patrail recipe which has the leaves stuffed with minced meat. The leaves are also used in fritters known as vadis or wadis. The famous palusami of the South Pacific uses taro leaves as wrapping for coconut cream. These are usually baked. *Kaikai Aniani* by R.J. May has leaves stuffed with corned beef and coconut which sounds similar and Sri Owen, in her *Indonesian Food and Cookery* gives a recipe called Buntil which are equally packages of taro leaf and coconut but much spicier and these are steamed.

Other names:

Old cocoyam, Dachine, Dasheen, Eddo(e), White eddo, Chinese potato, Egyptian ginger, Tanyah tuber, Poi plant, West Indian kale[1], Elephant's ear[1],

France	Taro de Chine, Chou caraïbe[1], Rou-rou[1]
Italy	Aro di Egitto
Portugal	Alcoleaz, Taro
Spain	Tayoba
Cyprus	Kolokassi
Egypt	Kolkas, Qolqas, Kulkas
Hindi	Arvi, Arwi, Arbi, Arwi ka patta[1], Arwi sag[1], Patra leaves
Tamil	Seppan kizhangu, Seppamilai[1], Seppam ilaigai[1]
Marathi	Alu kanda[2], Alu cha pan[1], Alva-chi-pane[1]
Bengal	Kochu, Kochu sag[1]
Telugu	Chama dumpa, Chama akulu[1]
Oriya	Saru
Gujarati	Alvi, Alu na patra[1]
Malayalam	Chembu, Chembu-ila[1]
Kannada	Sama-gadde, Sama-gadde yele[1]
Sri Lanka	Kandalla
Thailand	Peu-ak
Vietnam	Khoai au nu'oc trang
Malaysia	Keladi, Talas, Tallas, Taloes
Indonesia	Talas, Daun talas[1], Wu
Philippines	Abalong, Amalong, Dagmay, Gabi, Gablos, Gahula, Linsa, Lubiñgan, Pising
Cantonese	Oo t'au, Yu tao, Woo tau[6], Ya
Japan	Sato-imo, Ara-imo
W. Africa	Bari, Koko, Ya beré
Hausa	Gwaza

Ghana	Eddo, Eddy-root
Malawi	Koko
Hawaii	Kalo, Poi spinach[4], Luau[5]
Fiji	Dalo, Taro kalo, Taro
New Guinea	Anega, Ba, Biloun, Guarava, Hemar, Ifen, Jefam, Kemb, Keu, Kom, Kukun, Mabo, Niang, Nomo, Oema, O'ou, Sangni, Ument
Trobriands	Uri, Mwedu
New Caledonia	Dab, Di, Ekengad, Io, Inagad, Kening, Moa, Moe, Ne, Nere, Ta, Tari, Waela, Walo, Wamo, Wamu, Wane, Weeo
Polynesia	Colulu, Talc, Ta'o, Taro, Tiee
Micronesia	Ioth, Kotak, Kukau, Mal, Ot, Oni, Ori, Sawa
Maori	Taro
S. Carolina	Tanga
Central America	Malangay, Malangu[3]
Cuba	Guagui
W. Indies	Barbados eddo, Chinese eddo, Chou bouton[1], Chou de chine[1], Coco, Cocoyam[3], Curcas, Madére Baddo, Callalou greens[7]
Guyana	Chinese tayer
Guatemala	Quisquisque
Venezuela	Danchi, Ocumo culin
Columbia	Chonque
Brazil	Taioba

1. These names are usually used for the leaves.

2. Which name means any edible root or potato in most Indian languages.

3. These names are also used for tannia, particularly the leaves.

4. See introduction.

5. Luau is a name for taro leaves but also for a traditional Hawaiian outdoor feast.

6. Small red budded varieties of the eddo type (Hung nga woo tau) are boiled in their jackets and eaten as a snack by the light of the moon at the mid-autumn festival in parts of China.

7. Callaloo is a famous soup. Other leaves may be used for it and in Jamaica callaloo is an amaranth, *(A. viridis)*.

Elephant Yam

Amorphophallus campanulatus *Araceae*

Elephant's foot yam, Elephant bread, Sweet Yam, Devil's tongue, Arum root

I used to have mild fantasies about the generic name *Amorphophallus*. Shapeless penis? Well, I was right. Some of these plants, of which there are about 90 species, produce a massive (up to a metre long) flowering structure/inflorescence (see all purpose arum drawing) with the spadix often blotched and malodorous. No comment. It is however the rhizome which is eaten. This is said to resemble an elephant's foot being some 25-30 cm in diameter, pouffe or footstool shaped with a corrugated rim, hairy, scaly and greyish brown. The flesh is usually white and very starchy. More often than not it is cultivated to produce a flour, konjac flour, which is used in Japan to make the noodles known as shirataki or processed into blocks which are sliced and added to soups and stews. In Northern India it would seem that the corms of one species, *A. rivieri*, are regarded as a delicacy usually grated, diced or sliced and boiled before being made into patties, curried with curds or fried with a sweet and sour sauce. These recipes describe the zaminkand simply as 'yam', ie it is the "yam" of the area, even though not a true yam. I have found these recipes only in Indian cookery books for Indians, not in books for Westerners. I found no evidence of the leaves being eaten but the young shoots may be treated like asparagus.

Hindi	Zaminkand, Jaminkand.
Tamil	Senai kizhangu
Malayalam	Chena
Telugu	Kanda dumpa
Kannada	Suvarne gadde
Oriya	Hathikojia aloo
Marathi/Gujarati	Suran
Bengal	Ol
Bangladesh	Ol kuchu
India general	Telinga potato, Arsagna, Balukund, Kidaran
Malaysia	Karak-kavanai
Indonesia	Ilis ilis, Kand godda, Sooweg, Waloor
Philippines	Oroy, Pañgapoñg, Puñgapuñg, Tigi
Cantonese	Mo yu
Japan	Konniaku, Konnyaku
Fiji	Daga
Tahiti	Teve
Polynesia	Koe

Giant Swamp Taro

Cytosperma chamissonis *Araceae*

This is the "taro" of Micronesia. There are about ten species. The plants have the virtue of being amenable to cultivation on coral atolls. Therefore they are a staple food on many islands in the South Pacific. They grow very large but take several years to develop.

Fiji	Tao kape, Viakana,
Tahiti	Maota, Ape de veo
Samoa	Pukaka, Puraka, Pula'a
Ponape	Muhang
Malaita	Kakake
Kiribati	Babai
Polynesia	Puna, Brokka, Opeves, Taa taa
Micronesia	Brak, Iaraj, Lok, Muang,
Philippines	Biha, Lok, Pananau, Palauan, Galiang, Gallan

Giant Taro

Alocasia macrorrhiza, Alocasia indica   *Araceae*

The first species is seldom cultivated today as it needs extensive preparation to destroy the toxins contained in it. The second is still grown in the South Pacific and parts of India and SE Asia. These plants are known by the general name of spoon lilies.

Fiji	Apai, Pai, Viamiloa, Viagaga. Viamila
Samoa	Talanu, Ta'amu
Tahiti	Apé
Tonga	Kappe
Polynesia	Ape, Kape, Ta'amu, Uvea
Melanesia	Awere, Kape, Koe, Pia, Twowe, Wawe
Micronesia	Bisech, Fole, Ka, Lai, Oht, Piga, Wot
India	Alavu, Alooku, Manaka. Mankachu, Mankanda
Sri Lanka	Habara
Assam	Boro
Malaysia	Birah, Brak
Indonesia	Senté
Philippines	Biga
Japan	Doku imo
Brazil	Inhame gigante, Toyoeu
Guyana	Hog tannia

Tannia

Xanthosoma spp.   *Araceae*

Tanya, Tannier, Malanga, New coco yam, Pomtannia, Yautia

There are some forty five species, generally known as spoonflowers. They are very similar to the taros but have a brighter green leaf with a distinct intramarginal vein and a longer central root. They also grow considerably taller than taro plants. They are native to the West Indies, Central and South America and were cultivated in pre-Colombian times. The main rhizome is not normally eaten. (It is fed to the pigs). The lateral tubers and the leaves are of most interest, with some varieties being cultivated only for the leaves. Most must be very well cooked as they contain calcium oxalate which stings the mouth and throat. They are of most importance in Cuba, Puerto Rico and the West Indies. In Shepherd's Bush Market they are sold simply as coco, the notion of old and new varieties not being understood. The following varieties are most commonly cultivated. The common names below are those of Puerto Rico.

Xanthosoma atrovirens	Yautia amarilla	Dark leaved tannia
	Yautia vinola	
Xanthosoma brasiliense	Yautia belembe	Belembe
	Calalu	

These two are grown almost exclusively for their leaves. Belembe is popular as it contains very little calcium oxalate.

Xanthosoma sagittifolium	Yautia blanca, Malanga	Arrow leaved
Xanthosoma caracu	Yautia horqueta	Caracu (a type of cattle)
	Yautia mavola	
	Yautia viequera	
Xanthosoma violaceum	Yautia guayamera,	Primrose malanga
	Yautia morada, Oto,	Blue taro, Red coco
	Yautia palma, Yautia prieda	

Other names

West Indies	Tayobe, Tayonne, Tayo tayo, Tata(e), Coco
Jamaica	Badoo
Costa Rica	Tiquisque
Antilles	Chou caraibe, Mafaffa, Mananga(y)
Brazil	Mangareto, Rascadera, Taioba
Venezuela	Ocumo cumán
Guatemala	Quequeque
Fiji	Dalo ni tana
New Guinea	Kong kong taro
Malaysia	Keladi telur, Keladi saratah, Keladi sarawak
Kampuchea	Macabo[1]
Indonesia	Kimpool[2]
Tanzania	Maduma[1]

1. I have these names but can't find the source so I am more than a bit dubious about them.

2. *X. violaceum*.

Sources of Information

Herklots, G.A.C. *Vegetables in South East Asia*, George Allen & Unwin, © 1972.
Kokua, Na Lima. *Taro (Kalo) Uses and Recipes*, Helping Hands of the Pacific Tropical Botanical Garden, Hawaii, © 1977.
Kay, Mrs D.E. 'Root Crops' *Crop and Product Digest 2*, Tropical Products Institute, Crown Copyright, 1973
Bazore, Katherine. *Hawaiian and Pacific Foods*, M Barrows & Co., New York, © 1949.
Skinner, Gwen. *The Cuisine of the South Pacific*, Hodder and Stoughton, New Zealand, © 1983. (with acknowledgement for names to *Food Plants of the South Sea Islands*, South Pacific Commission, 1956)
Lovelock, Yann. *The Vegetable Book*, George Allen and Unwin Ltd., © 1972.
Phillips, Thangam E. *Modern Cookery for Teaching and the Trade*, Orient Longman, New Delhi, © 1965.
Bisen, Malini. *Vegetable Delights*, Wilco Publishing House, Bombay, © 1970.
May, R.J. *Kaikai Aniani, A Guide to Bush Foods, Markets and Culinary Arts of Papua New Guinea*, Robert Brown and Associates (Aust) Pty, Ltd., NSW, © 1984.
Olaore, Ola. *The Best Kept Secrets of West and East African Cooking*, W. Foulsham and Co Ltd, Slough, 1980.
Anthonio, H.O. & Isoun, M. *Nigerian Cookbook*, Macmillan Publishers, London 1982.
Dahlen, Martha & Phillips, Karen. *A Popular Guide to Chinese Vetables*, Frederick Muller Ltd., London, 1983. First published in Hong Kong, South China Morning Post, 1983.
Zee, S.Y. & Hui, L.H. *Hong Kong Food Plants*, The Urban Council of Hong Kong, 1981.
Stobart, Tom. *The Cook's Encyclopædia*, Cameron and Tayleur Books Ltd., London, 1980.
Parkinson, Susan & Stacey, Peggy. *A Taste of the Tropics*, Mills and Boon Ltd., London 1972.
Owen, Sri. *Indonesian Food and Cookery*, Prospect Books, 2nd. edition, 1986.

Unfair Game

by Carolyn McCrum

For the Sioux, Cheyenne, Dakota and Comanche Indian tribes of North America, the buffalo was a primary source of food, shelter and clothing. These were migratory tribes, traversing the uncompromisingly rough plains to the west of the Mississippi River, carrying their homes and possessions with them on dog-pulled *travois*. By the time the first western explorers penetrated the Great Plains, the Indians had developed a way of life which depended heavily on the buffalo. Buffalo hide, scraped with the creature's shinbone and softened by rubbing in its brains, was cured and made into clothing, teepees, shields, saddles, harness, lassos and bowstrings. Indian arrowheads and feathers were attached to the shaft with glue made from buffalo hooves. Tendons and sinews were split to make thread. The hair was braided into rope. The stomach was used for cooking pots and water buckets; the bladder for raft floats. Horns were softened by steaming, and shaped into spoons and ladles. Buffalo chips provided fuel where no trees grew. The heart membrane was dried and used as a nursing bottle; the ribs became runners for sleds, and the scrotum was dried and filled with pebbles to serve as a baby's bottle.

Unfortunately for the Plains Indians (and for the buffalo), westward expansion of the Frontier was to put an end both to the hunter and the hunted. Before the white men came, even the most wasteful Indian methods of trapping the buffalo (setting fire to the dry prairie grass, or stampeding herds over a cliff) had barely dented the huge numbers of animals roaming the plains. Early white explorers reported their astonishment at the sight of the vast herds; estimated at between sixty and a hundred million, said to be the greatest population attained by any large animal at any time in the history of the world.

Before the Civil War, the only Americans apart from the Indians who showed much interest in eating buffalo were the pioneers crossing the plains in wagon trains bound for the Rocky Mountains. They called it "humpbacked beef", and there is evidence that they considered it at least as good as domestic beef. Wanton killing of buffalo was carried out by plains hunters who had discovered a market for the tongues, which could be smoked and shipped to luxury restaurants on the East Coast and were sought after even in Europe. Animals were killed by the thousands; their tongues cut out, and the carcasses left for the coyotes. A vogue for buffalo carriage robes accelerated the carnage, and thousands more were killed for their pelts, with the rest of the animal once again wasted.

Cattlemen were hostile to the buffalo on the grounds that the grazing land could not support both wild and domesticated herds. Railwaymen found buffalo herds a menace, as the animals obstinately refused to give way to locomotives. The railroads encouraged the hunting, as buffalo hides were profitable items of freight. Even before the coming of the railroads, the wasteful slaughter of buffaloes was decimating herds; in 1840, St. Louis received the pelts for 67,000 buffalo robes, and in 1848, 25,000 buffalo tongues found a market in that city.

But with the westward expansion of the railroads, the process of extermination accelerated. The Kansas Pacific ran special trains for buffalo hunters, who could fire from train windows into the herds grazing alongside the tracks. Railroads even hired professional hunters to kill the buffalo; Buffalo Bill was employed by the Union Pacific, and is claimed to have shot 4,280 animals in twelve months. The carcasses were not brought on board for meat then, as refrigerator cars did not come into service until 1875.

Why were the buffalo sacrificed and the Longhorn cultivated, one wonders? Buffaloes can live on two-thirds of the acreage of grass needed by the same number of Longhorns, and grow heavier. The sheer size of the beast was a consideration. Before the arrival of refrigerated railway cars, the transportation of the meat from a 2,500 pound animal to the nearest potential market would have been impossible. Longhorns could be driven and herded into cattle cars; the buffalo herds were considered wild and intractable and incapable of domestication.

But chiefly, the buffalo was eliminated by the white man in order to eliminate the Indians. This intention was admitted by General Philip Sheridan in addressing the Texas legislature, "Let [the hunters] kill, skin and sell until the buffalo is exterminated, as it is the only way to bring about lasting peace and allow civilization to advance." President Grant was of like mind with Sheridan. In 1875, the United States Congress passed the first bill in its history aimed at preserving wildlife, to protect the buffalo. Grant vetoed the bill, and the destruction of bison went on unchecked.

Deprived of their natural food, the Indians not surprisingly took to looting white settlements for an alternative. Indignant whites convinced the Army (who probably needed little convincing) that the raiders were savages worthy only of eradication, and that the only easy way of accomplishing this was the killing of all the buffalo. After the Civil War, there remained in 1865 only about fifteen million buffalo, and in that year alone, a million more were killed. By 1872, half of those remaining were gone, and by 1879, there was hardly a bison left on the southern prairies.

In 1880, when the railroads penetrated the North, large-scale slaughter was perpetrated there also, and in 1883, the largest remaining northern herd was wiped out in days by sharpshooters stationed at every known buffalo waterhole: an operation jointly carried out by white hunters and Crees, who unlike the Plains Indians, did not rely on the buffalo for their livelihood.

A national census taken in 1894 counted 1,090 bison in the whole United States. According to Root and de Rochemont in *Eating in America*, some local inhabitants tried to protect a herd of bison in the Lost Park of the Colorado which numbered one-fourth of all the American bison left in the world, but poachers relentlessly mowed the animals down, and in 1897, the last four were shot on behalf of a taxidermist who wanted specimens to stuff.

At the opening of the twentieth century, there were no wild bison left on the entire North American continent except 21 in a protected game reserve in Yellowstone Park, and a small herd near Lake Athabasca in Canada, which was under the protection of the Northwest Mounted Police. The deliberate eradication of the American bison had served its purpose, and ultimately cleared the way for the construction of railroads and cities, but the conscience of the United States was not altogether clear.

Dale Brown in *American Cooking: The Northwest* recalls the words of Charles Russell, the eminent painter of the wild west, to a crowd that had gathered to honour him, "I have been called a pioneer. In my book, a pioneer is a man who comes to virgin country, traps off all the fur, kills off all the wild meat, cuts down all the trees, grazes off all the grass, plows the roots up, and strings ten million miles of bob wire. A pioneer destroys things and calls it civilization. I wish to God that this country was just like it was when I first saw it, and that none of you folks were here at all."

In 1905, Theodore Roosevelt set up a buffalo reservation near Wichita, Kansas, stocked with animals donated by the New York Zoological Society, and this was followed by the creation of a national bison reservation in Montana. Under protection, the herds increased rapidly, as while a bison cow gives birth to only one calf a year, she is fertile for 40 years, and the young animals are hardy. The buffalo had escaped extinction by the skin of its teeth, but only through relegation to reserves: numbers of wild bison are restricted by the limited amount of grazing land left in the United States, and the herds are periodically thinned out to satisfy a small but growing market for buffalo meat.

Some private breeders have become convinced that providing grazing land for buffalo is a profitable operation. Dale Brown encountered one of these in his travels in the Northwest: a Mr. Roy Houck, part-owner of 50,000 acres of land in South Dakota, with a buffalo herd of 1,500. He acquired the unpromising land in 1959, and by building ponds to conserve rainfall and rotating herds, by the 1970's had turned it into a thriving ranch, chiefly devoted to raising buffalo. His success and that of others like him seems to confirm that buffalo are eminently suited to life on the plains; that they require less pasturage per head than cattle, and that they are natural survivors. The white hunters of the plains, Ulysses S. Grant and Buffalo Bill have a lot to answer for, it would seem.

Traditional Table Manners in Dagestan

By Dr. Magomedkhan Magomedkhanov & Sergi Luguev.
Edited by Robert Chenciner

Background

Customs and traditions connected with nourishment, as well as norms of etiquette and decent behaviour at the dining table[1], appear to be an essential part of the cultural life of any people. At the same time, different peoples' genetic and cultural attitudes towards eating seem to be more alike than different[2]. Our research shows that this is also true of the (at least) 33 linguistically different peoples of Dagestan.

This approach is supported by the following selected evidence gathered from the authors' notes on annual ethnographic expeditions to different regions of Dagestan over the past eight years.

Restraint in eating

People in Dagestan usually had three meals a day, the largest for supper. It was thought healthy to eat at regular hours, but never to satisfy the appetite fully. At breakfast and supper if no strangers were present, the whole family sat together at one table. In large families, the head and the older sons might eat apart from the women and children, but in the same room. The choicest pieces were usually served to the head, who gave small portions to the rest.

As in all the Caucasus[3], restraint in eating denoted rank and importance in people, especially in men - as shown in the popular sayings of different peoples in Dagestan. "A wolf, even when hungry, pretends to be full up" (wolves personified strength, will-power, manliness and bravery) - Kumykh. "A real man, even at home before his wife, does not overeat" - Avar. "Insignificance in a man is fed through his stomach" - Dido. "A person who has hunger in his bones" - Laki. "...to look on food with longing" - Kumykh. "Greedy glutton" - Dargin. "Hungry as a locust" - Avar. "Greedy as a clothes moth" - Lezghin. "Where there is bread, that is where his head is", "A hungry man has a full stomach - he has no eyes", "a shameless man will eat even what is not put out for him", "although the 'kazan' is open, even a dog should feel ashamed" - Kumykh. "Although shame is on his face, his belly is pleased", "bottomless barrel" - Lezghin.

When invited to taste an ordinary dish, such as wheat & beans flavoured with fat from a sheep's tail or feet, or thin patties with melted butter *(churek)*, or slices of pie with cheese[4], the guest must taste but not fill himself, and then quietly say a prayer of thanks. At meals it was customary to eat sufficient for physical need, finishing with a slight feeling of hunger.

Conversation at meals was edifying, but joking was discouraged as food was a serious matter. "If you don't work, forget about eating - you will wither away" - Avar. "No need to feel sorry for one who married on time and had an early breakfast", "Without food, there's no life, without work - no food" - Lezghin. "If going to a wedding start off well fed, for you will be starving before you return." Restraint in eating, characteristic of highlanders is eloquently described in Tolstoy's Hadji-Murat[5].

Prohibited Foods

It is well known that Islam prohibits eating pork, wild boar and related products, also 'unclean' wild and domesticated animals - foal of ass, cat, dog, bear, wolf, fox and more. Eating horseflesh was rare though not forbidden - only if a horse broke its leg, or was otherwise useless for riding. But in western Dagestan - Akhvakh, Karata, Dido, Bejta - the flesh of a riding pony, if it had to be killed was taboo for the head of the household and his male relatives. Also it was taboo to serve guests horsemeat - furthermore guests might only be given clean meat killed in the correct Muslim way - the animal's head facing south-east to Mecca, with the Muslim formula in Arabic "In the name of Allah, gracious and merciful. Allah the Great". Other bans were on food come by ill-gotten gains - stolen or by cheating; dishes prepared by women or men who had not completed head to foot ablutions after sex, or by menstruating women; and food licked by a dog.

Where and how to eat

It was incorrect to eat in the streets, or walking, or in a crowded place, or on horseback, or lying down, standing or when sleepy, with a person having a contagious disease - or next to a sickbed, or in a barn, hayloft or stable. It was good manners to chew food thoroughly, not hurrying, but bad to stuff your mouth, blow on food to cool it, take large bites and eat with your mouth open. It was rude to be excessively gay, joke constantly, laugh loudly or talk with undue animation. It was impolite to ask for anything, implying discontent with what was served. Children were taught that worthy people do not grab leftovers. "A bite of bread is appetite" - Lezghin. "My granary is empty, but my heart is full of honour" - Avar. "Only a dog stays where appetite is heard" - Kumykh. Food could give satisfaction if obtained through hard work. "Work, not meditation, provides bread" - Lezghin. "Free food is not to be had", "He who has no work, has no bread" - both Kumykh. "If you don't make your brains boil in summer, your pot won't boil in winter" - Laki.

When a family gathered at table, the eldest started first after he had recited a prayer, then taken a piece of bread or *churek*[6] and eaten a couple of spoonfuls of soup - *kasha*. He finished last taking care that all had had enough. First to finish had to be the youngest of the family. Eating started with bread taken with the right hand, using only the thumb and third finger.

Rituals of the meal

A meal began with washing hands. One of the household - often the youngest daughter - brought an ewer or basin of water. The head man washed first, symbolically inviting the others to table. After the meal, as a hint that it was time to leave the table, the host was last to wash.

You started eating from the near edge of the plate, lifting spoon to mouth without inclining the head. When food was on a common dish, the nearest was taken. But if that was the biggest or best apple or apricot (say) it was impolite to choose it and it was then offered on to an elder or a child. A cough or sneeze at table was covered by your hand while turning away your head from the table. It was impolite to dip a bitten piece of bread into a common dish. "If you lick it first, then gulp it" - Avar. Away from home, it was rude to eat more than others from a common dish, but worse to comment on another's manners.

Hospitality and the community

Journeyman

In Dagestan when a man agreed to do some building work for the master of the house, he was invited to a good supper every day. But however much he was asked, he never accepted and it was never taken as payment for work done.

Reputation

News of improper behaviour or lack of personal integrity often spread beyond the village and groups of villagers with bad reputations settled in the far corners of Dagestan. To eat at the same table with a person of low conduct who rudely intruded, violated the rules of honour and shame, was considered humiliating. If a man, met on the road, took advantage of politeness to receive shelter and food and assume the status of guest and got drunk, the host would refuse to sit at the same table.

Traditional custom required that each member of the village council *(Djamat)* should have a benevolent attitude to relatives and countrymen. Often when a family had cooked a specially delicious meal of sheep or ox, it must invite all relatives, neighbours and other members of the council to supper, or else earn the mocking title of "Shameful family". This counted as the severest censure in Dagestan where rumour and community opinion drew the line between honesty and dishonesty, dignity and disgrace, nobility and baseness.

Avoiding quarrels

Sitting together at table could also have a symbolic meaning - for two quarrelling relatives it showed they were reconciled. In contrast if someone categorically refused to have a meal with another, he was given to understand that the former was secretly offended. If a man heard a sharp or unkind word of a friend, he only had to say "I ate bread with him (in his home)" for the argument to stop.

It is usual in Dagestan for a proposal of marriage to be made by relatives or trusted friends of the youth, who come to the home of his intended with gifts of refreshments - festive dishes, sweets and fruits. If the host tastes something of the gifts, he approves of the arrangement - if not, he does not. The hostess sets another table for the guests and after some time invites the girl to join them. If she does not sit at their table, she shows she is not interested in a polite and tactful way, whereas an abrupt spoken refusal was rude and could have serious results - agreements on the side were unethical.

Guest ritual

The hospitality of Dagestan people is often written about[7]. Guests were invited into the best room in the house, to sit first before the host, men and boys, and served the best dishes. The hostess only sat down if repeatedly invited by the host and guests. Women and children were not invited to sit at the table.

The guest had to taste a little of every one of the numerous dishes on the table. To mention the quantity of the food or compliment a dish was bad form. The host constantly reminded the guest that he should not be shy and eat more. The guest never refused directly which was rude, but answered "Don't worry, I'm eating..." or "Yes, everything I could need is on the table". If someone drops in to greet the guest after everyone is already at table, he is introduced to the guest, but remains standing while the host gives him a drink which he drains before formally greeting the family and guest. The host might invite local singers or a storyteller for a special guest, who (or the host in the guest's name) would present each performer a small gift in appreciation. There was seldom a bustle of activity on the arrival of a guest. The host was always prepared to invite home relatives, neighbours and countrymen to eat and drink without loss of face or embarassment.

The ritual drink

Until about 1900 the most popular alcoholic drink was *buza*, made from millet as well as home-made beer. It was bad form to drink until merry and drunkenness was condemned by the community. The horn goblet of drink had more of a symbolic meaning and was not necessarily part of the table setting. After light refreshment the host would stand to propose the health of his guests. After drinking from the goblet (formally a polished horn decorated with niello silver ends), the host offered the refilled goblet to the guest who in turn drank to the host, household and any senior member with "may this horn reach you with prosperity", "I drink to your health and happiness". The goblet then passed to him and so on. The recipient should stand to hear the kind words. Between 1850 and 1900 more goblets were placed on the table and then the toastee did not raise his glass or drink from it.

The ritual sheep

Often a sheep was killed for the arrival of the guests. It had to be fresh, but all preparations had to proceed without the guest noticing, otherwise he would object to the trouble he was causing and beg to be given something simple. Before the guest arrived, the host would check the quality and prestige of the sheep dish - stewed shoulder, leg, pelvis bones and breast. The cooked head was also placed before the guest, who would cut off bits and offer them to the others. However hungry he was, the guest ate small portions slowly - it was impolite to show appetite.

Today

Many of the old customs have entirely disappeared. Those which we hear about today are usually from old people. Some manners are now thought of as old fashioned morals.

Notes

1. European style tables began to appear at the end of the 19th century. Traditionally meals took place on thick felts and carpets spread on the floor - sitting on low cushions or low 'mountain' stools. Sometimes the head of the family and guests dined on small tables (40-50 cm high). So the expression "at table" came into use long before tables did.
2. Bromlei YU V, *Ocherki Teoria Etnosa*, M, 1983, p 23.
3. "To look upon food with longing" mentioned in Gardanov V K, *Obshchestvenuy Stroy Adigshii Narodov (XVIII- per. pol XIXvv)* M, 1967, p 296.
   Mambetov G H, 'O Gosteprinmstve i Zastolnom Etikete Adigov' / *Uch. zap. Adigeiskovo NII yazika. lit. i ist. T. Ush.* Maykop, 1968, pp 242, 243, et al.
   Magometov A KH, *Kultura i buit osetinskovo naroda*, Ist.-etnog. isledovanie Ordzhonikitze, 1968, pp 448-452. *Obshchestvenui stroy i buit osetin (XVII-XIXvv)* Or., 1974, pp 296-7 et al.
4. Special dishes were eaten on special occasions, such as at rain-invoking ceremonies, on Fridays, at thank-offerings for the dead, on celebrations of happy events, or when warding off auguries of misfortune.
5. Tr "The wife of Sado brought a small round table, on which stood tea, 'pilichish', pancakes with butter, cheese, 'churek', bread and honey. A girl brought a basin, jug and towels ... although Hadji-Murat never ate more than a cheese, he ate a little bread and cheese and, taking a small knife out of his pocket, spread honey on the bread. "Our honey is good. This year and every year there is honey: lots and good", said the old man, pleased to see Hadji-Murat eating honey. "Thank you," said Hadji-Murat and left off eating. Eldar wanted to eat more, but he, like his 'murshid', moved away from the table and gave Hadji-Murat the basin and jug."
6. Tr Serdzhinutovski A K, *Poyezdka v Nagornii Dagestan.* Petrograd, 1917, p 289.
   Many mountaineers did not eat bread every day, preferring *khinkal*. "There is no bread around Andi. The Didoy have *churek* - like unleavened bread for special occasions and important guests."
7. Gadjieva S SH, *Kumiki*. Ist.-etno. isled, M, 1961, pp 290-292.
   Gadjieva S SH, Osmanov M O, Pashayeva A G, *Mat. Kultura Dargintzev,* Makhachkala, 1967, pp 124, 268 et al; *Mat. Kultura Avartev*, M'kala, 1967, pp 168, 290 et al; Egorova V P, 'Iz Narodnukh Traditzii Dagestana '/ *Vapros Ist. Etno. Dag.* t. 2, M'kala: Dagosuniv, 1970, pp 83-89.
   Luguev S A, 'Gosteprimstva i Kunachestvo i Laktzev (vtor. pol. XIX-nach. XXvv)' / *Semeinuiy buit naradov Dagestana v XIX-nach. XXvv*, M'kala, 1980, pp 67-74 et al.

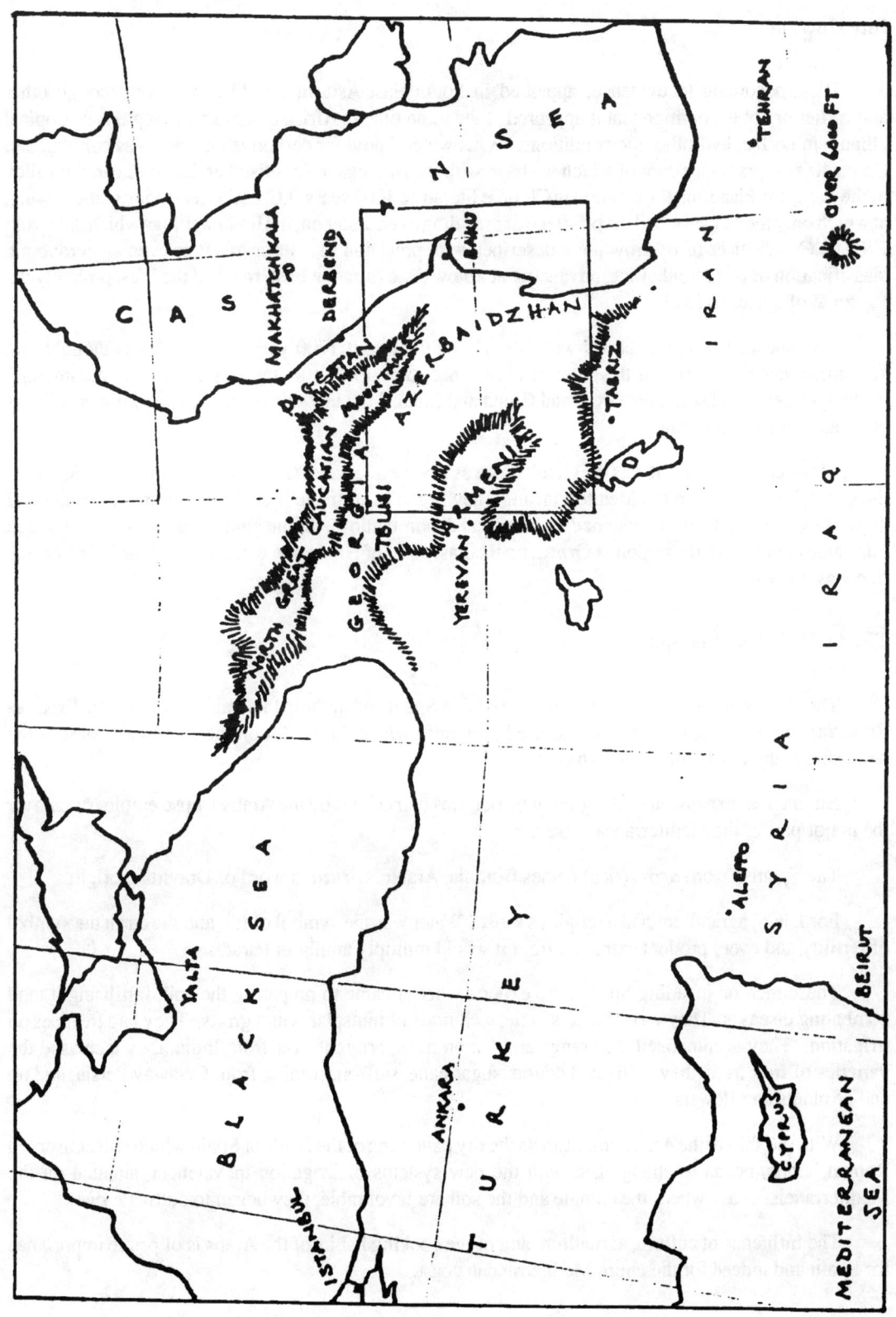

Rice, a Staple Food in Spain

Lourdes March

Introduction

Rice, according to evidence, appeared in South East Asia in a wild form, even though other authorities are of the opinion that it appeared at the same time in Africa coinciding in regions of tropical climate in perfect hydrothermic conditions. What we do know for certain is that rice was cultivated in China 4000 years B.C., proof of which we have with the findings of fossilised grains of rice in the valley of the Yang-tse-kiang and the mention in Chinese literature 3000 years B.C. of the ceremony of the sowing, at which only the Emperor had the privilege of presiding over. Later on, in Hindu writings which date from 2700 B.C., practices of its growth are described with precision like a transplant or even an agronomic classification of this cereal. Rice, advancing at a slow pace from the East, reached the West probably via the route of spices and silk.

It is known that the Hebrews were already cultivating it 1700 years B.C. and it is thought that accompanying the Hicsos in their invading advance, they took it as far as Egypt. The commercial exchanges between Darius the Great and China and India show that this cereal became known and was cultivated in Mesopotamia.

Many centuries later it was the Greeks who as a consequence of the invasion of India by Alexander the Great, brought rice to the Mediterranean. They began to cultivate it in Greece giving it the name of *Oryza* (a word which has maintained its botanical denomination). Some authors are of the opinion that this name comes from the region of Orissa, on the East coast of Hindostan where, since time immemorial, rice was grown.

The Arrival of Rice in Spain

There are some doubts as to how rice arrived in Spain and authors like Paul Gaubert in *'L'Espagne Bizantine'* are of the opinion that it was the Byzantines who cultivated it on the south-east coast of the peninsula at the beginning of the 6th century.

But the real expansion of the growth of rice was carried out by the Arabs whose empire ruled over the major part of the Mediterranean coast.

The Spanish word *arroz* (rice) comes from the Arabic *Al-ruza*, a word of Dravidian origin.

For Islam, agriculture had a religious sense. Water was the symbol of life and the earth the symbol of fertility and every product extracted from it would multiply amply in Paradise.

These men on invading Spain were experts when it came to preparing the soil, fertilising it, and combating diseases. They were masters in the technique of transplants and grafts. They had treatises on irrigation. They acclimatised the orange and lemon trees, brought over from India, they increased the varieties of fruit trees, they cultivated cotton, sugar-cane, saffron, coming from far-away Persia, and no end of plants and flowers.

With the skill of the Arab agriculturists the dry panorama of the fields of Spain which still conserved Roman habits, began to change and with the new systems of irrigation in Valencia, situated on the Mediterranean coast, where the climate and the soil are favourable, they began to cultivate rice.

The influence of culture, agriculture and of the cooking habits of the Arabs is of prime importance for Spain and indeed for the entire Mediterranean coast.

With the passing of time, the areas of rice growth began to extend towards other coastal zones, especially in the delta of the Ebro and in the marshes of the Guadalquivir near to Sevilla.

Gastronomical History

In Spain the historical-gastronomical references about rice date back to the thirteenth century and in an anonymous script on Hispanic-Maghreb cooking, many recipes are given of this cereal in the majority of cases prepared in the modern style of rice with milk.

In the twelfth century, in the *Book of Agriculture* of Abu-Zaccaria it is said that: 'they grow two types of rice, the aquatic and the dry and that before eating it, they wash it seven times in hot water, then they cook it with milk'. It is also said that with rice and pulses they prepared a type of bread which they had with milk or oil or accompanied by meat as well.

In the *Libre de Sent Sovi* of 1324 we also find a recipe of rice boiled in almond milk and perfumed with cinnamon. There is no question of doubt as to the Arab influence which we can clearly appreciate in this book and other recipes of the fourteenth and fifteenth century and even with a hue of Hebrew cooking, difficult to define on many occasions.

In the books of that era, many of the ingredients, seasonings and spices that the Arabs used are enumerated: dried fruit, almonds especially which combined with onion were and still are the main base colour and taste to so many stews and casseroles, aromatic herbs which exquisitely perfume the dishes, rose water which gives the characteristic touch to desserts and cinnamon appreciated since times remote which gives an aroma, not only to sweets, but also to stews and rice dishes.

The European recipe books of the Medieval and Renaissance era give us the formula for *menjar blanc*, an exquisite dish composed of a mixture of ground rice, breast of hen or capon, almonds and sugar. With the passing of time the recipe began to change and doing away with the meat, it became a dessert to which milk and orange blossom and rose water were added. As the years went by, rice became the principal and main dish. In the *Libre del Coch* by Ruperto de Nola (in the year 1520) you can find a rice dish prepared in the oven, very similar to what we prepare today. Juan de Altimiras, a Franciscan friar, in the *Nuevo Arte de Cocina* (1745), offers his recipes to prepare 'Eel with rice' and 'Rice with almond milk'.

The historical-gastronomical references about rice increase as we approach modern times.

Between the eighteenth and nineteenth centuries it was gradually becoming important as the basic dish of Spanish cooking and in cookery books of that period the recipes for preparing rice are numerous.

Ways to Cook Rice in Spain

The varied Spanish orography covers several natural zones where, in very few kilometres, from the coast to the interior you can come across fruit and vegetable orchards and mountainous zones.

This circumstance of the soil, the climate and the different produce is what gives such character to its varied cooking. The numerous rice dishes which enrich the Spanish kitchen have one common denominator which is the intuition and ingenuity of the people who prepare it, along the coast or in the interior, because they know how to combine and adapt this cereal with the ingredients they have at hand or those they can acquire easily.

Although the 'Valencian Paella' is the most famous of all Spanish dishes, there are many and varied forms of preparing rice. Without exaggerating at all, we can safely say it is possible to cook a different rice dish for each one of the 365 days of the year.

Rice is normally served as a first course, even though on occasions it is used to garnish or as a filling, while it is also served sweet, like in desserts or as a drink.

Ingredients

Ingredients which accompany rice: meat, fish, vegetables, etc. vary and can be combined endlessly. Normally the typical rice of every region is boiled with:

- poultry or meat stock (with or without vegetables or pulses)
- pork
- game
- fish or fish stock
- dry cod (with vegetables or pulses)
- vegetables
- pulses
- vegetables and pulses
- dried fruit

It is very important for the rice to be cooked tender but firm, not too soft, and never so that the grain doesn't open.

Olive oil is the basic vegetable oil. Saffron, paprika and salt are ever-present condiments.

Tomatoes, garlic and on occasions onions are important ingredients for the *sofrito*.

The rice used is of medium grain.

Methods of Cooking

Rice, irrespective of the ingredients which accompany it, can be cooked in two ways

a) In abundant stock called 'soupy rice'.

Soupy rice is suitable for the winter months and usually consists of meat and pulses or fish and vegetables. First the ingredients are cooked and then the rice is added for the last 17 to 20 minutes.

b) In a lesser quantity of stock called 'dry rice'.

'Dry rice' is cooked in two ways:

- in a flat pot, not very deep (generally earthenware) which you can put in the oven. The main ingredients are normally meat stock or game with pulses and vegetables and sometimes dry fruit.

- or in a 'paella-pan' where the different types of rice called 'paella', whose ingredients can be very varied, are cooked.

In Spain rice is also cooked as a dessert with milk, sugar and, depending on the different recipes, adding some fruit - fresh or dry - or cream too.

Rice because of its astringent properties has been used for centuries as a drink or infusion.

The water in which the rice was cooked is often given to children. To prepare it, after filtering this water, you add lemon peel, grains of aniseed and sugar.

The *horchata* of rice as a refreshing summer drink is prepared by soaking rice in water and lemon peel for 24 hours. After beating, filtering and adding sugar, once dissolved, it is put into the refrigerator to cool.

The Most Popular Spanish Rice Dishes

Without question the *Valencian Paella* is known world-wide and is identified with Spain, but let us not forget the many types of exquisite rice dishes which are traditional in Spanish Cooking.

1) Among the rice dishes from the coast, generally of fish, we can pick out:

   - Rice with seafood, cooked in the paella;
   - Rice *a banda*, cooked in the paella, too;

2) In the orchard areas, rice is prepared with vegetables, meat and poultry,

   - 'Valencian Paella', cooked in a paella (of course);
   - Rice with vegetables, cooked in the paella;
   - Rice with chard, cooked in a stew pot;
   - Rice with dried beans and turnips, cooked in a stew pot;

3) As we move away from the coast and the orchards in very few kilometres, the soil and climate change and the ingredients for rice are more substantial. These rice dishes are normally prepared with meat stock or stock from poultry or pork or dried cod, and also pulses and a selection of cured meats, for example:

   - Crusty rice, baked in an earthenware casserole in the oven with beaten eggs, thus achieving the crusty top.
   - Rice with meat balls, baked in the oven;
   - Rice with partridge and rabbit, cooked in paella;
   - Rice bullfighter's style, cooked in a casserole whose basic ingredients are rice, dry cod, chick-peas, fresh red peppers (which look like a bullfighter's cape) and garlic.

This brief summary of traditional rice dishes, taken from my book on rice dishes and paella, is only an eye-opener to the immense variety of Spanish rice recipes.

We can see that rice, which arrived centuries ago from the Orient, made its home in the Mediterranean, thus becoming a staple food for many of the surrounding countries and especially for Spain.

Healthy - or a Health Hazard

Yesterday's Diet, Today's Disaster

Harry Morrow Brown MD FRCP

We are what we eat - but what we eat does not always agree with us. Severe reactions to food were known long before Hippocrates first reported it in 460 BC.

It is true that we must adapt continuously to our environment from birth to death, for death is really the final failure of adaptation. Partial failure of adaptation is the cause of allergic disease, which is over-reaction to the environment. As many as one person in five may not adapt normally to substances which we inhale, such as pollen or dust, substances we contact, such as nickel, or to the foods which we must eat to support life. An allergic person is one who reacts excessively to something which he inhales, touches, or swallows, which does not upset normal people in the slightest. An allergic reaction is an abnormal phenomenon, so violent as to be self-damaging, thus producing hay fever in the nose, asthma in the bronchi, colitis or irritable bowel in the gut, eczema in the skin, arthritis in the joints, hyperactivity and behaviour disorders in the nervous system and so on.

Foods as a cause of a wide range of illness were clearly defined by many American investigators during the first half of this century, when there were no effective drugs to relieve or to suppress symptoms. Thus the incentive to find a causative food, the avoidance of which could result in cure, was very great.

Since the fifties the advent of many powerful drugs, such as steroids, has unfortunately created a tendency to concentrate on finding the pill for every ill rather than looking for a cause which could be avoided. Today it is often forgotten that drugs seldom effect a real cure, and that it is a lazy doctor who does not at least try to find the cause of a problem.

Allergic reactions can affect any part of the body, therefore the nature of the reaction depends entirely on which part of the body has become sensitised. The reaction may occur at the point of first contact, such as in the gut, but on the other hand it may be absorbed into the blood stream and distributed through the blood and body fluids to be presented to every cell in the body within minutes. Thus an allergy victim may react in different body systems in different ways, and hence may attend several specialised hospital departments instead of the non-existent allergy department.

Food Allergy and Food Intolerance

Before we can begin to understand what is happening when a food reaction occurs, it must be defined clearly that there are distinct types of reaction to foods.

Food allergy is present where tiny amounts of the foods to which the individual is sensitive can cause severe and sudden reactions, accompanied by specific antibodies in the blood and positive skin tests. The victim usually becomes aware of the problem at an early age and has to take great care to avoid that particular food because the reaction can sometimes be dangerous and life threatening.

Food intolerance is quite different as the reaction is delayed for at least three hours, large amounts of food are required to trigger it off, no antibodies or skin reactions are found, and after avoidance for some time tolerance can often be re-established. At the present time the diagnosis can only be achieved by dietary manipulation. The intolerance reaction is not properly understood, is much commoner than allergy, and is a subject of controversy within the medical profession. For example, some physicians regard food intolerance without demonstrable antibodies as being all in the mind, and their approach tends to be bigoted and dismissive. Their attitude closely resembles that of the medical profession towards those

American physicians who first reported allergy to foods in the twenties. The laboratory tests which proved their findings were developed many years after they had died.

At the present time excessive media exposure of bizarre cases of allergy must confuse both doctors and patients and encourage those who consider it to be all in the mind. However, much attention has been paid in recent years to the possibilities of dyes, preservatives and other chemicals used as food additives causing allergic reactions. My experience is that reactions to chemicals are less common than to staple foods. My impression is that intolerance is becoming more common, but frequently fails to be recognised. Perhaps chemicals, additives, and exotic materials in animal feed could be having some subtle influence.

The Dairy Revolution

Allergy to cow's milk was recorded by Hippocrates as causing gastric upsets and urticaria, Galen described allergy to goat's milk in 130 AD, and in 1745 Charles Edward Stuart, the pretender to the English throne, was reputed to have the "bloody flux" due to milk, which could have been allergic colitis.

Until about one hundred years ago, the usual infant feed was human milk either from mother or from a wet nurse. However, with the expansion of the dairy industry at the beginning of the century, infant formulæ based on cow's milk were soon being mass produced as an alternative to breast feeding, which became less and less common until revived in the last ten years.

Advances in animal husbandry and nutrition, directed at maximum milk yield, plus advances in food technology such as pasteurization, sterilization and so on, have resulted in tremendous expansion of the dairy industry. Today enormous quantities of cow's milk products are consumed as compared with one hundred years ago, yet even the most intensive marketing failed to prevent milk lakes and butter mountains until production was cut back. The first major foreign protein which we encounter is undoubtedly cow's milk, and through life we are exposed to much larger quantities of milk products than our grandparents ever were. It seems reasonable, therefore, to suggest that adverse reactions of either allergy or intolerance have become more common since the dairy revolution. There are still difficulties with family doctors and with pædiatricians in recognising allergy or intolerance to milk, perhaps because constant promotion of milk as a dietary essential and a healthy food may have influenced the profession in this way. Perhaps it is difficult for both public and profession to accept that milk is not always good for you and can sometimes cause disease. Cow's milk is meant for baby cows, not for baby humans. In fact there is a close similarity between the proteins and the composition of all animal milks, but human milk is quite different. The promotion of goat's milk as a healthy food may not be correct as the goat is not designated as a dairy animal and is not subject to the strict regulations as for cow's milk.

Babies allergic to milk will react to traces of cow's milk protein in their mother's milk, and this applies also to egg, wheat and peanut and other substances. Hence the diet of the mother can be important to an allergic baby, supporting the view that what the cow eats can be important. Cow feed is a very highly developed branch of food technology, and includes many surprising items, even to adding colour to stimulate the cow's appetite.

Milk can be associated with arthritis, but it is not possible to find out how often this is true unless every case stops taking milk products to find out. Unfortunately many specialists do not believe in this, but I certainly have seen a trickle of cases, usually self-diagnosed, by observing that when they cannot eat their joints feel better.

Eggs

Eggs usually cause straightforward allergy, but this can quite often remain unrecognised. I have never seen intolerance.

Allergy to egg white can be extreme. Cases have been known where a patient could react sixty feet away, and is often necessary to forbid eggs altogether in the house of an egg sensitive child to get a good result.

Hen food can be important, as in a man who began to get asthma at his work, which was found due to Spiramycin, an antibiotic used in hen food. He improved when he gave up his job, but did not recover properly until his wife, who only worked in the office, also left the factory. Some time later he began to get attacks again, which coincided with eating eggs which contained traces of this antibiotic from the food of the hen.

Viruses are often grown on chick embryo, and after an influenza injection one man began to have asthma every Sunday when he ate chicken, but not when he had his ham and eggs every morning.

Potatoes

It has been known for a long time that the aerosol produced by scraping new potatoes can produce allergic symptoms. A few years ago I saw a lady with eczema whose skin had become very thin and fragile from applying steroid ointments for about thirty years. The clue to the cause was that her hands itched badly and got worse when she prepared potatoes, and with removal of potato from the diet her skin is now as near normal as it ever will be.

This lady was the first of a succession of patients with allergy to potato, often diagnosed by pricking through a little raw potato juice to produce an obvious reaction.

I have now found that potato can cause a wide range of adverse reactions, the most remarkable of which was Christopher, a five year old whose mother refused to accept that his dreadful behaviour and the development of asthma could be her fault for being a fussy mum. She pestered her doctor until he referred him to me, when I found a large pot bellied boy, with heavy circles under his eyes, who could not sit still and would not submit to examination or tests. This is the sort of behaviour which always suggests food intolerance. His pot belly had been put down to obesity, but on questioning he had the typical large very smelly fatty stools, typical of Coeliac disease or malabsorption, along with excessive thirst and bed-wetting. The crucial observation was that if for any reason he was quite unable to eat for a few days he suddenly became a totally different personality and a very lovable little boy.

I put him on a very limited diet of lamb and rice for a trial period, and mother and father also went on the same diet in order to encourage him. The result was that within two days his pot belly began to disappear, his dreadful smelly motions stopped, and he had a wonderful personality change to a little angel. Thirst and bed-wetting also ceased, and his schoolwork, writing and drawing improved beyond recognition.

This was wonderful, but the bonus was that his father's personality also changed for the better, and on adding foods one by one to their diets to see what would happen, it was found that every time Christopher or his father ate potatoes they became aggressive, difficult and nasty within half an hour. Introducing wheat produced dreadful smelly motions but not on gluten free products, so he did indeed have Coeliac disease. It turned out that father did not usually have potato during the week until Friday lunchtime when he had chips, and he was invariably in a filthy temper when he came home. Milk or cheese would also make him miserable within half an hour.

Both father and son are now completely new people, with new personalities and no illness at all.

This unusual family case history showed how food intolerance can be inherited and how dietary manipulation can revolutionize a family.

Bread and Wheat

In my experience true allergy to wheat is uncommon, but intolerance is often associated with the so-called irritable bowel syndrome, usually attributed to emotional factors, but which can be relieved by dietary manipulation.

Coeliac disease is due to gluten sensitivity, but is not usually classed as an allergy. However, the lining of the gut is destroyed by the reaction to the gluten and avoidance will give a complete cure.

Wheat flour today contains many so-called improvers, including potassium bromate, chlorine dioxide, enzymes derived from various fungi, and soya and milk powder are often added without being declared.

It is not known how often people are sensitised to the additives in the flour, but allergy to yeast is quite common, but there are no laboratory tests which work as yet.

It seems that both the national beverage (apart from tea), and the staff of life may often be involved in intolerance and allergy.

Allergy to Sex

Severe allergic reactions to seminal fluid, even anaphylaxis and collapse, have occasionally been recorded over the last thirty years, but food or drugs taken by the male partner can also be transmitted to a sensitized female during intercourse. For example, one patient broke out into a rash every time her husband took Penicillin. A report on one of my cases where traces of milk protein in the husband's semen caused a reaction in his wife's joints is attached.

A similar type of case was recently reported from Switzerland, when a lady who was very keen on horse riding, but was not allergic to horses, complained of intense allergic reactions in the genital area after horse riding especially if she was wearing thin pants. The allergist suspected, and confirmed by skin testing that the polish she was using for her saddle contained a resin called balsam of Peru. This is not an uncommon ingredient in various shampoos, mouth washes, perfumes, soaps and sometimes ointments.

After she stopped using this polish, she did not get attacks after horse riding, but repeatedly had intense local reactions following intercourse with her boyfriend. Testing her skin with his seminal fluid and two other unrelated samples of semen gave inconsistent results, being sometimes positive, sometimes negative, for no apparent reason.

On intensive questioning as to the habits of the couple, it was revealed that the boyfriend was a very heavy Coca Cola drinker, and that reactions after intercourse appeared only when he had been drinking large quantities of Coke. Obviously the use of condoms eliminated reactions completely

The Swiss investigators reported, and I quote, 'it has not yet been possible to persuade the pair to proceed to indirect sexual coke provocation test, which would definitely prove the causal role of coke in this affair and its relationship to balsam of Peru sensitivity'. The mind boggles at how this could be done.

Diagnosis of Food Allergy and Intolerance

Skin tests and blood tests are helpful in definite food allergy, but are useless in diagnosing intolerance, where dietary manipulation is the only way. The methods applied are outlined below.

How to find out if your symptoms are due to food.

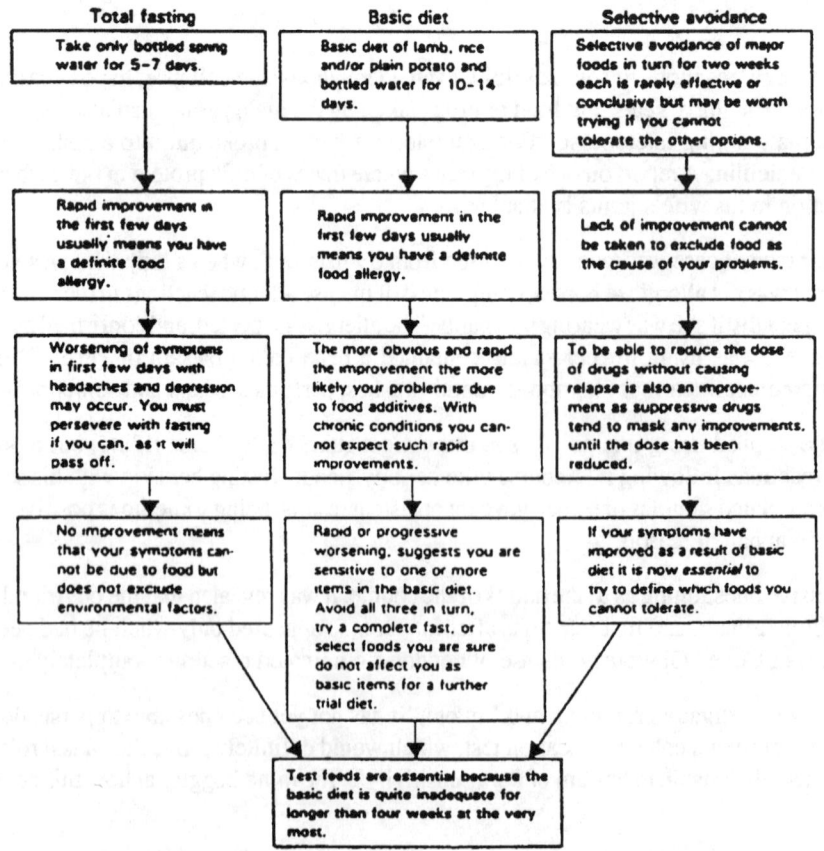

Total fasting	Basic diet	Selective avoidance
Take only bottled spring water for 5-7 days.	Basic diet of lamb, rice and/or plain potato and bottled water for 10-14 days.	Selective avoidance of major foods in turn for two weeks each is rarely effective or conclusive but may be worth trying if you cannot tolerate the other options.
Rapid improvement in the first few days usually means you have a definite food allergy.	Rapid improvement in the first few days usually means you have a definite food allergy.	Lack of improvement cannot be taken to exclude food as the cause of your problems.
Worsening of symptoms in first few days with headaches and depression may occur. You must persevere with fasting if you can, as it will pass off.	The more obvious and rapid the improvement the more likely your problem is due to food additives. With chronic conditions you cannot expect such rapid improvements.	To be able to reduce the dose of drugs without causing relapse is also an improvement as suppressive drugs tend to mask any improvements, until the dose has been reduced.
No improvement means that your symptoms cannot be due to food but does not exclude environmental factors.	Rapid or progressive worsening, suggests you are sensitive to one or more items in the basic diet. Avoid all three in turn, try a complete fast, or select foods you are sure do not affect you as basic items for a further trial diet.	If your symptoms have improved as a result of basic diet it is now *essential* to try to define which foods you cannot tolerate.

Test feeds are essential because the basic diet is quite inadequate for longer than four weeks at the very most.

How to find out which food is causing your symptoms.

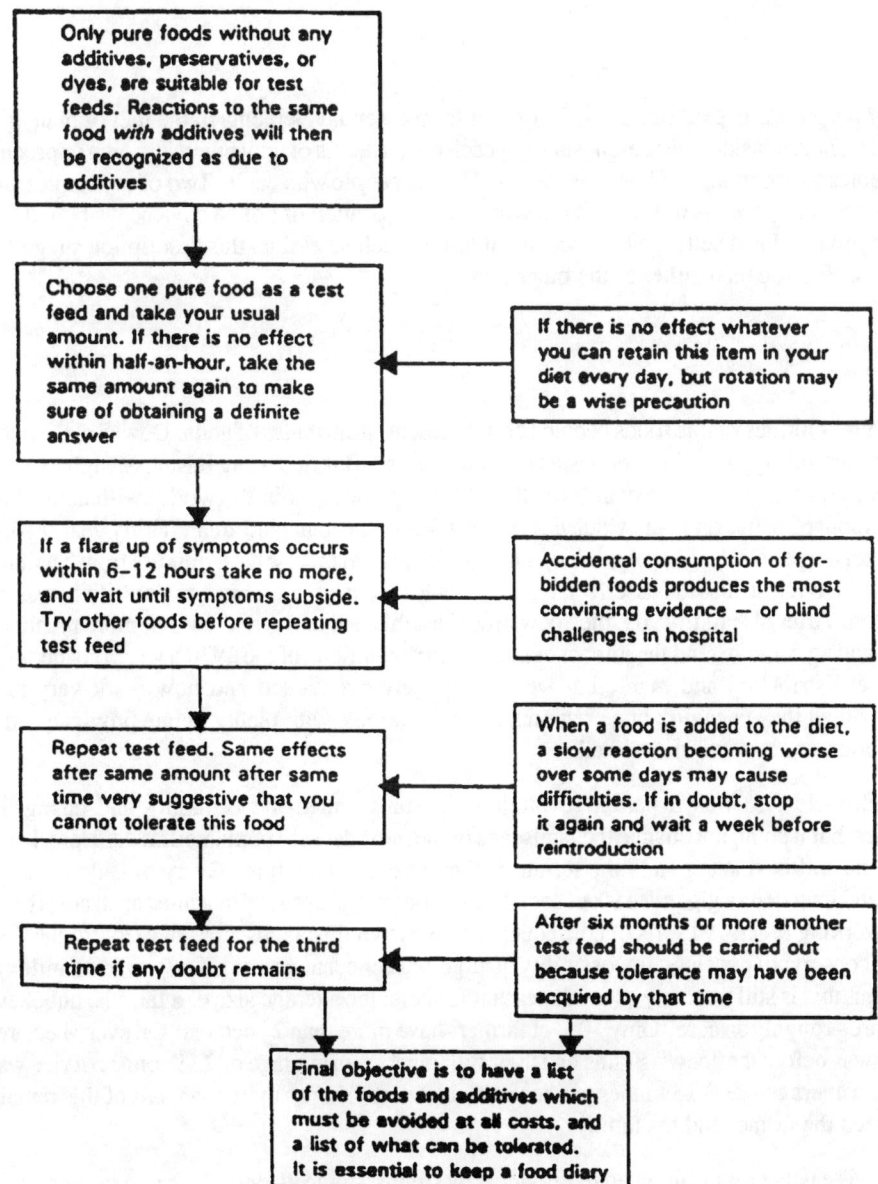

(Diagrams above from *All About Asthma And Allergy*, by Dr. H. Morrow Brown MD FRCP, to be published June 1990.)

Three Staples of Indonesia: Rice, Coconuts, Tempeh

By Sri Owen with Roger Owen

My purpose is to examine three foods that form essential elements in the diet of many millions of people. I want to consider the present status of each one, in terms of how much is grown or produced, what it contributes to nutrition, and how it is regarded by the people who eat it. Two of my staples are familiar to everyone, but tempeh is not so well known; it is a preparation of boiled soya beans bound together by an edible mould. I had better say at once that it tastes much nicer than this description suggests, and it is much better for you than either of the other two.

Rice

In most cultures, staple foods become gods or essential attributes of gods. Dewi Sri, the rice goddess, is still a powerful figure in the Indonesian consciousness[1]. But rice is big business, high technology and a sensitive political issue. The country is the fifth most populous in the world, with about 170 million people. About 60% live on Java, which has less than 7% of the total land area[2]. Thirty years ago, President Sukarno set his people the goal of growing enough rice to feed themselves; imports from Thailand and the USA were having a disastrous effect on his balance of payments, and malnutrition was not at all uncommon. After his fall in 1965, the "new order" established less grandiose but more practical policies for trade and agriculture, and the impact of the "green revolution" of the 1970's was considerable[3]. At the same time distribution and marketing were completely overhauled and now work very much more efficiently than they once did; huge regional price variations, with plenty in one province and dearth in its neighbour, are rare and short-lived.

High-yield rice was a partial revolution. It transformed the economy and changed farming techniques, but it could not change the landscape of the most densely populated areas. Here, land is fertile because the soil is volcanic, and the terrain is therefore mountainous. Every hillside is painstakingly terraced and irrigated - a gigantic task of social co-operation and a legacy that must at all costs be preserved in good working order. All sorts of consequences flow from this: landscapes of extraordinary beauty; an intensely conservative society; impossibility of large-scale mechanisation[4]. Despite some influx of capital investment, this is still a society of small peasant farmers: the average size of a farm is, unbelievably, 0.5 of a hectare - roughly an acre. Only 10% of farmers have more than 2 hectares. On every hectare, in 1977 (and that was before the "new" strains had their full impact), an average of 2.79 tonnes of rice was grown; in Java, the average was 3.13 tonnes per hectare[5]. Not surprisingly, more than half of this remains on the farm to feed the farmer and his family.

Rice is easily cross-pollinated to provide new strains[6] - indeed, one of the rice farmer's problems is that his cultivated rice so easily crosses with wild rices[7] growing around the edge of his field. For this and other reasons, the new high-yield varieties depend on specially-grown seed that has to be distributed to the farmer for every sowing. Fertilizers and pesticides are also needed for the new rice, and because the rice-stems are shorter it is harder for harvesters to cut them with the tiny knife - the *ani-ani* that the women tuck into their hair to carry to the fields, and that they conceal between their fingers as they cut, because they do not wish to offend the goddess. There are many questions still unanswered about the long-term future of rice. The latest that I have come across raises the spectre of methane gas, which rice-growing produces in large quantities and which contributes to the greenhouse effect. But there is no denying what was achieved in the five years after 1977: near total self-sufficiency in the one food that every Indonesian, unless he or she is seriously poor, must eat three times a day.

From the farm, the rice is taken first to be milled. In the old days, it was pounded by hand, by women using heavy wooden "pestles", up to 5 feet long, and stone troughs or mortars. Right up to the 1960's there was a strong demand for *beras tumbuk*[8], hand-pounded rice; fewer grains were broken, and a little of the husk remained, so that the rice had more flavour and food value. Beras tumbuk has now virtually disappeared, and even farmers' families send their rice to be mechanically milled. Mills are encouraged to be local, small-scale businesses, in line with the Government's policy of encouraging low-investment, labour-intensive projects.

Overall, Indonesians get about 50% of their total calories and almost 50% of their protein from rice. However, one of the big changes in Indonesian society over the past 20 years has been the growth of a large and fairly prosperous middle class, made up partly of successful businessmen (it is a myth that Indonesians cannot run a business profitably) and partly of civil servants (it is, alas, all too true that one of the most tenacious legacies of colonialism is bureaucracy). These people can afford to buy all the rice they want, and having done that they can go on to add variety to their diet by buying imported or locally produced western-style foods: thus, one can now buy passable butter and cheese in Jakarta stores, apples and other temperate- zone fruits, even sliced bread which is almost as flabby and tasteless as Mother's Pride.

This move away from rice has been encouraged by the Government as part of its third five-year plan, and in 1985 it even published a book of non-rice recipes[9] - partly to cut imports, partly to improve health. But if high-priced wheat and refined white flour take the place of rice in Jakarta and Medan, it is hard to see what has been gained. Having more money and less time, people want to buy packeted goods in shops rather than haggle in the market-place, and like all of us they feel sure that expensive imported goods must be classier than cheap local ones.

Rice itself has a complex price structure in the market. The study made a few years ago by Leon Mears and colleagues high-lights the price variations dictated by quality and, to a lesser extent, by season. (Rice can be sown and reaped at any time, and differences are not as great as one might expect; but wet and dry seasons make it easier to plant in certain months than in others, and the cycle of holidays and religious festivals strongly affects demand.) This brings us to the question of how Indonesians perceive quality in rice.

Rice comes in several different colours and many grades, all of which have their uses; but as the basis for a meal it should be pure white and rather soft and sticky. The whiteness is a status symbol, just like the whiteness of bread; milled rice has always been more expensive. Milling removes the outer layer of the grain and with it most of the nutritional value and the fibre. Indonesians know this. They are as conscious of the importance of diet and nutrition as westerners are; but the cultural insistence on whiteness is too strong for them. In Leon Mears' study selected prices (per kilo) are quoted for different types of rice in the city markets of Jakarta in July and December 1977. The cheapest rice was imported from Korea and sold for Rp116. Not far above it came the lowest-grade "new" rice, locally grown, at Rp120 (July)/ Rp131 (December) and US Medium Grain at Rp127/Rp135. Thai rice with 25% broken grains was down at about the same level. At the opposite end of the scale was *Cianjur Sliyp Kepala*, top-grade polished rice from Cianjur in West Java, at Rp205/Rp227. Mears noted that the "new" high-yield rices were lower in quality and price, but were often more profitable for the farmer. 58% of the "new" crop, against only 14% of the "traditional" rice, comes to market - which suggests that farmers know what they like and, however poor they may be, keep the best for themselves and send the city folk the stuff that brings the profits. Much low-grade domestic rice is brought to market with a high (sometimes illegally high) water content. Cheap rice, incidentally, often expands most when cooked because it absorbs most water; it is therefore chosen by manual workers because it makes them feel fuller.

According to Sapuan's BULOG report, the most sought-after cooking quality is what the Javanese call *pulen* - the quality of softness and succulence - which he associates with rice that is low in amylase. The cooked rice grains should be just sticky enough to make it easy to eat with the fingers, rolling the rice grains into a ball. Premium rices tend also to be fragrant and tasty (hence, I suppose, the name "Thai Fragrant Rice", which puts off some English shoppers, especially when they see it translated and printed on the bag in French - *riz parfumé*. It is not, of course, "perfumed" at all; it just smells nice while it is being cooked). Really glutinous or "sticky" rices are highly prized and highly priced; *ketan hitam* and *ketan putih* (respectively, black and white sticky rice) came third and fourth in Mears' Jakarta market survey. But these are used for making sweet rice cakes, or as part of an invalid diet.

The part which rice plays in ritual and festivals, as well as in the formality of daily life; its importance as a symbol of health and well-being; its central place in the mental as well as the physical landscape of Indonesia - these are topics about which someone, I feel sure, must have written a book, though I have not yet come across it. In any case, the changes in society and technology during the past twenty years have certainly brought changes in people's awareness and use of rice, and these changes ought to be charted and recorded.

Coconuts

Coconuts are much less satisfactory to write about. To begin with, there is (relatively) much less literature on them[10]. And though everyone recognises their value, their inevitability as one of the bases of tropical life, they are somehow not self- evidently a staple food, as rice pre-eminently is. I know of no coconut-hero or goddess, no generally-held beliefs or superstitions regarding them. Coconut palms are the ideal furniture for desert islands, but when the islands become populous the coconut palm, however intensively cultivated, remains at the level of furniture - useful, perhaps beautiful, but nothing more.

I suspect that most westerners think coconuts grow wild, being self-seeded and needing no maintenance. This is quite wrong; some palms may grow wild, but the vast majority, even on the remotest and most idyllic beach, have been planted and are cared for by someone. Coconuts need a light soil that allows air to circulate around their roots, and moist winds to keep their fronds from drying out in the sun. Hence their presence on sandy beaches cooled by the seawind; but inland they are planted among the village houses to provide shade, fresh nuts, decorative fronds, raw materials for making household implements and weaving containers of all sizes, juice from the flowers to be fermented and distilled into a potent liquor, and eventually fence-posts and fuel. In a landscape of wet rice fields, you may pick out the villages from afar as islands of shady palm and bamboo.

Isaac Henry Burkill, author of the massive *Economic Products of the Malay Peninsula*[11], explains why coconuts are something of a botanic freak: they collectively make up one member of a family of four, the other three all being natives of the Americas. How did they cross the ocean, and at what stage in their evolution? And *Cocos nucifera* is virtually the only member of its branch of the family, with relatively few races: very different from the hundreds of varieties and races of the rice plant. Burkill continues gravely:

> The coconut palm is one of Nature's greatest gifts to man. Nature gave it to him ready-made, whereas whatever excellence is possessed by most of his other important food-plants has been his reward after ages of semi-conscious effort to ennoble them by selection. It is true that he now possesses a number of partially-isolated races of the coconut, and that he has obtained a few relatively dwarf and slightly precocious races, so that he reaps a crop with reduced trouble, but this is little to his credit as an agriculturist.

The value-judgments implied in Burkill's remarks are fascinating and I think relevant to an enquiry into staple foods and people's attitudes towards them. Burkill's praise of the coconut reminds me a little of William Cobbett's invectives against the potato, which he called "Ireland's lazy root"[12]. However, speculative digressions would be out of place here.

Coconut plantations are big business in most tropical countries. Indonesian government figures[13] show that in 1985 the country's estates produced 1.2m tonnes of palm oil (more than a third of which was exported) and 1.8m tonnes of copra. Plant breeding and hybridisation are actively pursued, and *Indonesia 1987* has a photograph of a new type of coconut palm which appears to have shrunk into the ground so that its trunk is only about four feet high - no need to climb up to pick the nuts or to wait until one falls on your head. What would Burkill have said to that?

Apart from its usefulness, the coconut is also a luxury - the only one of the three foods investigated here which could be so described. Most Indonesians, and notoriously the Javanese, love sweet things, and coconuts contain a lot of natural sugar. One of the pleasures of living in the tropics is to stop by the roadside on a hot afternoon and drink the juice of a young coconut, which is then roughly chopped open so that you can scoop out and eat the soft flesh. This coconut water is considered to have no commercial value, and most of it goes to waste; but it contains some vitamin A as well as proteins and sugars. Very ripe nuts usually do not have much water left in them, or if they have it is salty and unpleasant to drink. Much more valuable, especially for the cook, is *santen*, the milk that is extracted by squeezing the shredded flesh of the older nut in water. This contains the same proteins, sugars and salts, but in stronger concentration. Most important, it also contains some coconut oil as an emulsion. This oil is essential in cooking a great many Indonesian dishes, and even - as in *rendang* - allows you to start a long cooking process by boiling (in the milk) and end it by frying (in the oil, after all the water has been driven off or absorbed by the meat). You can, of course, make your own coconut oil by boiling a pan of santen until all the water has been driven off. This process produces another luxury, a semi-solid residue called *belondo*, which is used in various recipes to impart a characteristic and delicious flavour; mixed with peanuts, for example, in a kind of peanut sauce, or with spices in cooking chicken.

Indonesians are health-conscious people, many of them with tendencies towards faddiness, and they are well aware nowadays of the risks that lurk behind the attractions of the coconut: a high level of cholesterol and saturated fats as well as a great deal of sugar. Maybe we shall see the luxury consumption of coconuts diminish over the next 10 or 15 years. But, like butter in the cuisines of France, coconut oil will hold its necessary place in Indonesian cooking, because there is really nothing which can adequately replace it - certainly no western-style dairy product, since the local dairying industry is still in a very early stage of development and is never likely to produce milk in very large amounts. Whether it is true (as I have seen asserted somewhere) that the great majority of Asians lack the enzymes needed to digest dairy products, I do not know; but it is certainly true that they regard milk and butter as acquired, unusual and expensive tastes.

Tempeh

However differently people in Indonesia may think of rice and coconuts, they treat them both with respect; nobody was ever sneered at, or felt he had to apologise, for eating them. The soya bean, however, has in the past been less well regarded, though it contributes much more of nutritional value. It has been grown in large quantities in Java and the other islands for a very long time (it was brought here from India), and it has always been known as a healthy food. The trouble with soya beans (*Glycine max*) is that they look, and taste, boring. They are also rather indigestible.

I have described elsewhere, and so have many other writers[14], one of the ways in which Indonesians have long overcome this problem, fermenting the soya bean to make it both palatable and digestible while actually enhancing its food value. In brief, soya beans are boiled, dehulled, boiled again, inoculated with a mould known as *Rhizopus oligosporus*, and incubated (usually in plastic bags pierced with pinholes to let air pass through) at 88 degrees F for about 28 hours. As the mould develops, it breaks down the outer layers of the beans, converting them into nutrients which the human system can readily digest; its myriad tiny fibres bind the beans (which remain recognisably whole) into a solid cake, covering its surface with a whitish nap which has been well compared to the fluff on a tennis ball. It needs to be cooked before you eat it; Indonesians usually fry or boil it.

As a food, tempeh is rich in protein (about 43% by weight) and contains all the amino acids which the body cannot synthesise for itself. It is claimed to be the world's richest vegetable source of Vitamin B12. It has a good proportion of natural fibre (which tofu, of course, does not). 100gm of soya bean tempeh contains only 157 calories. It is low in saturated fats, high in polyunsaturated. It contains no cholesterol and is even said to lower existing cholesterol levels. It is digestible, a good source of other vitamins and minerals, mildly antibiotic, non-toxic and cheap[15]. In the west, it should be an obvious food for vegetarians and almost obligatory for vegans, but its agreeable texture and mildly nutty taste make it an excellent basis for almost any kind of savoury dish, either replacing meat or complementing it.

Tempeh is certainly one of the great natural bonuses of life in a warm country. *Kecap* - soya sauce - is of course also made from fermented soya beans and is a great gift to cooks, but you cannot eat or drink it by itself. Tempeh is a solid food that needs only very simple cooking, although it can be prepared most elaborately if the cook or the customer so desires; it can also take the place of meat in a great many oriental and western meat dishes. There are numerous accounts of how tempeh saved lives in Indonesia during the Japanese occupation of 1942-45, and it is beginning to be written about in the USA and (to a much lesser extent) Europe as both a major new healthfood and a partial solution to the long-term planetary food crisis that we seem to be wishing on ourselves.

You would certainly expect Indonesians to have a very high regard for tempeh indeed. Not really. Until very recently it was looked on as food for the poor, a mark of backwardness that newly-independent Indonesia should shake itself free of. Sukarno, the first President of the Republic, frequently attacked the "tempeh mentality" in his speeches in the late 1940's and 50's. One reason for this food snobbery may have been that tempeh had done its job during the war too well - it had become associated with experiences that everyone wanted to forget. Sukarno's attacks fortunately did nothing to reduce the quantity of tempeh being made and eaten, but you will find little mention of production figures in the official statistics for those years; even today the official attitude seems to be one of indifference. The IMF, after all, is not likely to be impressed by this peasant foodstuff that needs to be explained every time it is mentioned. *Indonesia 1987* does, however, record that in 1985 the country produced over 39,000 tonnes of MSG.

In 1979, according to William Shurtleff, there were 41,000 small businesses making and selling tempeh in Indonesia, producing 169,000 tons of the stuff each year with a retail value of US$85.5m. (This works out at roughly 50 cents a kilo, say Rp200 at the then rate of exchange). More recent figures are obtainable from a report published in 1987 by the CGPRT Centre in Bogor, West Java[16]. Unfortunately they cover only one region, and we cannot extrapolate to get an estimate of production for the whole of Indonesia because by no means all Indonesians like or will eat tempeh; in West Sumatra, for example, it is relatively little known.

The CGPRT report shows small farmers in the hills of West Java taking crops of soya beans between crops of rice, thereby improving their rather poor upland soils and restraining weeds and pests. Most farmers sell about two-thirds of their crop; much of what they retain they make into tempeh themselves, while much of what goes to market is processed in small factories into *tahu* (the Javanese equivalent of familiar tofu or beancurd). In 1985-86 soya bean prices in West Java markets fluctuated around Rp600-Rp700 per kilo. A kilo of beans makes at least 1.7 kilos of tempeh, since the beans absorb a lot of water when boiled. Ready-made tempeh was being sold in chunks of about 90gm at Rp40 - Rp50 each, i.e. around Rp500 a kilo. (The US dollar was at that time worth Rp1126, so in dollar terms Shurtleff's 1979 price was little changed.) No comparison with the 1977 rice prices given above can be made, because all prices rose sharply between 1977 and 1985; unfortunately, at the time of writing I have no data on 1985/86 rice prices.

The eye-opening figures in the CGPRT report relate to soya bean imports. Far from Indonesia becoming self-sufficient in this area, the report states that in 1975 she imported 18,000 tonnes of soya beans; by 1984 this had risen to 401,000 tonnes of beans and 206,000 tonnes of pressed soya bean cake. Imports had thus risen until, according to the CGPRT, they were roughly equivalent to domestic production. *Indonesia 1987* says total soya bean production in 1985 was 865,000 tonnes; it gives almost no figures for imports of any kind.

Clearly, soya beans in general and tempeh in particular play a large and by no means diminishing part in feeding a sizable proportion of the Indonesian nation. My own impression, when I visited Java, Bali and Kalimantan in 1987, was that tempeh is eaten by most people, at any rate in Java, on most days; it certainly was not treated, by my modestly prosperous friends and relations, as food for losers. Maybe they just knew I was fond of tempeh; at any rate, it seemed to turn up on the menu more often than not. They did not make their own tempeh, but bought it from small shops or market stalls. Its manufacture is still a cottage industry. From somewhere I acquired a little booklet on tempeh-making, published in Bandung as long ago as 1961[17]. Its author launches confidently into his subject - perhaps a little too confidently, as though he were not sure of his reception:

Nothing Unusual About It Any More - It's Everywhere!

Look at any cooked-rice stall. On any stall you're sure to find tempeh. Usually it's fried to make Tempeh Goreng. And every day in the market there are people selling uncooked tempeh... Tempeh is familiar to people in Java, in Sumatra, in Kalimantan. Yes, people everywhere know tempeh. From mountain-top hamlets to the big cities and wealthy homes, no one forgets to serve tempeh as the companion of rice.

You spoon steaming hot rice onto your plate. The tempeh is freshly fried. A little salt and tamarind to season it. That's the way. So simple! But that dish of rice and tempeh, by itself, is a guarantee of good eating. Really scrumptious...

The author presses on, assuring us in sub-headings that this is "A really healthy food!" and that "It's so easy! Anyone can make tempeh!" - statements which are certainly true, as long as you have an uncontaminated culture of the right starter mould[18]. He has a pleasant habit of making encouraging noises at the start of a paragraph - *O, Ah, Nah,* even the occasional *Hhiih!* He goes on to tell the reader how to make *oncom*, a speciality of West Java, from peanut or coconut presscake. Indeed, tempeh is only one of a range of fermented foodstuffs found in different parts of Indonesia, all of which can be made from a variety of ingredients.

I draw no general conclusions from this brief survey, except to suggest that the way in which people perceive their staple foods and relate them to the value-system of the world they live in is interesting in itself and may reveal something about individual psychology, the effect of social pressures, and the give-and-take between tastes and market forces. There are of course other staple foods in Indonesia: sweet potatoes and cassava are especially important. Neither of these has the depth and complexity of cultural meaning that rice has. Neither is anywhere near as attractive or as versatile as the coconut, or as nutritious as tempeh. It seems to me curious that the food we regard most highly is that whose natural goodness we have taken the greatest pains to remove. Geography and economics presumably play the biggest parts in deciding what we shall eat, but just what choices can we exercise, and how?

Notes and References

1   Clifford Geertz, *The Religion of Java*. University of Chicago Press 1976, p. 81n; Sri Owen, *Indonesian Food and Cookery*, Prospect Books, 2nd ed., 1986, pp. 13-29.

2   *Indonesia 1987: an Official Handbook*, Dept of Information, Jakarta, pp. 9 sqq. It is important to keep in mind the extraordinary range of population densities in Indonesia. Of its 13,000 islands, over 12,000 are said to be uninhabitable; large areas of most islands are arid and infertile. People therefore huddle together on a small percentage of the total land area. The overall density (per square mile) was given by Shurtleff and Aoyagi (note 16 below) in 1979 as 217, but that of Java as 1,533, making Java the most densely-populated region of the world. Quite large parts even of Java are very sparsely populated, so the problems of feeding the populous areas (central Java, the north coast, Jakarta and its neighbourhood) are indeed formidable.

3   Leon A. Mears, *The New Rice Economy of Indonesia*, Gajah Mada University Press, Yogyakarta, 1981.

4   For a brilliant and very readable account of rice cultivation and water distribution in the rather special conditions of Bali, see Clifford Geertz, *Negara*, Princeton University Press, 1980.

5   For comparison, Taiwan produced 4.4 tonnes/hectare, Japan 5.6, Bangladesh 1.83. Mears, op. cit., gives annual per capita consumption and income figures for several countries:

	rice: kg/head/yr	income: $US
Bangladesh	181	109
China (People's Rep.)	99	-
Japan	102	4026
Malaysia	110	716
Thailand	209	323
Indonesia	113	200

Needless to say, sources disagree in detail; one cause of confusion is that some do not state whether figures are for unthreshed newly-cut rice or finished grain.

6   Sapuan, *Laporan Survey Klasifikasi Jenis Beras*, BULOG (Bureau Logistik, the government body responsible for overseeing the distribution of foodstuffs), November 1978.

7   Really wild rice, of course, has nothing to do with the so-called "wild" rice which is sold in expensive delicatessens and which is actually a type of rare edible grass seed found only in a few North American lakes (though Waverley Root says there is a Chinese species, kaw-sun; see *Food*, p. 416).

8   For the record, *beras* = rice grains, as sold to the consumer; *padi* = harvested rice, still in the husk or on the stalk; *sawah* = wet rice fields. The boiled rice that comes to table is *nasi*.

9   *150 Resep Makanan Rakyat Non-Beras*. compiled by Departemen Pertanian (Dept. of Agriculture) and published by Indradjaya, Jakarta/Bandung 1985. The new generation of Indonesian women's magazines are as much taken up with diet, slimming and healthy exercise as their western counterparts.

10   From my own library, only Taufik Rusdi, *Kelapa: Menanam dan Menolah Hasilnya*, Karya Bani, Jakarta/Bandung, 1986; from the public library, Jasper Guy Woodroof, *Coconuts: Production, Processing, Products*, AVI Publishing Company Inc., 1979 (Taufik Rusdi's title is not a bad translation of Woodroof's); from the basement of SOAS Library, University of London, a charming book by Smith and Pape, *Coco-nuts: the Consols of the East*, "Tropical Life" Publishing Department, 1912. Smith and Pape perform dazzling calculations to show that coconuts must be

what the smart money will go into now that the market for rubber is finished.
11   Burkill's two vols. (1935) were reprinted by the Min. of Ag., Kuala Lumpur, in 1966.
12   William Cobbett, *Cottage Economy*, C. Clement, 1822; Oxford University Press (paperback ed.), 1979, p. 52. Cobbett was in fact quoting a certain Dr Drennan.
13   *Indonesia 1987*, p. 135.
14   William Shurtleff and Akiko Aoyagi, *The Book of Tempeh*, Harper and Row 1979; preferably the Professional Edition. This excellent book (which I have drawn on heavily) contains a full bibliography. Sri Owen, *Tempe: A Javanese Health Food Comes West*, PPC no. 5, May 1980 (though later experience of making and using tempeh would lead me to make some changes in that article).
15   All the statements about nutritional values of tempeh in this para. are from Shurtleff, op. cit. Since I gave this paper at the 1989 Symposium, I have had some correspondence with a commercial tempeh maker in the West Country who has cast doubt on this claim for the high nutritional value of tempeh, particularly for the assertion that it contains a useful amount of Vitamin B_{12}. She is not allowed to state on her packaging that her product contains this substance.
16   Hayami, Kawagoe, Morooka and Siregar, *Agricultural Marketing and Processing in Upland Java*, CGPRT Centre, Bogor, 1987. CGPRT stands for Coarse Grasses, Pulses, Roots and Tuber Crops in the Humid Areas of Asia and the Pacific. There is a copy of the report in SOAS Library.
17   Soedigdo E. Muljokusumo, *Tempe dan Oncom*, Penerbit Tarate, Bandung, 1961 (5th printing, 1978). This useful little book is one of a series called "Kita membuat sendiri" - "Let's make it ourselves".
18   Anyone wanting to make tempeh at home can obtain a starter kit, price £5 inclusive of p. & p., from Murphy and Son Ltd, Wheathampstead, St Albans, Herts. AL4 8QY.

Couscous And Its Cousins

by Charles Perry

Couscous is a North African staple as far east as Tripoli, and particularly in Morocco and Algeria, where the local name for it is sometimes identical to the word for 'food' in general. It is also widely known in neighboring African countries from Chad to Senegal[1] and has footholds in Europe (the *cùscusu trapanese* of Sicily), Syria-Palestine (where it is called *moghrabiyye*, 'the North African dish'), and somewhat surprisingly in southern Brazil, where it is made from maize[2].

Dried couscous looks rather like a small soup noodle, but it is made by a process of its own that does not involve kneading: A bowl of flour is sprinkled intermittently with salted water as the fingers of the right hand rake through it in sweeping, circular movements, causing balls of dough to coagulate. The granules are also rubbed between the palms or against the side of the bowl to shape them, and when complete they are dried.

The contents of the bowl are sieved several times to obtain granules of uniform size. Couscous proper (called *kuskusu* in Arabic and *seksu* in Berber) is about 2mm in diameter, but everywhere in North Africa a 3mm or larger variety is also made, usually known as *berkukes*, and sometimes there is an additional grade somewhat larger or smaller than *berkukes*. Along the Mediterranean coast of Algeria and Morocco an ultra-fine variety with 1mm granules is known as *seffa* or *mesfuf*.

Couscous also differs from noodles in that it is not boiled but steamed, in a metal or earthenware colander sealed with paste to a pot in which, as a rule, the stew with which it will be served is being cooked. (*Berkukes*, however, because of the difficulty of steaming the large grains thoroughly, is usually cooked directly in the stew for a while after being steamed.) Steaming goes hand in hand with the stirring and rolling process by which the granules are made; if couscous is boiled it tends to have a slippery, porridgy consistency far inferior to that of noodles and particularly to the fluffiness of steamed couscous.

Sophie Ferchiou argues[3] that the stirring and rolling process by which the couscous granules are formed is a way of preserving grain. Couscous is traditionally made from freshly ground whole grain, which in fact is much better suited to the purpose than bolted flour, because starch readily accumulates around the larger and harder particles of bran and germ, much as a pearl forms around a grain of sand. The resulting granule is, in effect, a grain turned inside out, with the starch on the outside and the germ, the most perishable part of the grain, on the inside sealed off from the air. Now it can be stored for months or years without danger of staling. This would explain why *berkukes* is the couscous traditionally carried by travelers on long journeys - it has a thicker protective envelope of starch.

However, I suggest that the wide spread of couscous is due to a combination of economic and aesthetic reasons as well. Although the best-known couscous is made from wheat flour, one of the attractions of the couscous technique for an African cook is that is produces a light, elegant grain food which, unlike pasta or leavened bread, is held together not by gluten but by the weaker proteins found in all grains, and so may be made with grains that otherwise are suitable only for porridge or coarse unleavened bread.

For instance, in parts of Morocco and Algeria barley couscous (most often called *abelbul*), maize couscous (*abaddaz*) or couscous of a mixture of grains is more common than the wheat variety. A specialty of the Kabyle Berbers is *ahethut*, made from barley, bran and ground acorns. In West Africa, couscous is made from millets, the native grain *fonio* (*Digitaria exilis*) and a number of minor grains, some of them wild rather than cultivated. The Tubu of the eastern Sahara sometimes harvest a form of goosefoot which makes a black couscous.

It should be noted also that other things than couscous are sometimes steamed in North Africa, such as *dshisha* (cracked wheat or barley) and the usual Arab noodles *reshta* (rather like orzo) and *she'riya* (vermicelli). One medieval Spanish Arab cookbook[4] actually gives a description of steaming breadcrumbs like couscous.

Algerian folklore has it that couscous was invented by the Jinn. Certainly its early history is obscure, but the evidence does not suggest that it dates from remote antiquity. In the forties, H. Pérès published in the *Bulletin des Etudes Arabes* a compilation of the earliest literary mentions of couscous then known, and all were from the 15th century or later. The only citation that even claimed to be earlier was a 14th century anecdote related in the 17th century book *Nafh al-Tib*, which told how the mysterious illness of a North African visitor to Damascus was cured by making couscous for him.

Since the forties we have become aware of 13th and 14th century Arabic cookery books which contain couscous recipes. But altogether, the suspicious silence about couscous in sources from before the 13th century, coupled with the evident Berber origin of the Arabic word *kuskusu*, suggests that couscous arose among the Berbers of northern Algeria and Morocco during the obscure period between the 11th century collapse of the Zirid Kingdom and the triumph of the Almohads in the 13th[5].

A peculiarity of the way couscous is described in the 13th century cookery books also argues that it was a relatively recent invention, if we see in it an explanation for how the unique process of creating the couscous granules originated. Several of the books show a connection between couscous and a small soup noodle.

The Syrian cookbook *The Link to the Beloved*[6] gives two basis recipes for what it calls couscous, one ('couscous of the North Africans') consisting of granules made by stirring and then steamed. The other, however, instructs the cook to knead a stiff dough and twist it into short noodles to be boiled. A book written in Moorish Spain[7] describes a soup noodle called *hummais* (literally, 'small chickpeas') as resembling coriander seeds and connects it with the area around Bougie (that is, the country of the Kabyle Berbers). Another Spanish Arab book, *The Superfluity of the Table*[8], describes essentially the same noodle under a different form of the name, *muhammas* ('made like chickpeas').

This second book remarks offhandedly that *muhammas* could be 'spun in the kneading trough as *zabzin* is spun, for him who wishes to accomplish it quickly'. The book has already described *zabzin*, and it is a dish made with *berkukes* (which it calls *barkus*); the recipe has given a clear description of the couscous technique, which the text refers to as 'spinning'.

This accords with the double meaning of the word *muhammas* in North Africa today. *Berkukes*-like balls of dough which are made in noodle fashion, by kneading rather than stirring, are known today in Algeria - Kabylia and the Mzab oases - although now they are steamed, not boiled. They are called *tihemzin*, a Berberized form of the *muhammas*. Meanwhile, elsewhere in Algeria the same word (now pronounced *mhammsa* in Arabic, *tikhemzin* in Tuareg) means a large steamed couscous like *berkukes*.

The simplest explanation for this is that *kuskusu* was originally a small noodle, and that the peculiar stirring technique was originally a hasty way of making a noodle. Its resistance to staling and its unique lightness when steamed would have been discovered later.

References

1. E.g., Tubu *wasa*, Hausa *kuskus*, Mandingo/Bambara *basi*, Wolof *tyere*, Susu *kusikusi*.
2. Ortiz, Elizabeth Lambert, *Latin American Cooking*, Time Life Books, New York 1968, p.129-130. Curiously, the *cuscuz* of northern Brazil is a sweet made by mixing coconut meat and milk, tapioca, sugar and boiling water.
3. Ferchiou, Sophie, "Conserves céréalières et rôle de la femme dans l'économie familiale en Tunisie", in *Les Techniques de Conservation des Grains à Long Terme*, edited by Marceau Gast and Francois Sigaut, Editions du Centre National de la Rècherche Scientifique, Paris 1878, pp.183-185.
4. Huici Miranda, Ambrosio, *Traducción Española de un Manuscrito Anónimo del Siglo XII sobre la Cocina Hispano-Magribi*, Imprenta y Editorial Maestre, Madrid 1967, p. 204.
5. "Le couscous *(kuskusu ou kuskusî)*, attesté à l'époque hafside, ne figure dans aucun de nos textes zirides". Idris, Hady R., *La Berbèrie Orientale sous les Zirides, X-XIIe siècles*, Adrien Maisonneuve, Paris 1862, t.II, p.589. The Hafsids, local agents of the Almohads, began their rule at the beginning of the 13th century.
6. *Al-Wuslah ila al-Habib*, British Library Oriental 6388, pp.56a-56b
7. Manuscrito Anónimo, p.207.
8. *Fadalat al-Khiwan*, Staatsbibliothek Preussischer Kulturbesitz, 5473 (Wetzstein 1207), p. 24b.

Wine, Women and Song: the Staples of Life

By Graham Pont

The standard dictionaries do not recognise a specifically gastronomic usage of the adjective "staple": industries and commodities have long been described as staple in a purely economic or statistical sense, as those "having the chief place among the articles of production or consumption" *(Shorter Oxford)* or "for which the demand is constant and not dependent on variable factors (as season or fashion)" *(Websters)*. The concept of staple foods, therefore, must be a fairly new one. The term is not indexed, for example, in Anne Wilson's *Food and Drink in Britain from the Stone Age to Recent Times* (1973) and it is not employed in the catalogue of the British Library, nor in the classification of the *British National Bibliography*. But the term is in current usage; and I shall be interested to observe how it is used during this symposium. In particular, I will be looking for evidence of its philosophical connotations, assumptions and presuppositions, for these will surely reflect the present state of gastronomic thought in this country.

I take it as self-evident that a symposium on food and cookery is a gastronomic enterprise, but will not be surprised if I encounter some British resistance to the concept of gastronomy itself. My aim in this paper is to offer a small contribution to gastronomical philosophy - or, more accurately, the history of ideas in gastronomy. I hope thereby to bring out some of the dangers inherent in the concept of *staple foods*.

Brillat-Savarin's *Physiology of Taste* is the bible of modern gastronomy, yet this famous work has inspired only the most rudimentary exegetical literature, even in its native language, and it lacks a much-needed concordance to help the reader grapple with its encyclopaedic learning. Such deficiencies might partly explain the curious position of Brillat-Savarin in the world of English learning. His book is well-known, having gone through several translations, but it has attracted hardly any sustained discussion since the pioneering articles of Abraham Hayward in the *Quarterly Review* of 1835-6. A similar intellectual vacuum can be discerned in the United States, where the new *Journal of Gastronomy* shows as yet little evidence of historical or critical interest in the men and the writings responsible for the creation of the field. In the *rezeptionsgeschichte* of gastronomy in the English-speaking world, the anomaly is Australia, where during the 80's Brillat-Savarin might have seen the sudden appearance of much that he had hoped for, including gastronomy's academicians, professors and prizes. And the talent assembled at the Symposium of Australian Gastronomy - the fourth meeting was at Sydney in October 1988 - could well make a new kind of university.

If gastronomy were to become an academic discipline, as distinct from a field explored by other disciplines, it would require a fundamental text-book, and the *Physiology of Taste* would be the first candidate. From several years' teaching experience at the undergraduate level, I know that it is an ideal introduction to the subject. But I also realise that, with the dearth of academic learning in classical gastronomy, a subject without a well-developed philosophy and critical tradition, a little bit of Brillat-Savarin can be a dangerous thing.

Brillat-Savarin's charm is indefinable and inimitable. He is disarmingly persuasive, and that persuasiveness is due not only to his eloquence and address but also to his quick imagination. As well as being a brilliant raconteur, he has a wit which is both sharply pictorial and endlessly suggestive: his vivid and striking images are often accompanied by allusions to what is unstated or cannot be stated and yet is picturesquely delineated. He is a master of parody. These gifts which make Brillat-Savarin such a delightful companion in the dining room, the music salon and the lecture class also conceal his intellectual weaknesses - his dilettantism, his eclecticism and his occasional lapses into sophistry.

He makes no secret of the fact that his book is a hasty assemblage of notes and memorabilia collected during a long life. Modelled on the ideal of a dinner, the composition takes the form of a diverse collation improvised by a genial and generous host who makes all his guests feel at home. The entertainment is complete and the subsequent expressions of satisfaction almost entirely unanimous; the memories are fragrant, evocative and alluring; and the spell has remained unbroken for more than a century and a half of enchanted devotees. At such a splendid table, one is hardly inclined to notice the bread-and-butter issues of staple foods.

But imagine if we were to catch our host in one of his more professional moments and insist on a straight answer to a straight question, a judicial ruling on a point of philosophical consequence: *what are staple foods?*

Brillat-Savarin's first reaction might be to reach for his own copy of the text-book, which did not have time enough to become well-thumbed; or, more characteristically, to recall some of its memorable passages that might be suitable for quotation in the present circumstances. But, sensing the seriousness of the interrogation, he might then hesitate: the ready witticisms of the salon are not the cautious argument of the courtroom. His reputation as a gastronomical philosopher is now on the line.

Brillat-Savarin, I imagine, would continue to hesitate between two possible ways of answering our question, both of which he would like to offer, in spite of their being apparently incompatible. The different answers depend on different conceptions of food and nourishment, and these in turn go back to two conceptions of humanity and ultimately two different world-views that Brillat-Savarin inherited and wished to reconcile. The unifying concept he chose was *le goût*, taste.

Criticising Brillat-Savarin is like poking a finger into a perfect soufflé or being the first to disturb the arrangement of an attractive buffet. But if we must lay his ideas out on the critic's cold dissecting table, then we have to disentangle the two quite distinct world-views which he tries to amalgamate: one is the mechanical world-view, the other is the magical or musical world-view.

The mechanical world-view is the one so easily adopted by the fashionable and well-educated lawyer who grew up in the full flowering of the Enlightenment and knew some of its greatest authors by heart. Enthused by Newtonian mechanics and the analytic successes of the new chemistry, the *philosophes* and their followers envisaged a future in which science would eventually reduce all phenomena, including human behaviour, to the mechanical simplicity and rational certainty of Newton's astronomy. In the same spirit, Auguste Comte went on later to project the reduction of human psychology and the new social sciences to a mechanical, deterministic physiology which would be fully embraced by the fundamental sciences of physics and mathematics. Comte and Brillat-Savarin were widely-spaced contemporaries, but their world-views have much in common. The first to conceive biology as a systematic science, Comte would have entirely approved of Brillat-Savarin's first Aphorism: "the world is nothing without life and everything that lives must eat". Brillat-Savarin too would have sympathised with Comte's evolutionary derivation of the sciences, through which, as the unsociable positivist had failed to observe. "Gastronomy arose in her turn, and all her sisters gathered round to receive her". But, as we shall see later on, Brillat-Savarin rightly detected a very different world-view, from the prevailing scientism of his day, in his neoclassical image of Gasterea, the goddess of gastronomy, joining the hallowed circle of the Muses.

After the formalities of the hors d'oeuvres, the physiologist's feast opens in the full confidence of the Enlightenment with a highly speculative *mécanique du goût*. As the logical foundation for a mechanistic physiology of taste, the nature and operation of the five standard senses are enunciated; but then Brillat-Savarin adds a new, sixth sense, *le génésique* or the sense of physical desire. *Le génésique* does not have a specific organ - it is the sense which brings a couple together to perpetuate the species. For Brillat-Savarin this bold addition to the bodily endowment turns out to be something of an Achilles Heel.

In the encyclopaedic spirit, Brillat-Savarin reflects on the gradual improvement of the senses through time, both in the natural development of the species (he does not have a word for evolution) and the artificial extension of sensory powers with tools like the lever, the telescope and the microscope. Here the wit of the dilettante soon outstrips the narrow programme of mechanical science as he launches into a rather Comtean classification of the various arts produced by the improvement of the senses. Sight gives rise to painting, sculpture and all kinds of spectacle (architecture and gardening are unaccountably omitted); sound produces melody, harmony, music and even the dance (though it is also a visual art); smell gives rise to the art of perfumery; taste to the production, choice and preparation of food; touch affects all the arts; and physical desire produces the arts of sexual union, including romantic love and coquetry. This last refinement, according to Brillat-Savarin, is an exclusively French invention; but, as he warms to the praise of that Parisian specialty, his genealogy of the sciences begins to lose its classificatory neatness: physical desire involves all the organs of sense and consequently influences all the sciences. As Brillat-Savarin expands his argument, *le bon goût* also begins to exhibit transcendental tendencies.

Brillat-Savarin is unable to claim gastronomy in general as an exclusively French creation, but he rightly groups the practice of gourmandism with coquetry as two of the arts which had reached their modern state of perfection in Paris. Grimod de la Reynière had noticed certain analogies between gastronomy and sex but, in Brillat-Savarin's thought, these become two arts which are complementary means to human survival: practical gastronomy, or gourmandism, ensures the survival of the individual and coquetry ensures the survival of the species. Having both reached a high state of refinement, these twin arts of survival might have merited equal importance in in his meditations. But Brillat-Savarin suppressed his essay on coquetry. It would surely have drawn the parallels and connections of coquetry with the elaboration of the gastronomic arts in those superior dinners and banquets which cater for every desire and sensibility of the complete gourmand. The analogy of coquetry and gourmandism is central to Brillat-Savarin's thought; but, in exploring the parallels, he quietly abandons the restrictive premises of the mechanical world-view, the original foundations of his gastronomic science. This transition can be followed in Brillat-Savarin's changing conceptions of his principal subject, *le goût*.

In his preliminary exposition, *le goût* is simply one of the sense organs, the apparatus of taste and ingestion, including teeth, tongue, palate and stomach. This narrow primary meaning is confirmed in his list of the arts which minister to the sense of taste, that is, the production, choice and preparation of food. But, having drawn the analogy between the senses of taste and physical desire, Brillat-Savarin sees them both as capable of enjoying all the accessories of luxurious society. Expanding on *la puissance du goût*, he writes as if the field appreciated by *le goût* now embraces all the luxuries of sumptuous dining, including pretty women - who would, of course, be appropriately coquettish. Not only do *le goût* and *l'amour physique* have similar capacities for pleasure and satisfaction, but one sense now seems to include the other! The cause of this remarkable synesthesia is the intrusion of quite another kind of taste, the sensibility that enjoys *les jouissances du goût*. These are no longer just the pleasures of food in the narrow sense of nourishment, but the pleasures of the table in general. This sophistic confusion pervades the book, which might just as well have been entitled "The Physiology of Dining" or "The Pleasures of the Table", but at the risk of making the physiological speculations irrelevant. In his preface to the book, Brillat-Savarin expresses his interest in *le plaisir de la table sous tous ses rapports* and this is the subject that finally comes to dominate the argument at the end of the theoretical part, in the "gastronomical mythology" of Meditation XXX. His vision here belongs to what I call the magical or musical world-view, although room is also allowed for the scientific study of food, in the laboratories beneath the Baconian temple of Gasterea.

The inconsistency I have identified in the *Physiology of Taste* can be stated quite simply: Brillat-Savarin wants to use the same word "goût" to refer to a specific bodily organ or apparatus and also to aesthetic awareness in general. But if *le bon goût* appreciates the beauties of all the arts, then the organ of this taste is more than a single bodily sense. Brillat-Savarin would like to have it both ways: but it is obvious that the particular organ he calls "le goût" in the narrow mechanical sense cannot be the same as the entire organism that is delighted by the panoply of *les jouissances du goût*, the domain of Gasterea who charms or disarms all the senses simultaneously.

What I have distinguished as the magical or musical world-view is not so explicitly presented in the *Physiology of Taste*. but its position in the landscape of Brillat-Savarin's ideas is hardly obscured by his dabbling in the fashionable science of his day. The musical world-view is no other than that personified in the classical Greek Muses, the mythology that still remains the foundation of the western conception of culture. In creating Gasterea as the tenth Muse, Brillat-Savarin is not only admitting gastronomy to the inner circle of the civilising arts, but also locating his new science in a magical, musical and ceremonial tradition immeasurably older than the Age of Science. Musical magic appears to be a necessary ingredient of ritual feasting the world over, especially of sacred feasts like the Eucharist. The origins of such practices are certainly prehistoric.

In Brillat-Savarin's final vision of the new gastronomical mythology, the central observances of the Christian religion are to be replaced by a new cult of rational gastronomy based on an up-to-date scientific technology of food and embellished with all the refinements of modern taste. The worship of Gasterea - that is, the religion of epicurean gourmandism - is to be celebrated in an operatic synthesis of "all the arts", a phrase that is regularly repeated throughout the book. In the charming image of Gasterea, we discern a thinly-disguised Madame Récamier, the object of Brillat-Savarin's hopeless love and the dedicatee of his book. In this celebrated beauty Brillat-Savarin saw the personification of *le bon goût*, a figure transcending the delights of the individual senses and providing the ideal focus for the pleasures of the table in their totality. Thus the worship of Gasterea is no other than a reformulation of the *festin* or musical banquet, the ancient form of ritual dining in which wine, women and song have usually been considered necessary parts. In this new Muse we also recognise an epicurean Mary, perhaps not so generously endowed as her predecessors, the ancient earth goddesses.

While it is not acceptable to think of modern western women as commodities or chattels, they are nevertheless in regular demand and, without stretching the language unduly, could be described as staples in a diet. Brillat-Savarin loved women - too much to marry any one of them - and his book eloquently demonstrates his appreciation of feminine beauty, arts and accomplishments. There is no denying the indispensable role of women in Brillat-Savarin's theory of taste, especially when the most solemn observances at the table are to be celebrated in honour of a female goddess.

The role of music in Brillat-Savarin's gastronomical philosophy has long been one of my enthusiasms, but I shall restrict myself here to a new insight gained while preparing this paper. In his first Meditation, Brillat-Savarin mentions the belief of certain *docteurs allemands* that those who have a feeling for musical harmony are in possession of an extra sense. A very tempting hypothesis for a philosophical musician and amateur medico! In the following paragraph, Brillat-Savarin glosses this dictum with some speculations on the physiological state of those who are insensible to music or "tone-deaf" in the extreme sense. I very much suspect, however, that his thoughts on this subject have been edited here to fit in with his foregoing analysis of the senses and their mechanisms; for it surely must have occurred to him that, if physical desire can be regarded as a sense, then so might the musical sensibility itself. However, the very next paragraph reveals the hopelessness of such an analogy. Here he lists the recent improvements brought to the sense of taste by the discovery of sugar, alcoholic liqueurs, ices, vanilla, tea and coffee; but, having just discussed the radical improvements in music since the fifteenth century, Brillat-Savarin could see that the modern sciences of counterpoint, harmony and tonality far exceed anything comparable in the gastronomy of his day. So, for all his love of music, Brillat-Savarin, I believe, made a deliberate choice to abandon any hope of developing a theory of musical sensibility that would allocate the appreciation of music to a special harmonic sense and thus place it on a physiologically equal footing with gourmandism and coquetry. In the end, he had to be content with a very traditional role of music among the pleasures of the table, but this did not stop him from borrowing or adapting several concepts of music theory to create his philosophy of gastronomy.

Women, then, occupy a central role in Brillat-Savarin's larger world-view, in the actual as well as the ideal pleasures of the table; and music also remains in very much its traditional role as a necessary embellishment of both sacred and secular feasting. It is a great pity that Brillat-Savarin chose not to preserve some of his own music in notation, but he has at least left us the words of some of his songs. As we would expect, these belong to the ancient tradition of the *chansons à boire*, which could be sung from memory or improvised by the true sons of Bacchus. But wine is not just an accessory in Brillat-Savarin's magical or musical world-view, which owes much to the cult of Dionysus: in this the magical association of music and wine was consecrated and, in the old worship of Dionysus, women played an unusually liberated role. Even today, an orgy without wine, women and song would be unthinkable.

Critics have rightly pointed out that wine connoisseurship is not one of Brillat-Savarin's strengths: he simply takes good wine for granted and he accepts without question the established social, ritual and medicinal uses of alcoholic drinks. As with many other parts of his book, Brillat-Savarin's observations on wine are not particularly novel in themselves, but they often provide food for thought by placing a familiar experience in a new context. The service of the wine at the feast of Gasterea, with the discreet analogy between the breast and the bottle, is a memorable example. In this ancient image, gourmandism and physical desire are subsumed in a higher spiritual taste, which is the real subject of Brillat-Savarin's book. The basic organs of taste are found in every animal; but the taste that can appreciate *les jouissances du goût* is something of a quite different order. The delights of wine, women and song are only part of that; but, in its higher reaches, *le bon goût* transcends the merely epicurean enjoyment of the bodily pleasures and approaches a more spiritual or platonic love of the Good. The physiology of this transcendental taste defeated Plato himself and, knowing his limitations, Brillat-Savarin bowed elegantly before the banner of Epicurus. The master parodist could have easily written a purely materialistic gastronomy, but the book would have been dull and ephemeral - like his essay on political economy.

How, then, would Brillat-Savarin respond to the concept of *staple foods*? To what might have appeared to be a simple question, I doubt he could give a straight answer. If he were tempted to pursue the mechanistic assumptions of his Enlightenment science, he would soon realise the dangers into which modern medicine and nutrition science have so often fallen, through treating human beings as if they were just machines needing so much fuel. He realised that, in addition to the organism which needs regular feeding - *l'homme physique* - there is also *l'homme moral*, the spiritual being whose needs, tastes and desires have not yet been greatly illuminated or well served by mechanical science. Had Brillat-Savarin been able to overcome the old dualism of body and soul and bring these two conceptions of humanity together in a unified theory, he might have salvaged the argument of the book and remedied its fundamental eclecticism. But here he shows his age - something that rarely happens in this ever-fresh masterpiece.

In his meditation on the pleasures of the table, Brillat-Savarin rightly distinguishes between the pleasures of eating, which we share with other animals, and the pleasures of the table, which are peculiar to the human species. The recognition of commensality as an early distinguishing characteristic of humanity is one of Brillat-Savarin's most profound insights. He also recognises that the pleasures of the table do not depend directly on the basic pleasures of eating, since the former do not arise from the satisfaction of any simple hunger or appetite like that which makes the pleasure of eating so immediate and natural. But his demonstration of this fact shows how much manners have changed. At the beginning of a typical dinner, he says, people are so preoccupied with eating that they ignore each other and any conversation that might occur. Only when the physical needs have been satisfied do the moral and spiritual faculties come into play and the company is enlivened with thought, conversation and conviviality. Such poor manners, which remind one of stories about Dr Johnson's behaviour at dinner, can still be observed today, but they are no longer typical of polite society. It seems that Brillat-Savarin accepted the gross manners of his time as reflecting the natural order. He did not foresee the possibility of modern dinner parties, which proceed on the expectation that the guests are not desperately hungry and assume a high level of conviviality and conversation from the very beginning. The affluent society is not alone in separating survival from social eating. In *Consuming Passions* (1980), Farb and Armelagos point out that the Trobriand Islanders regard eating as a social rather than a biological necessity.

In a society where communal eating is not always an urgent physical need, but sometimes rather a necessary duty, it makes more sense to separate the notion of staple consumer goods from the notion of basic or necessary nutrition. By so doing we can proceed to consider what might be the staple foods or nourishment, not of man-the-machine (if there is such a thing), but of the complete moral and spiritual being. Viewed in this context, the traditions of wine, women and song are an important part of gastronomy - that is, if gastronomy is held to include the pleasures of the table.

Some might object that this is taking too broad a view of gastronomy and that wine, women and song are really parts of good living in general, rather than food and sustenance in particular. There would certainly be advantages in adopting a broader view of the subject, for which the Australian gastronomer François de Castella proposed the alternative expression "eubiotics", meaning "the art of good living". But it would still be necessary to distinguish the notion of staple foods and other eubiotic commodities from the notion of what is absolutely essential for the maintenance of life itself. For millions of starving people, there is no practical difference: the staples are the foods they survive on, if they can get enough of them. But, for more fortunate millions, the staples include tea, coffee, sugar, marmalade, milk, bacon and eggs - none of which is absolutely necessary for survival. Here we must be wary of nutritionists and dietitians whose science is based on the mechanical world-view and whose catering, as a result, is directed almost exclusively to *l'homme physique* and almost invariably dismal. It was a real break-through when hospitals realised they could attract patients out of their beds by putting on good food in a pleasant dining room - and thus save money.

If we were to undertake the theoretical reconstruction of gastronomy, and try to bring Brillat-Savarin up to date, what we should scrap is not the holistic conception of the pleasures of the table enjoyed by *l'homme moral*, but rather the pseudo-scientific machinery of *l'homme physique*. The fashionable conception of humanity adopted by Brillat-Savarin owed its paradigmatic formulation to La Mettrie's *l'homme-machine*, a model that fascinated the scientific minds of the age but is now as quaint and antiquated as the ingenious automata which were built to demonstrate the theory. The final absurdity was Vaucanson's mechanical duck which appeared to both ingest food and defaecate.

Freeing itself at last from the rusty shackles of scientism, the Technological Society looks in a more kindly way on its real predecessors, the magicians, the song-men, the shamans, the priests, the cooks and the courtesans who knew that their art could touch the human soul and transform the human body. Their magical world-view, so long dismissed as "unscientific", is now being sympathetically re-examined by the human and social sciences, which did not exist in Brillat-Savarin's day. That research, I believe, should be extended to one of our most ancient and valuable institutions, the feast or musical banquet.*

Brillat-Savarin was right in sensing the significance of the sacred feast; his mythology was better than his science. And, even though what he had to say on the subject was hardly novel, he knew that the magic of wine, women and song is essential to the pleasures of the European table. Being fond of the curiosities of the English language, he might even have agreed that wine, women and song are the staples of life.

---

* See Graham Pont, 'In search of the *opera gastronomica*', forthcoming in *Food in Festivity; Proceedings of the Fourth Symposium of Australian Gastronomy*, ed. A. Corones *et al.* (Sydney, 1990).

Fishery and the Utilization of Fish Products in Russia and the USSR

T.S Rass
Institute of Oceanology, Academy of Sciences, USSR

Fishery and the utilization of fish for food radically changed in the USSR after the First World War and the Russian Revolution, as a consequence of the sociopolitical and administrative structure of the country.

The beginning of the 20th century presented a continuity from the 19th century: fishery was mainly based on anadromous and freshwater species of fish, which were widely distributed in the numerous rivers and brackish seas and lakes of Russia. The cookery books of the period mention only 25 to 30 species of fish, principally of the sturgeon, salmon, carp, and perch families, with from three to five species of each family. Freshwater pike, eel, smelts, and burbot were also mentioned. Marine species of fish were only represented by cod, White Sea herring and navaga (*Eleginus* spp), anchovies and plaice.

In the first decades after the October revolution the situation began to change, with major changes coming after the Second World War. The number of species of fish mentioned for culinary use increased to 55-60, mainly from the same families but adding the herring family and lampreys. In the 1970s and 1980s more than three hundred species of sea fish, suitable for culinary use, are listed as being caught by Soviet fishermen.

There was also a sharp change in the size and character of the catch: from 0.9 million tons before the revolution, to about 1.6 million tons in 1936-37, 9 million tons in 1977, and 10.6-11.3 million tons in 1975-87. Today the Soviet Union holds the second place in total fishery yield, just after Japan with over 12 million tons, out of a world annual catch of 85 million tons.

In the late 19th and the early 20th century, the catch in Russia was based on the anadromous fish of various families, which ascended the rivers to spawn, as well as on several freshwater fish and a few coastal marine species. The remainder was represented by several freshwater and coastal marine species.

Anadromous fish are very high in food value, because their bodies are up to 15-30% fat, necessary for their subsistence during migration. The abundance of species of anadromous fish in Russian waters was the result of the geological history of the area, as the melting of the ice from the last glaciation occurred only about ten thousand years ago, giving the fish the possibility of descending the rivers to feed in the salt water and develop their anadromous life-cycle.

The development of fishing techniques first enabled the fishermen to exploit coastal waters, and then allowed them to fish in the open seas. Small open boats were replaced by seiners, side trawlers, stern trawlers, and finally the newest super-trawlers with an unlimited geographical range. In the beginning the fishing fleet obtained fish from the coastal waters of distant continents. After that came the exploitation of pelagic and deep sea fish of open and oceanic waters, e.g. sardines, silver smelts, mackerel, jack mackerel, pelagic gadoid fish and grenadiers, and the fish of Antarctic seas.

A substantial diminution, in some cases the near extinction, of the valuable anadromous fish was caused by the industrialization of the territory of the Soviet Union. The construction of many hydro-electric stations led to the damming of many rivers, transforming them into a series of reservoirs connected by streams. Great harm was done by the extensive use of mineral fertilizers and pesticides, leading to the run-off into the rivers of polluting and poisonous substances.

The development of fishing vessels capable of unlimited cruising demanded a substantial change in the technology of fish processing, a passing to the industrial preparation of frozen fish, filleting, canning, and the transformation of part of the catch into fodder for fur-bearing animals. Valuable smoked and salt fish products, limited in quantity, were displaced by industrially prepared products, from less valuable but more available fish species, including some not formerly exploited, and also forage species.

At present anadromous and freshwater species of fish constitute about 0.8 million tons of the catch, less than 8% of the total. The rest, more than 10 million tons, is provided by marine fish. Out of the total production, 7.5 million tons are distributed as follows:

> chilled and frozen fish: 3.3 million tons
>
> salt, dried and smoked fish: 0.8 million tons
>
> marinaded and spice-salted fish: less than 0.08 million tons
>
> canned: 1.1 million tons
>
> non-food fish production: about 1.9 million tons, including 0.8 million tons of fish meal.

Thus the industrialization of the country, the intensive use of the rivers for hydro-electric energy, the changes in the character of agriculture, and the development of marine fisheries have led to a radical change in the utilization of fish; a change which reflects the shrinkage of the supply of anadromous and freshwater fish species, balanced by the present enormous predominance of marine species of fish from the world's oceans.

The Importance of Herring in the Daily Life of the Coastal Population of Norway

By Astri Riddervold

Her Ligg Vår eigen gilde by.

by Lars Meling

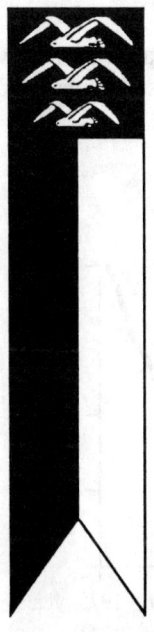

Her ligg vår eigen gilde by
med utsyn yver vide havet,
der båra syng ved kveld og gry
sin song, som alltid høyrest ny.
Derute gjeng det sild i kavet,
    halli, hallo, halla,
    og sildi ho er bra.
    Ja byen her
    den syng og lær,
når sild i not og garn dei fær.

På silda-bein er byen lagt
med sildatunnor i sitt våpen.
På alle buer er der vakt
— høyr diksla slær i jamne takt! —
Med sild gjekk byen vår til dåpen,
    halli, hallo, halla,
    då byens fedrar sa:
    Vårt Haugesund
    hev fulla grunn
til kjøpstadsrett frå denne stund.

So great was the importance of herring in the daily life of Haugesund, a town on the west coast of Norway during the last century, that a prominent citizen of the town, Lars Meling, wrote a song in its honour, a song which pays especial homage to the herring. Here is a translation of the song:

    Here stands our beloved town
    facing the open sea
    where the waves are singing their song
    all day and all night long,
    a song which is ever new.
    In this sea the fine herring abound.
    The whole town rejoices
    when the herring catch is landed.

> The town is built on herring bones
> Its flag emblazoned with three herring barrels*.
> Along the coast the people keep watch
> and listen for the sounds of the barrelmaker at work.
> For our town was identified with the herring
> and the burghers of the town could say:
> Our beloved Haugesund merits being honoured this way.

Behind this song is an interesting story of the life and activity which developed all along the west coast of Norway at the beginning of the 19th century. At that time the herring returned in large shoals to spawn after an absence of more than twenty years, and continued to do so regularly, every winter, for some 70 years.

The immense growth of the population, better living conditions, increased productivity, and commercial changes in the economic system, and the creation of three new coastal towns, Haugesund, Ålesund and Harstad, all seem to have owed their existence to the herring.

DEVELOPMENT OF THE POPULATION IN NORWAY c. 1000 - 1975

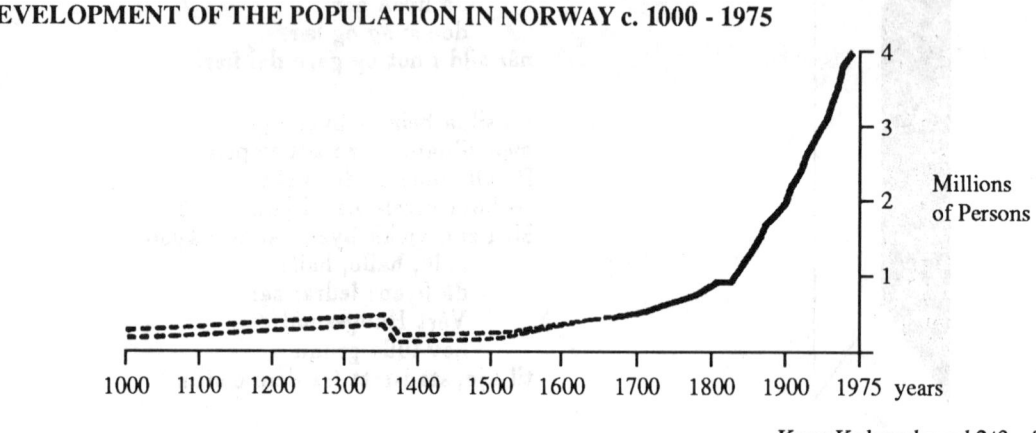

Knut Kolsrud mad 2/3 - 8

The nutritional importance of the herring

In the 19th century, the potato was introduced into Norway. It became very popular, as both the soil and climate proved ideal for its cultivation all along the coast to within the Arctic Circle. The staple cereals, oats and barley, often did not ripen before they were ruined by the frosts of autumn.

Norwegians usually had four meals a day. Oats and barley were used for porridge or gruel, often made with whey. Norwegian bread, the crisp, thinly rolled flatbread, was also made from oats and barley. It was usually baked only two or three times a year, as it can be stored for months. Every meal included either porridge, gruel or flatbread. After the introduction of the potato, it replaced cereals in the mid-day meal. Herring and cereals were eaten up to four times a day by most of the population, both rural and urban. Salt

---

* The banner illustrated at the beginning of the paper shows seagulls but the original design had barrels.

herring was always at breakfast, and often at one or two of the other meals as well. Inland, herring was usually served only at breakfast; at other meals, freshwater fish was used, either salted, dried or fermented. On the coast, salted, dried or slightly fermented saltwater fish was used; before the beginning of the 20th century fresh fish was seldom eaten. Such fish was regarded, like meat, as unhealthy. (Riddervold, Ropeid:1981) The usual Sunday midday meal consisted of a soup made from saltmeat, pearl barley, and locally grown vegetables such as turnips, peas and potatoes. Pieces of salt herring could be used instead of salt meat, and the soup was then called herring soup.

Herring in the history of Norwegian food

The importance of herring in Norwegian nutrition can be traced back for over 1000 years. Archaeologists have found herring bones dating from about 600 A.D. in settlements all over the country. In certain domestic sites on the coast, which seem to have been occupied exclusively by men, large quantities of fish bones have been found. Herring bones account for about 80%, cod bones for 15% and saithe (coley) only 5%. Digs at these sites have yielded bones from reindeer, deer, cow, ox, sheep, goat and pig. This suggests that men living in these localities from time to time came from farms at the head of the fjords to fish for herring during the season. (Magnus 1979, p.47)

Some evidence has been found which indicates the importance, during the Viking period around 800 A.D., of seasonal herring fisheries. From the early Middle Ages Norway was divided into four main areas, each with its own legal system, the oldest of which were *Frostatingslag* and *Gulatingslag*, dated from about 900 A.D. Together they applied to the western part of the country, whilst *Eidsivatinglag* and *Borgartinglag*, founded a hundred years later, covered the eastern part. These laws were recorded by the end of the eleventh century, and they provide interesting proof of the importance of the herring fisheries. (Holmsen : 1961, p.145, p.185)

A papal exemption from the general prohibition of fishing on Sunday was included in the *Frostatingslag*. It was granted by Pope Alexander III, 1159-81, and stated that on Sunday and on any other day on which God leads the shoals towards the coast, herring may be caught. A hundred years later this exemption was repeated by Cardinal Wilhelm of Sabina, who visited Bergen in 1247. (Grøn : 1926, p.116) (N.G.L.l., p.139) In 1273, the four regional laws were merged into one national law of Norway. Here we find detailed regulations concerning the herring fisheries, tools and tackle, distribution of the catch and punishment for those who disobeyed the law. That law also stated that every lawsuit should be postponed until after the herring season, a decision which stressed the importance of the fisheries. The law referred to the spring herring, which belong to the Atlanto-Scandic variety, and which enter the coastal waters in large shoals in the spawning season.

Preserving herring by drying and light fermentation

The law also prescribed how the herring was to be preserved. It laid down that when the herring were caught at Christmas, fishermen were permitted to take them on board, bring them to the shore and hang them on previously prepared wooden rods to dry. Cutting and preparing the rods was prohibited between Christmas and New Year's Eve. (Grøn : 1926, p.117), (N.G.L.I., p.140) Hence we know that herring were dried by hanging them on wooden rods, but the law was silent as to how and where the drying was done. Later, however, large purpose-built drying houses are reported as having existed on both the east coast (Berg : 1979, p.180) and the west. Herring can still be seen today hanging on rods, which almost cover the northern walls of houses and outhouses on farms. Both methods keep the drying herring out of the sun, to prevent the fat becoming rancid. Salt also encourages rancidity and the dried herring was originally unsalted. Today herring may be rubbed with salt or dipped in brine before being hung up to dry. As the fishing season is between Christmas and Easter, the low temperature makes salting unnecessary for preservation of the herring. The result of this process will be semi-dried, semi-fermented herring which can be eaten as it is, boiled, or roasted in hot ashes. Sometimes it will go mouldy on the surface. I experienced this as a child when we visited a farm and, together with the farmer's children, we were allowed to pick some of the herring from the wall and eat them. The mould was scraped off with a nail or a knife before we ate the fatty flesh by biting along the backbone. The rest we threw to the animals. We liked the taste, but when we got home afterwards, our cheeks shining with herring fat and mouth and fingers smelling from the partly fermented fish, we needed a thorough wash before being allowed into the sitting room.

In the sagas, the love of this dried herring is described in a story about the famous god Thor. He asks a man to take him across a fjord in his boat and in return offers him a meal of dried herring, which he is carrying with him: "There's no better food; I ate some myself with oaten bread before setting out, and I am still not hungry."

In this story, herring is described as food fit for the gods. It has always been - and still is - loved and eaten by the highest and lowest in the land.

Dried and smoked herring

When the aged are nowadays asked to reminisce about their childhood, almost everyone who lived on the coast speaks of smoked herring. Roasted over a low fire or in hot ashes, often purloined from the storehouse while parents were out of the way, this fat, smoked, warm herring was favourite. No-one knows how old the smoking tradition is, but it is believed to be very old. When the herring was hung from the ceiling to dry above the open fireplace, it acquired a fine smokey flavour. The smoke prevented it from going mouldy and lightly salted smoked herring will keep in the storehouse throughout the winter. It has been a popular food for centuries, with or without added salt. It was fried in its own fat on hot stones in the fireplace, or it could be eaten raw, boiled, or fried in butter and simmered in beer, a popular dish in my own home, served with mashed potatoes.

Fermented herring in light brine

Fermented herring also seems to have very old traditions in Norwegian nutrition. It was called "Bondegods", - a farmer's commodity, and is described as herring of varying size, only lightly salted and having a characteristic, strong, sour smell. It is favoured by the coastal population and eaten daily by the farmers. (Pontoppidan : 1754, p.235) Home-burnt Norwegian salt from seaweed or evaporated seawater was used, while imported salt and fully salted herring of regular size were used for export. These official rules were given in 1683 and 1753. The famous author and clergyman, Peter Dass, describes in the poem "Nordland's Trumpet", written in 1739, how a mountain farmer from the head of the fjord travelled to the coast to exchange some of his produce for a barrel of herring. "The stinking goods he prefers, the sourest barrel tastes best to him. This appeals to his stomach. It is smeared on the flatbread instead of butter...". Another official report from a region further south, the inner part of the Trondheimsfjord, from 1757 - 70, states that sour-smelling herring, lightly salted with locally produced salt is preferred by the local population who like the sour taste, find it easier to digest than fully salted herring and feel it is healthier. (Recent research has found this to be correct.) The sour-smelling, lightly salted herring seems to belong to the 17th and 18th centuries, preferred by the local population on the west coast. In the 17th century, imports of salt from Spain increased, and the price fell. The fermented herring, however, kept its position until the 19th century.

Salt and fresh herring

In this century, the fully salted herring, the "spekesild" seems slowly to have replaced the fermented herring as everyday food, along with smoked and dried herring which remain popular, whereas the fermented herring has disappeared from official reports and literature.

The return of spring herring in large quantities by the beginning of the 19th century, the immense growth of the population, and of the towns, brought fresh herring into Norwegian kitchens. It was slowly accepted, especially in the towns, along with salted herring, which now appear in a variety of qualities, and with sugar and spices added to the brine.

A survey of the cookery books of the 19th and 20th centuries illustrates this clearly. The famous book by Hanna Winsnes, published in 1845, running to 412 pages, contains only two recipes for fresh herring. In one of these, the fish is fried and eaten with scrambled eggs. In the other, it is also fried but served in a sour sauce. This dish may be kept for up to a week. Both dishes were suitable for breakfast or supper. The book has two recipes for salt herring: in the first a whole herring is fried, wrapped in paper and served hot in the wrapping: in the second, it is served in a salad. Both are recommended for breakfast or supper.

In 1845, the herring seemed still to be used for breakfast and supper in the traditional way. By the end of the century, however, recipes for herring pudding, herring cakes, herring soup, which appeared in the first cookery book published for use in the primary schools all over the country, indicate that herring had by then become a dinner dish. This referred to fresh herring. That book (Chritensen, Helgesen: 1897, p.103) gave seven recipes for herring, more than for any other fish. In the 20th century, the cookery book used in the domestic science school of the town of Haugesund has 21 herring recipes, 12 for fresh herring, 9 for salt. (Haugesund Husmorskole 1947, p.115) The leading Norwegian cookery book, Henriette Schønberg-Erken, 815 pages, first published around 1900, gives 39 recipes for herring, 21 using fresh for pudding, cakes, soup, etc., 18 using salt herring.

Some of the recipes are quite up to gourmet standard. In the period between the wars, men's parties where a variety of herring courses were served, accompanied by beer and aquavit, were very much in vogue

Figure 2. TRADITIONAL CONSERVATION METHODS AND TYPES OF HERRING IN NUTRITION

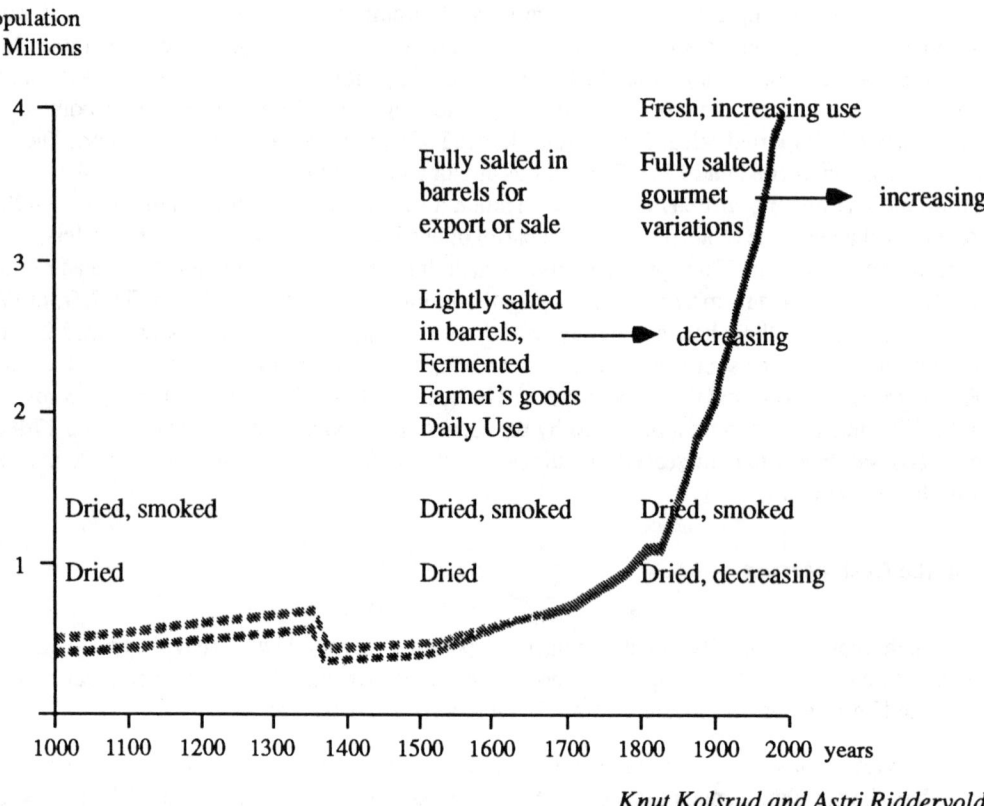

Knut Kolsrud and Astri Riddervold

in higher social circles and gave much credit to the housewife. In this period, fresh, salt and smoked herring were served for dinner at least once a week in the towns.

The ordinary salt herring kept its position in rural areas, served for breakfast and for dinner as well all over the country. In the towns pickled herring made its appearance on the breakfast table and as such it has maintained its position. During the Nazi occupation, Norwegians largely survived on herring, eaten morning, noon and night, dried, smoked, pickled, fried, boiled, marinated. At that time, "God sent the shoals close to the shore" according to the Frostatingslag 900 years ago. But in the late nineteen-fifties, the spring herring forsook inshore waters again for 20 to 30 years. As we know from historical records, this has happened several times, following a half-century or so of regular annual arrivals.

Small, immature herring, however, are always caught along the coast, but not in large quantities. The herring disappeared as part of everyone's daily diet but pickled herring survived, and is now commercially produced. In a variety of sauces it appears on breakfast and lunch tables in every Norwegian hotel, and in many private homes also.

The herring party, with herring pickled at home according to old and new recipes, is more fashionable than ever. Spring herring are now returning to the Norwegian coast, and according to recent research the

Figure 3. ECONOMIC IMPORTANCE OF SPRING-SPAWNING HERRING DURING THE 19th CENTURY

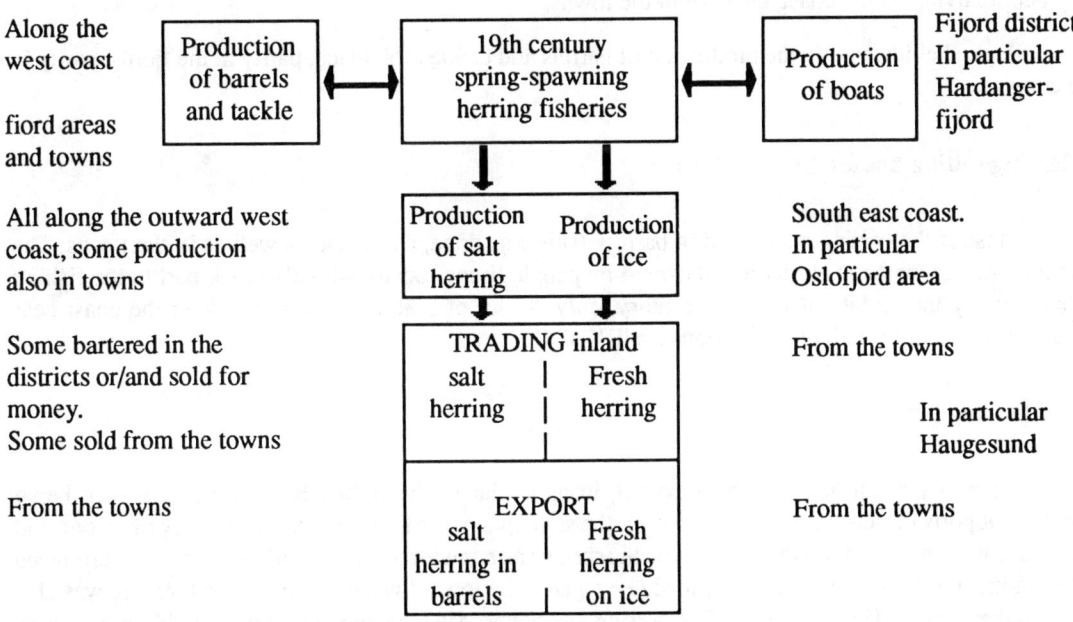

old tradition of eating herring daily makes good sense. Scientists suggest that we might vary the old proverb to "Some herring a day keeps a heart attack away."

Immature herring have always been caught all along the Norwegian coast. Such fishing is on a small scale and has never involved such intense activity as occurred when the spring-spawning herring returned to the coast at the beginning of the 19th century.

Growth of population

The growth of population in coastal areas was larger than the average growth in the country as a whole. Records show that this growth was greatest in those areas where and at times when the herring came close to the shore. Official reports from those areas stated: "Everybody, professional people and farmers alike, the head of the household and the servants participated in the herring activities." (Solhaug 1979, p.64) The population of Haugesund trebled between 1855 and 1865, from 1066 to 3221. The growth was permanent and, later in the century, resulted in the port acquiring the status of a town.

Shipbuilding, barrel- and tackle-making

Haugesund is situated between two fjords, the Ryfylke and the Hardangerfjord. It is important in timber, shipbuilding and commerce. The ethnologist, Bjørn Fjellheim, has collected a wealth of material relating to shipbuilding in the south-west part of Hardangerfjord. Between 1815 and 1875, some five hundred to six hundred ships were built there. Most of those built before 1820 were ordered by customers in Bergen. After 1820, a great change took place, in that the ships were then ordered by local residents or by people living on the coast, away from the towns.

A parallel increase in the production of barrels and tackle took place, partly in the fjords, partly in the towns.

Herring salting-houses

Most of the herring was salted in barrels. This was done, coastwise as well as in the towns. The salting-houses were mostly built and owned by people living locally who also took part in the fishing industry. By the middle of the 19th century, only 20 out of 800 salting-houses along the coast near Haugesund were owned by townspeople. (NRK 1982)

The herring trade

Most of the salt herring was exported, in particular to the Baltic, Riga and Kønigsberg being important ports for this trade. The shipowner, the sea captain, the herring fisherman - very often one and the same person - would sail to those ports in the spring and summer season, to sell the salt herring produced during the winter. This activity was mostly engaged in by town dwellers, while inland trading was also undertaken by people living nearby. The herring was bartered in local markets and in neighbouring areas, or sold to merchants. A new market for fresh herring developed after the Napoleonic Wars, which was accelerated by the general growth of population and towns all over Europe. This trade, both inland and export, generated great activity in ice production, in particular in the Oslofjord area. The cold winter climate, and the steep hillsides where small brooks could be dammed up to create small lakes where the ice was cut, were ideal for ice production. The plentiful supply of sawdust from the sawmills in that area provided ideal material for insulation of the ice during storage and transportation. Together with the many good harbours all along the fjord, these factors made it possible for people living in areas on the east coast to participate, to some extent, in the herring industry.

The harvesting of the catch

The yield of all this herring-based activity along the Norwegian coast was harvested both by the urban and the rural populations, in particular by the fishermen from the agricultural farms. We can trace the improvement in their economic position in the archives. During the 19th century, they bought the land they previously rented. This happened contemporaneously with the arrival of the herring shoals on the south-west coast between 1825 and 1855. By 1855, only 9% of these fishermen still rented their land. On the northern part of the coast the same thing happened about 1870.

New and better houses were built on the farms, and iron stoves replaced the open fireplaces. (Engen 1978) (Riddervold 1980) It has not been established that the herring was solely responsible for these changes and improvements, but recent research has shown that the herring played a very important part in bringing them about; so important, indeed, that the author of the song which started this thesis, Lars Meling, a member of the Stortinget (Norwegian Parliament), included the herring in his song of praise to his home town.

BIBLIOGRAPHY

Berg, Gøsta,1962:'Gravlax och surstrømming'. *Gastronomisk Kalender.* Stockholm.
Berg, Gøsta, 1979: 'Sill som føda'. *Norveg 22,* Oslo.
Christensen, Dorothea, Helga Helgesen, 1897: *Kogebog for folkeskolen og hjemmet.* 3.utg. Sandefjord.
Davidson, Alan, 1979: *North Atlantic Sea Food.* NewYork.
Devold, Finn, 1979: 'Sildefiske i Nord-norge', *Norveg 22.* Oslo.
Dass, Petter, 1763: *Nordlands Trompet,* Oslo 1980.
Grøn, Fredrik, 1926: 'Om kostholdet i Norge indtil aar 1500'. *Skrifter utgitt av Det Norske Vitenskapsakademi i Oslo.*
Grøn, Fredrik, 1941: 'Om kostholdet i Norge fra omkring 1500 tallet og opp til vår tid'. *Skrifter utgitt av Det Norske Vitenskapsakademi i Oslo Nr. 4.*
Haugesunds kommunale husmorskole. *Kokboke, 1947,* 11utg.
Holmsen, Andreas, 1961: *Norges Historie. Fra de eldste tider til 1660.* Oslo.
Hoven, Bjarne, 1954: *Ole Torjesen Hovens slekt.* Kristiansand.
Lind, Grøtteland, Kristine, 1978: *Daglig brød, daglig dont.* Oslo.
Magnus, Bente, 1979: 'Sild og samfunn i vestnorsk jernalder'. *Norveg 22.* Oslo.
Norges Gamle Lover I (NGL).
Norsander, Gøran. 1976: *Salt sill.* Lund.
Olsen, Kåre, 1980: *Endel trekk ved oppkomsten av borgerskapet i Haugesund i perioden ca. 1840-1900.* Hovedoppgave i historie, Universitetet i Oslo.
Pontoppidan, Erich, 1752-54: *Forsøg til Norges naturlige Historie I.* Copenhagen.
Rasmussen, Holger, 1979: Mennesket og silden. *Norveg 22.* Oslo.
Riddervold, Astri, 1978: *Konserveringsmetoder for kjøtt, fisk, ville baer og urter.* Magistergradsavhandling i etnologi. Universitetet i Oslo.
Riddervold, Astri. 1980: 'Jakten på den nordlandske røykovnen'. *Nordland fylkesmuseum årbok.* Bodø.
Riddervold, Astri, Andreas Ropeid, 1981: 'Popular Diet in Norway and Natural Science during the 19th Century'. *Ethnologia Scandinavica,* Lund.
Schønberg Erken, Henriette, 1942: *Stor Kokebok,* 15. utg. Oslo.
Solhaug, Trygve, 1976: 'Kulturformer, økonomiske og sosiale mekanismer i det førindustrielle "Fisker-Norge" 1790-1890'. *Heimen XVII,* Oslo.
Solhaug, Trygve, 1979: 'Vårsildfisket på Vestlandet 1808-1873'. *Norveg 22,* Oslo.
Vollan, Odd, 1971: *Sildefisket gjennom 1000 år.* Oslo.
Winsnes, Hanna, 1845: *Lærebog i de forskjellige Grene af Husholdningen.* Oslo 1985.

TYPICAL NORWEGIAN HERRING RECIPES

1. Pickled Herring

    3 salted herrings - *spekesild* (preserved in brine - 10/20%)

Marinade:
- 1 dcl (3½ fl. oz.) white wine vinegar
- 1 dcl (3 oz.) sugar
- 2 dcl (7 fl. oz.) water
- ½ tsp black peppercorns
- 1 onion

dill weed

Gut, wash, clean and fillet the herrings. Soak the fillets in water or milk to get rid of excess salt. Dry them and then cut them slantwise into 2½ cm (1 inch) pieces. Peel and slice the onion. Place the herrings in a glass jar layered with the onion and peppercorns. Stir together the vinegar, sugar and water and pour the liquid over the herrings. Set aside for a few hours or, if possible, overnight.

Serve garnished with dill sprigs.

2. Sherry Herring

    2 spiced herrings - *kryddersild* (preserved in spiced brine - normally purchased as fillets in tins)

Marinade:
- ½ dcl (7 tbs.) red wine vinegar
- 1 dcl (3 oz.) sugar
- ½ dcl (7 tbs.) water
- 4 black peppercorns
- ½ dcl (7 tbs.) dry sherry

    2 tbs. chopped almonds

Prepare the sherry marinade. Slice the drained fillets into bite-sized pieces and put them in the marinade. Preferably leave in the refrigerator for a day or two.

Sprinkle the dish with almonds prior to serving.

3. Baked Omelette with Smoked Herring

- 4 smoked herrings - *røkesild* (kippers may be used)
- 4 large tomatoes, sliced
- 4 eggs
- 1 dcl (3½ fl. oz.) milk
- salt, if required
- chopped dill

Skin and bone the herrings and place in a layer in a buttered fireproof dish. Cover with tomato slices and chopped dill. Whisk together the eggs and milk and pour the mixture over the herring. Put into a hot oven at 180C/350F/Gas 4 for about 20 minutes.

Serve immediately.

4. Herring Salad

    1 salted herring - *spekesild*
    1 apple - cored, peeled and diced
    100 g (4 oz.) diced boiled potato
    150 g (6 oz.) diced cooked beetroot
    3/4 tbs. finely chopped onion
    2 tbs. white wine vinegar
    2 tbs. sugar
    1 pinch pepper
    3/4 dcl (5 tbs.) whipping cream

Garnish:
    hard boiled egg and parsley

Rinse and clean the herring and cut into small cubes. Mix the solid ingredients with the herring in a bowl. Stir together the rest of the ingredients with the exception of the cream. Pour the mixture over the diced ingredients. Stir and season to taste.

Whisk the cream and fold into the salad just before serving.

Garnish with hard boiled egg and parsley.

5. Rollmops

    1 Kg (2-2^{1}/2 lb.) fresh herring - *fersk sild*
    salt
    chopped parsley
    dry bread crumbs
    margarine or butter
    1 dcl (3^{1}/2 fl. oz.) white wine vinegar
    1 dcl (3 oz.) sugar
    2 dcl (7 fl. oz.) water

Clean and wash the herring. Remove the backbones, fins and heads without separating the fillets from the skin. Sprinkle the flesh sides with salt and chopped parsley. Roll up the herrings from the tail towards the head (skin outside). Place the rolls tightly together in a greased fireproof dish. Sprinkle with the bread crumbs and dot with the fat. Bake for 25 minutes at 225C/425F/Gas 7. Mix together the vinegar, sugar and water and pour over the freshly baked herring.

Serve hot or cold.

Sylvester Graham and the Origins of the Breakfast Cereal Industry

By Elizabeth Riely

The graham cracker - that brown, crisp, sweetish rectangle familiar to millions of Americans - is Sylvester Graham's monument to posterity. Yet it stands for everything he opposed. Mechanically processed by industrial giants, it contains two different kinds of flour, three forms of sugar, salt, and sometimes palm oil. Bicarbonate of soda lightens its crumb and vitamins and minerals supplement its nutritional value. Commercial products such as the graham cracker have revived interest in health food and brought Graham's efforts at diet reform full circle. Perhaps it is not such an inappropriate symbol for this man of contradictions after all.

Let no man "put asunder what God has joined together," proclaimed Graham of the milling process. He believed that removing the kernel from the grain was the basis of society's ills, and he spent a lifetime preaching this gospel. His *Treatise on Bread and Bread-Making* gave detailed instructions on how to grow wheat (or select flour if people could no longer grow it themselves), store, and grind it. They should mix their unsifted wholewheat flour with nothing added but water and yeast and bake the bread themselves in their own ovens at home to be eaten cold. Graham directed that only the wife should bake the bread, the mother who "rightly perceives the relation between the dietetic habits and physical and moral conditions of her loved ones."

Graham is an easy target for our ridicule, but we should consider him in the context of his period and, at the same time, see more clearly our own. In the early 19th century, the rapid growth of industry and commerce drew many people from country farms to the cities, where they found it cheaper to buy bread than make it themselves. The new machines revolutionized milling. By leaving out the germ, processors could make a highly refined and less oily flour that kept well without a thick protective crust. The vitamins, minerals, and fiber that they eliminated from the diet were not yet understood. By the 1830s, the typical loaf of bread - one that we easily recognize - had become white, soft, thin-crusted and quick-baking. Unscrupulous bakers often adulterated it with alum, chalk, plaster of Paris, and other additives. Graham scolded them for altering it from its natural state, remembering nostalgically "those blessed days of New England's prosperity and happiness, when our good mothers used to make the family bread."

Graham had no such mother. Born in rural Connecticut in 1794, he was the youngest of 17 children. His father died intestate at the age of 74, when Graham was two years old. His overburdened mother soon became insane, and he soon showed signs of the poor health and emotional neglect that pursued him the rest of his life. The ideal of a happy home, with a good mother baking the family bread, was a romantic fantasy he never knew. Graham grew up unable to make friends and establish himself in a career. Soon after entering preparatory school at 29, he was unfairly expelled. A nervous breakdown followed, and upon his recovery he married his nurse. After studying for the Presbyterian ministry like his father and grandfather, he failed to find a parish. His evangelical sermons against alcohol alienated his congregation, but made him welcome within the temperance movement. Here he found his calling. During the 1830s Graham lectured in Philadelphia, New York, and Boston, broadening his crusade to include unbolted bread, vegetarianism, and sexual abstinence.

Digestion was to the 19th century what the psyche is to the 20th - the latter-day seat of the humors. To Graham and his contemporaries, who knew nothing of modern science, gastrointestinal irritation was the main cause of disease, The digestive tract was the battleground between organic and inorganic forces - good and evil - much as, on another scale, man struggled against the artificiality of modern city life. When Graham was beginning to lecture in 1830, he came across the writings of François Broussais, a French pathologist who theorized on the effects of stimulation. He believed that nerves spread the irritation of the digestive system and endangered the whole body. Graham was deeply impressed and incorporated these principles into his lectures without acknowledgment. Graham preached that his diet could oppose the dreaded threat of stimulation. Such excitement would lead to debility, both physical and moral.

The cholera epidemic of 1832 came at the right time for Graham. In the panic he preached that his regimen could prevent the disease. One should eat only the smallest amount necessary for "the vital economy of his system," he wrote, "knowing that whatsoever is more than this is evil." A subsistence diet of vegetables, Graham bread, and pure water could establish optimum health. Audiences flocked to Graham's lectures, gave testimonials, and brought notoriety if not acclaim. Ever the flamboyant evangelist, he basked in his growing fame and thrived on the controversy. Several times he was mobbed. In 1837 a group of angry bakers in Boston tried to disrupt his talk by throwing slaked lime on his followers.

More than a few noticed the contradictory side of his nature. One contemporary criticized Graham for his intemperance: his regimen was the "result of a most unruly passion for temperance." Ever more distrustful and self-pitying, he suffered another nervous collapse before retiring to Northampton Massachusetts in 1839 with his wife and two children. Financial losses and poor health filled his last years: embarrassment turned to rancor. His wife, who rejected his strict diet, tempted him with a "table luxuriously spread" with the food and drink he despised. At the end he accepted drugs and meat from a doctor, but they too failed. Broken and bitter, Graham died in 1859 at the age of 57.

We may wonder how Graham succeeded as well as he did, but undoubtedly his regimen did help many people. In this period of social change and upheaval, many felt threatened by the demands of their increasingly urban existence. Most of his followers were small craftsmen from New England villages who went to the city to ply their trade in a newly commercial culture. The Graham regimen gave a measure of order to their lives. Even if they could not control the outside world, they could regulate their own bodies to reduce the confusion they felt around them. Graham was the first to offer a system for coping. He succeeded so well that instead of turning them back to their old way of life, his followers learned to compete in their brave new world.

Hygiene was one of many 19th century American reforms, along with abolition, women's emancipation, and college education, among others movements. Grahamites often embraced them all, practicing them together in communal experiments. New York and Boston had Graham boardinghouses where men rose at 5 a.m., went to bed at 10 p.m., ate two small vegetarian meals, and drank only water. Students at colleges such as Wesleyan and Williams formed Graham clubs in their dining halls. In 1833 New England Congregational progressives founded Oberlin College in Ohio, the first coeducational college in the country. David Campbell, an abolitionist and manager of a Graham boardinghouse, was invited to become its steward. Campbell tried to control stimulation between the sexes ("amativeness") with a strict Graham diet. When a professor was fired for seasoning his food with pepper, a student protest forced Campbell's resignation. The Graham diet continued as an elective.

Brook Farm, started by transcendentalists in 1841, had a Graham table. Many intellectuals visited the Utopian colony near Boston, including Nathaniel Hawthorne and Ralph Waldo Emerson. In June of 1843, Bronson Alcott left Brook Farm to found Fruitlands, a farm farther outside Boston and based on stricter vegetarian principles. Besides dairy products and meat, also forbidden were oil lamps, wool, leather, silk, and cotton because they came from animals or were produced with slave labor. The problem was that, however learned in philosophy, they knew little of farming. "They look well in July," Emerson wrote in his diary after a visit. "We shall seem them in December." By early January the commune had failed, and the Alcott family barely survived the winter on barley, potatoes, apples, and water.

Until the 19th century no one had taken responsibility for public health. As orthodox doctors relied on drastic treatments to produce dramatic changes in their patients, violent therapies such as purging, bleeding, blisters, and emetics were common. Poisons such as calomel and mercury were used for everything from cleansing the bowels to keeping away bedbugs. Those looking for gentler methods had many to choose between, often overlapping with Grahamism. Phrenology, botanical medicine, homeopathy, and hydropathy were some of these alternatives. A doctor complained in 1848 that homeopaths "were equally full of transcendentalism, the year before of homeopathy, the year before of animal magnetism, Grahamism, phrenology." Hydropathy - the water cure - had many followers, including Graham himself who took it the day he died. Water-cure spas opened in the northeast where patients purified their bodies with brisk exercise, fresh air, cold food, and water, up to forty glasses a day.

In one water-cure establishment, a combined health spa, resort, and commune named "Our Home on the Hillside," James Caleb Jackson sold various health food products. Among these were Graham flour and "Granula" - wholewheat flour and water baked in thin sheets, then broken up, ground, and baked again. Jackson devised this early cereal around 1863 as a convenient breakfast food that could be stored on the shelf for long periods, to be mixed with milk the night before, and refrigerated. It was selling poorly and might have been forgotten altogether, were it not for a visit to "Our Home" by Ellen Gould White.

On Christmas Day in 1865, Mrs. White, the leader of the vegetarian Seventh-Day Adventists, experienced a vision telling her to build a similar facility, but without "the sophistry of the devil" - theatre, concerts, and socializing. The following September the Western Health Reform Institute opened in Battle Creek, Michigan, her Church's headquarters, on strict Grahamite lines: no meat, greasy food, caffeine, alcohol, drugs, tobacco, corsets, false hair, or "unnatural" sexual activity; plenty of Graham bread, vegetables, herbs, fruit, water, and fresh air.

The Institute struggled in its first decade until in 1876 Mrs. White hired a 26-year-old doctor named John Harvey Kellogg to head the medical department. Although the young Adventist happened to be the son of the treasurer and largest stockholder, Kellogg was also a recovered consumptive, Graham convert, and graduate of Bellevue Hospital Medical College in New York City. He had persuaded the Whites, in fact, to help pay for his orthodox medical studies, during which he regularly ate an apple with seven graham crackers for breakfast. His credentials were impeccable.

Kellogg immediately set about making changes at the Institute. By renaming it the Medical and Surgical Sanitarium - downplaying the religious affiliation without offending Mrs. White - he aimed for scientific respectability and broad popular acceptance. Both came quickly. Within a year he published the first of many books on "biologic living." Within two years he began making zwieback biscuits from wheat, oats, and corn which he "partially digested" by slow cooking, ground into breadcrumbs, and sold as "Granula." When Jackson sued, Kellogg changed the name to "Granola" and never looked back.

Soon the "San" (as the Battle Creek Sanitarium came to be called), with only twelve patients on Kellogg's arrival in 1876, attracted a steady stream of thousands each year. His younger brother handled the business and advertising end, which the doctor called the Extension Department. By the time of his death in 1943 at the age of 91, some 300,000 patients, including presidents, magnates, socialites, and celebrities, had stayed there. He enjoyed a success unknown to Graham.

"The breakfast food idea," Kellogg later said, "first made its appearance in a little third-story room on the corner of 28th Street and Third Avenue, New York City ... I was boarding myself while attending medical college, partly as a measure of economy and partly because I was making experiments in diet, and no boardinghouse in New York would have provided such a diet as I wanted. My cooking conveniences were very limited. It was very difficult to prepare cereals. It often occurred to me that it should be possible to purchase cereals at groceries already cooked and ready to eat, and I considered different ways in which this might be done."

Kellogg never stopped experimenting with grains, nuts, and legumes. In 1895, after trial and error, he came up with wheat flakes. A bushel of wheat costing 60 cents was transformed into "Granose" and sold at 10 ounces for 15 cents. In the first year he sold 113,400 pounds of them. Spurred on by competition with Shredded Wheat, Grape Nuts, Puffed Rice, and Ralston Purina, he continued his testing with a wide range of health foods. Encouraged by his wife - adhering strictly to Graham doctrine, he was married but apparently celibate - Kellogg learned in 1902 how to make corn flakes palatable by adding malt. Four years later, "Kellogg's Toasted Corn Flakes" made their appearance under the brilliant merchandizing of his brother, W. K. Kellogg. The Kellogg Company, as every American child knows, grew into an industrial giant.

The trick of making cereal was to separate the germ and bran from the starch. Indeed, the technique for processing corn flakes has changed little. First the yellow hybrid corn is milled: after the kernels are steamed and the germ and hull removed, they are cracked into grits. Then the grits are steamed again and flavored with sugar, salt, and malt, with vitamins and minerals added. After drying and further cooking, the grits are allowed to mellow for fuller flavor. Finally the grits are rolled into flakes and toasted.

If Sylvester Graham would have scorned Kellogg's method, what would he think of today's cereal manufacturing? S'Mores Crunch ("Crispy, chocolate-flavored graham cereal with bite-size marshmallows"), Honey Graham Chex, ("A Crisp, Sweetened, Rice Cereal with Honey and Graham Crackers"), Honey Graham Oh's ("Good Things In The Middle"), and a thousand other adulterations of his high-fiber wholewheat bread crowd the supermarket shelves. The irony is that in order to be accepted into the mainstream, Graham's ideas had to be modified. Americans have always tended toward extremism in diet and religion. Despite obvious differences, the parallels between his time and our own are many, and crusaders in our secular age share more with him than they might realize. The cult of whole-grain bread has come full circle.

Vilhjalmur Stefansson and the All-meat Diet

by Gillian Riley

"Unless it be religion, there is no field of human thought where sentiment and prejudice take the place of sound knowledge and logical thinking so completely as in dietetics."

Vilhjalmur Stefansson, 1921

A diet of nothing but meat and water is palatable and healthy and will sustain life over a considerable period of time. Vilhjalmur Stefansson proved this to his satisfaction during a twenty-four year period of exploration in the Arctic where he spent ten winters and thirteen summers, including an unbroken period of more than five years north of the Arctic Circle. From 1906 onwards Stefansson enjoyed the life-style and friendship of the "Stone Age" Eskimos of Alaska and Northern Canada, at a time when they were still relatively untouched by the benefits and disadvantages of Western civilisation. Later, in New York, he spent a year having his regime evaluated under laboratory conditions in what came to be known as the Cornell-Bellevue tests.

Clarence W. Lieb, the gastro-enterologist, a friend and collaborator of Stefansson on this experiment, described him as "an anthropologist by training, an arctic explorer by choice, who became a student of nutrition by necessity. Perhaps there is no other man living today whose experimental studies have been so well controlled and done on so large a scale. His laboratory was the Arctic Circle, his experimental subjects human beings and his experimental material, meat."

Stefansson loved the Arctic. He was born and brought up in an Icelandic community on the wide, rolling plains of Dakota and felt happy and at home in the wide open spaces of the Polar regions. Mountains, he felt, got in the way of the view.

He adapted happily to the Eskimo way of life as he found it in the first two decades of this century. Exasperated by the often-repeated question: "However did you endure the intense cold?" he would reply with pent-up irritation: "We did not endure it, we protected ourselves against it!" The Eskimo ways of doing this were elegant and comfortable. Layers of soft, beautifully made skin clothing kept out wet and cold and were light and decorative to wear. The traditional snow house was warm and dry, with sophisticated ventilation and heat control. The food, fat meat and fish, lightly cooked or raw, was varied and delicious.

Stefansson described how caribou meat was divided between a family and its dogs: "The children get the kidneys and the leg marrows nearest the hoof. All Eskimos known to me think the sweetest meat is nearest the bone; they boil the hams and round shoulder bones and the children pick from these the cooked lean that goes so pleasantly with the uncooked fat of the raw lower marrows. Perhaps the whole family and any visitors will share the boiled caribou head. The Eskimo likes the tongue well enough, and the brains; but what he prefers from the head is the jowl, and after that, the pads of fat behind the eyes. His next preference is brisket, then ribs, then pelvis. From the hams and shoulders he will peel off the outside meat as dog food, but will keep some of the inside meat for his family. The neck of the caribou is considered halfway between human food and dog food. Dog food, especially if the team is big, would consist first of the lungs, but not the windpipe or the bronchial tubes, for the Eskimo likes cartilage. The liver goes to the dogs and everything else from the body cavity except the kidneys. The heart is considered intermediate, not especially bad but not especially good. The dogs get the stomach and the entrails, but not the fat from the entrails. They also get the tenderloin and much of the meat from the backbone; but the Stone Age Eskimo likes to leave some meat on the vertebrae; he enjoys picking them when boiled.

"The four tidbits of northern meat eaters are all of them high-fat items and at their best must always be boiled but not overcooked. They are: beaver tail, caribou jowl, loche liver and moose nose. These are, in our farthest north, what the fan tail of their sheep was to the Old Testament Hebrews and still is to the tropical Arabs.

"When seals are eaten, heads are not the favoured parts and the dogs often get them. A seal's tongue, like the caribou's, is looked on as a food of intermediate quality, but inconsiderable because it is so small. Seal liver never goes to dogs, as the caribou's usually does, and is ordinarily eaten raw whenever it may be had frozen. Flippers never go to the dogs if there is time and fuel to boil them, but raw flippers are dog food. At a meal of boiled seal, the flippers are the tidbits, and after them the first joint of either pair of legs."

Fermented seal oil, prepared with meticulous care, was the equivalent of a cocktail or liqueur, or a carefully ripened cheese. "The colour was that of nut-brown ale and it had a slightly vinegary flavour and fragrance, to the connoisseur like the bouquet of a suitably aged wine". This oil was not drunk, but savoured in small amounts by dipping in a finger or a morsel of food.

A not inconsiderable advantage of the all-meat diet was the change in intestinal flora which produced odourless stools and a complete absence of flatulence. No farts, and relatively benign chamberpots. This must have improved the quality of life in the enclosed quarters of tent or snow house.

Although edible berries and roots can be found in the Arctic, nobody seemed much interested in eating them, and they did not, in Stefansson's time there, form a significant part of the diet. The delectable cloudberry was alright for little children, but not grown-up food.

The dogs, indispensable to an arctic explorer's mobility, were a vital part of the strategy of living off the land. Stefansson was not sentimental about his dogs. He treated them well and became fond of their foibles and characters. "Bones", for example, was a cheerful, willing beast who resolutely refused to move a paw when on short rations. Being a strong animal, he was kept in the team, but not included on expeditions where hard times might ensue. Game, set and match to Bones, I think.

The recycling of dogs was practised even by the humane and fastidious Scott. To let the poor creatures starve on short rations and then end up eating them, as a logical prelude to devouring one's human companions, seemed to Stefansson an unnecessary waste of resources. The plentiful supplies of seal and caribou he got on his hunting forays kept his animals, as well as his human comrades, robust and willing. He would even set his men to making boots for his dogs in conditions when the "needle ice" would have injured their paws. "It is a well-known fact that thirteen dogs have fifty-two feet," he wrote, "but I don't think anyone realises fully who has not had the task of making boots for dogs day after day." The boots were made of canvas. Sealskin ones would have been appreciated as snacks.

Stefansson applied what he had learned from the Copper and Mackenzie Delta Eskimos to his explorations. From 1913 to 1918 he was commander of the Canadian Arctic Expedition. His aspiration was to widen the extent of his discoveries by living off the land. Unencumbered by provisions his party of men and dogs could travel light, and fast, and further than ever before. He and his companions would eat the food they killed on the journey over the land and frozen polar sea beyond the arctic circle. On land they would find caribou and musk ox and polar bears. On the frozen sea would be seal. With sufficient rifles and ammunition they could get themselves a plentiful food supply and enough fat for cooking and heating. This was his theory. The problem was getting people to believe it. While he and his chosen companions were feasting off freshly-killed seal, lolling cheerfully in their shirt sleeves in the rugged warmth of their snow house, writing up their diaries and reading an article or two in the *Encyclopaedia Britannica*, their obituaries were being penned by the sceptical and unconverted.

Stefansson loved the rough and tumble of argument as much as he disliked the priggish certainties of vegetarians, dietitians, nutritionists and doctors. Their dogmatism about what we should and should not eat ran in the face of his own experience. When he got back to their world he was only too happy to set about routing their theories of a "balanced diet" and the Bellevue experiment was a pleasant way of doing this. He and a companion from one of his explorations, Karsten Andersen, lived for a year, under strict supervision and control, on fat meat, with no vegetables or fruit or salt or dairy produce or any of the things without which they would allegedly perish of deficiency diseases. The details of the Cornell-

Bellevue experiment, conducted at the Russell Sage Institute of Pathology in New York, were well documented in scientific journals of the time. Allowed out on parole, Stefansson lived the sedentary life of a New York businessman, working away at the writing, lecturing and publishing which financed his exploits, and surviving the oppressive New York summer and the flutterings of uncomprehending New York hostesses with his usual unfailing good humour. Both he and his companion spent the year in robust good health. Andersen got rid of his high blood pressure, and a colonic condition for which most authorities would have put him on a meat-free diet.

The experiment showed that, contrary to the received wisdom of nutritionists at that time, Stefansson and his colleague were fit and well, and suffered no deficiency symptoms, on a diet of fat meat and nothing else.

"Enough", said Oscar Wilde, "is as good as a feast. More than enough is even better than a feast." And there were days during the Canadian Arctic Expedition when Stefansson's advance group were obliged to sit and recuperate quietly from the abundance provided by the friendly arctic. And here, though they did not know it, was the answer to the disbelief and pessimism of his critics. Stefansson had established that he could lead an active, healthy life on a diet of fat meat. The vitamins without which it was alleged he would perish were somehow present in that diet. Since explorers suffering from scurvy could be cured by being given plentiful quantities of lightly cooked fat meat, it was assumed that vitamin C must be present in large quantities in this fat meat. To some extent it was, but not enough to have cured the scurvy single-handed. What Stefansson did not know then, but we know now, was that fat meat contains three essential fatty acids - linoleic acid (metyl linolate), linolenic acid and arachidonic acid. The delight in fat which sometimes made them poorly was the expression of a fundamental nutritional need. Stefansson and his companions and dogs flourished and prospered when they had enough fat meat. When times were hard and they had to get by on lean meat alone they became unwell with the symptoms of "fat starvation": diarrhoea, lassitude, and feelings of discomfort and misery which could have lead to death within a few weeks. As soon as they could eat even small amounts of oil or fat they got better fast.

Stefansson was right in claiming that he could be fit and healthy on a fat meat diet. He did not fully explain how, for he did not know. Having established that it could be done in some circumstances he made no claims that this was a desirable or appropriate diet for the modern city dweller. He agreed that it would be an uneconomical use of the world's resources, and had no wish to become as dogmatic and rigid as his critics.

Late in life, in his late seventies, Stefansson began to feel a little the worse for wear. Incommoded by the effects of a slight stroke and annoyed by pains in his joints he persuaded his wife to fill the freezer with fat mutton, his favourite meat, and embarked on a diet rich in eggs, bacon, cream, butter, chops and steak, and highballs. The expected improvements in health, "the decreasing weight, the lessened need for sleep, the growing energy, the near-euphoria which she (his wife) found near-obnoxious, especially before breakfast," were accompanied by the disappearance of all stiffness and aches and pains, and a "near-inability to feel gloomy and grouchy."

Perhaps the lesson for us in Stefansson's fat meat diet is to emulate his cheerful optimism and willingness to learn, rather than become embroiled in doctrines of prohibition and prescription, to avoid fads and dogmatism and remain open to advances in knowledge and understanding, like the discovery of the essential fatty acids which, unknown to Stefansson, was to vindicate the theory on which his achievements in Polar exploration were based.

A Medieval Staple.

Verjuice in France and England.

by Brenda S. Rose.

The staples of medieval cookery do not generally pose problems of identification. Some may be unfamiliar or little used but most of those basic to medieval cookery are still in common use and do not therefore require explanation. But there are a very small number of staples which are difficult to identify and which will require substitutes in modern adaptations of old recipes. One such staple is the liquid known as verjuice. That it is a staple is confirmed by the laconic way in which it is cited in the culinary section of the *Menagier de Paris*[1] written in about 1393. The author, an unknown bourgeois aged about fifty, says he is writing a treatise on cookery for his very young wife of fifteen. He is a man apt to explain everything unfamiliar and to describe in detail essential points about ingredients which his wife could not have known. Verjuice is merely noted in passing as an ingredient and certainly no basic recipe for it is given; the implication is that verjuice is an obvious store-cupboard staple and probably easily obtained commercially.

To the modern cook verjuice is unknown. Out of necessity he must find suitable substitutes for it if he is to reproduce as faithfully as possible the medieval recipes in which it is cited as an ingredient. It was when I turned to modern definitions and modern substitutes that I found it to be a subject requiring, unexpectedly, elucidation. As I shall suggest, modern writers on medieval cookery are by no means consistent in their treatment of verjuice. Definitions differ and cookery writers offer different, and sometimes surprising, alternatives. This paper is designed to bring some order. I propose first to look at verjuice itself, its actual constitution in a medieval context and in medieval commentaries. I propose then to turn to the way it is used in the recipes, hoping to discover how its function was perceived.

Etymologically verjuice could be defined simply as green (vert) juice (jus), but in order to obtain a full picture of its nature various French and English dictionaries[2] were consulted; all the French dictionaries state that the word was first recorded in the French language during the 13th century when it was cited in the *Roman de Renart*

> "Onc n'i ot savor de cuisine
> Verjus, sause, ne ail ne poivre"

Verjuice is defined primarily in all these dictionaries as

> "suc acide que l'on tire de raisins provenant de certains
> cépages ou de tout autre raisin cueilli vert"

An additional citation (1351) claims that verjuice was also the name of a

> "gros raisin qui mûrit imparfaitement dans nos climats,
> et qui sert à faire du verjus"

The word made its first recorded appearance in England at the beginning of the 14th century where it is cited in the historical rolls kept at Ely. The Oxford English Dictionary differs from its French counterparts in that it primarily defines verjuice as

> "the acid juice of green unripe grapes, crab apples,
> or other sour fruit expressed and formed into
> a liquor; formerly used in cooking, as a condiment,
> or for medicinal purposes."

The O.E.D. also makes reference to the figurative use in later centuries of the characteristic acidity of verjuice.

Verjuice as a staple was by no means limited to the cuisines of France and England. It was common to other European cuisines too. Elizabeth David writing on *Agresto*[3] (the Italian for verjuice) quotes the notes made by Castore Durante, a 16th century Italian doctor, in the herbal *Herbario Novo* published in 1585:

> "agresto is the juice of sour grapes especially obtained
> from the fruit of the vine called Agresto[4] on which
> the ripe fruit and flowers are sometimes seen simultaneously".

The origin of verjuice as a medieval staple remains obscure but it is possible that it was introduced into European cookery by the Crusaders returning from the Eastern Mediterranean, where they had developed a taste for Arabic dishes. The Baghdad Cookery Book of 1226[5] has a chapter on Sour Dishes which lists amongst its recipes those which demand the addition of the "juice of sour apples"[6] or that of "fresh sour grapes squeezed well in the hand"[7].

The constitution of verjuice, then, presents no major problems, since most culinary experts are prepared to accept dictionary definitions. It is only when one turns to these authorities for substitutes that this apparent simplicity becomes complicated.

Dorothy Hartley in *Food in England* recommends cider as a more authentic alternative than modern vinegar for those attempting to re-create the true flavour of the old recipes. She makes no reference to the verjuice of grapes. She considers the verjuice of crab apples to be "very best verjuice" and says that "verjuice was in constant use till the last century when its place was taken by a squeeze of lemon juice. It is really a sharp cider... not a vinegar..."[8]

Lorna Sass suggests, on the other hand in *To the King's Taste*, (a book of feasts and recipes in England at the time of Richard II), that modern cooks experiment with rose-hip syrup as a substitute. Verjuice is described in the glossary of this book as

> "the juice of green or unripened fruits such as
> grapes and (more commonly) crab apples;
> a popular ingredient in cookery which often
> replaced vinegar"[9].

"Jus de citron ou vinaigre de vin" are the two alternatives for

> "ce fond de sauce acide préparé à partir...
> de raisins verts d'où il tirait son nom"[10],

proposed by Jeanne Bourin in *Les Recettes de Mathilde Brunel*, a comprehensive collection of medieval recipes with modern adaptations.

A mixture of lemon and orange juice or cider vinegar used with discretion are suitable alternatives recommended by Professor Constance Hieatt in the culinary preliminaries to the Adapted Recipe Section of *An Ordinance of Pottage* where also Professor Hieatt gives the following account of verjuice

> "this was a tart fruit juice usually made from sour
> grapes. In southern Europe a special grape was
> cultivated for the purpose... Southern Europeans
> also used lemon juice and the juice of bitter (Seville)
> oranges and the British apparently made verjuice out
> of crab apples."[11]

The author of the MP indicates that there was certainly a special way of treating vines to ensure that grapes to give verjuice were available even at Christmas[12], and he confirms that orange juice[13] was used in French medieval cookery. Maître Chiquart in his culinary treatise *Du Fait de Cuisine*,[14] refers to "verjust de orenges" and "verjust de oyselle"[15] but these were not, in fact, variants, but rather side dishes of relish to be eaten as an accompaniment to other foods. They consisted basically of verjuice flavoured either with crushed oranges or pounded sorrel.

Most of the medieval cookery books provide little information on the various properties of verjuice, but this is not surprising since these books[16] were probably little more than aide-mémoires written by professional cooks for other professional cooks. The one exception is the *Menagier de Paris*. This treatise is designed to provide information of a very basic kind for the inexperienced or the ignorant and because of the greater detail that this demands we can often draw very useful information from a single recipe. A valuable case in point is the recipe given for *Lamproye*. After having advised using vinegar the writer goes on, in a useful aside, to suggest that

> "se le vinaigre est trop fort, si le
> actrempez de vin ou de vertjus"[17]

Now this has most useful connotations. The lamprey was an expensive and desirable fish. Great skill was needed not only in its initial preparation but also in the preparation of its accompanying sauce which was usually made from bread toasted until black, spices - a mixture of ginger, cinnamon, long pepper, cardamon, and nutmeg, the fish blood, and vinegar. From his comments we discover that not only was the author sensitive to the delicacy of this dish but more important, that even though we have no way of knowing just how strong medieval vinegar was, verjuice was seemingly seen to be a less pungent substitute. Verjuice however did vary in strength. In the chapter on *Saulses Boulyes* the writer of the MP asks the reader to note that

> "en juillet le vertjus vieil est bien
> foible et le vertjus nouvel est trop vert"

At that time of the year, he suggests,

> "le vertjus entremellé - moictié viel
> moictié nouvel - est le meilleur"[18]

New verjuice we are told reaches the height of perfection in January or February. There were, however, occasions when it was preferable to use the weaker, mature liquid. The author of the MP strongly advises its use in the recipe for *"Vertjus d'Ozeille"*, in the chapter on *Saulses Non Boulyes*,[19]. It seems the sauce was inclined to curdle, but this could be avoided if the verjuice used was "vieil blanc". This recipe also appears in a fifteenth century edition of the Viandier but here a crust of bread is added to prevent the mixture from turning[20]. To avoid this problem completely, Vertjus d'Ozeille could always be purchased ready-made from a professional sauce-cook.

Although basic verjuice was liquid, there was also a more solid variant, worthy of some clarification. It is described, variously, in French recipes as

> (a) vertjus de grain
> (b) verjus esgrené
> (c) grains de verjus
> (d) grappes de verjus[21]

Its true identity remains obscure, but it would appear to be a quantity of unripe grapes, loose or in a bunch, occasionally used in a raw state, but most frequently wrapped in a cloth and simmered gently in water "pour oster la premiere verdeur", indicating that only a measure of acidity was needed. This was then usually added to dishes just before or at the point of serving. For example, it is offered as a side-sauce with *Chaudun de Pourceau* roasted on the grill, or scattered on the top of each bowl if the same *Chaudun* is served as a stew, particularly in summer[22]. In this recipe and in one for *Rappé* common to both the MP and fourteenth and fifteenth century versions of the Viandier "groseilles" are listed as an acceptable alternative[23]. In either case the final overall effect would have been to add a tart fruit flavour to the finished dish.

To sum up then, from a careful reading of the texts, a comprehensive picture of the nature of verjuice has emerged which confirms dictionary definitions that it is a fruit-based, sour substance, presented most frequently in liquid form. But the dilemma facing the modern cook remains, since it is still not evident which of the suggested sustitutes most resembles this medieval staple. I propose, therefore, to turn to a study of those French and English medieval recipes which include verjuice in their list of ingredients[24], so that this may amongst other things help find the solution.

Taking the English recipes first, a principal fact to emerge is that verjuice appears in less than eight percent of them. This is rather surprising, and it seems counter to received opinion which is that verjuice was a popular, extensively used ingredient of English medieval cookery.

To judge from this small number of recipes, however, verjuice appears to have been used in three different ways. Firstly, as a seasoning; it was added at the end of cooking, to the cooking medium (water, wine, broth, or a combination of these), together with herbs and spices to form a covering or presentation sauce for the cooked dish. Interestingly, ginger seems to have been a frequent accompaniment. With the exception of no more than half a dozen of the recipes where spices are added along with verjuice, ginger is mentioned either specifically or as a component in a mixture of spices called "poudyr fort". Several recipes do state simply;

> "sesyn hit up with poudyr of ginger, a grete
> quantyte, and vergyes"[25].

Secondly, it was used as a cooking medium itself, especially for those meat or fish dishes encased in a pastry "coffin" or shell[26] and which required slow baking in the oven. Thirdly and finally, it was mixed with other ingredients (ground almonds, hard boiled egg yolk, herbs, and spices) to form side dishes of sauce[27] to be eaten with certain foods such as veal, capon, goose, bacon, trout, halibut, turbot, and mackerel.

There is then considerable consistency in the ways in which verjuice is added to dishes. Unfortunately, however, it is difficult to discern any pattern for its presence with particular foods or dishes. Some pointers are available. For example, it is named in some fish dishes, but by no means in all. It is used in recipes for pike, sole, gurnard, salmon, turbot, bream, skate, eel, mackerel, trout, and mussels, but equally there are recipes for these same fish where it is not present. The same too can be said for the meat dishes. Generally, verjuice seems to have been added to dishes of "light" meat such as chicken, capon, veal, kid, and rabbit, but the evidence is not conclusive since it is also included in the occasional dish of beef or mutton. But clearly, information is limited.

The English recipes tend to be short, often little more than precise lists of ingredients and cooking instructions are kept to a minimun, if given at all. Useful asides and comments (so helpful to the modern reader) are rare. We must therefore turn to the recipes of the French collections such as those in the MP, which are inclined to be more detailed and instructive. They reveal that verjuice was quite widely employed in French cookery during the Middle Ages; a good third of the recipes in the MP include it in their lists of ingredients. Almost all the recipes using verjuice which appear in fourteenth century versions of the Viandier reappear in the fifteenth century edition, indicating that culinary practice related to its presence in certain dishes had not changed drastically over a period of many years. There were, however, some apparent innovations, because several new dishes[28] not recorded in any other of the French collections under examination appear in this later edition of the Viandier, and verjuice is used in a few of them indicating that its use was also possibly being extended. Although verjuice is cited in many more French recipes, its uses are much the same as in the English ones. Again it is seemingly mostly associated with dishes of "light meat". Recipes for thick and thin stews of veal, pork, hen, kid, capon, and rabbit where verjuice is added to the cooking liquid (usually after the initial cooking process has been completed), are common throughout the various compilations. These same meats when roasted are served with verjuice to which may have been added any one of the following, or a mixture of several; garlic, onion, parsley, rosemary, sorrel, saffron, ginger, nutmeg, cloves, cardamon, mustard, mace, almonds, wine, and vinegar. This forms a side-dish of sauce to be offered separately and into which pieces of the cooked meat could be dipped. Verjuice employed in these various ways appears with such frequency in so many meat recipes in French cookery that it may be simpler to list some specific recipes where it is not included. For example, it was not apparently customary to eat certain roasted birds, in particular, peacock, swan, heron, cormorant, curlew, or pheasant, with any seasoning other than salt. Overall, almost all the white meat stews and broths had fish equivalents to which verjuice was added; these could be prepared for those days of the week when meat could not be eaten. Most of the fish recipes containing verjuice which are listed for the English cookery collections are recorded also in the French ones. In addition, there are others for barbel, carp, garfish, cuttle-fish, lamprey, and oysters.

With an examination of the recipes containing verjuice completed it seems permissible to say that verjuice was more popular in medieval France than in medieval England. The manner of its usage however, in the cookery of both countries shows a major similarity, namely that it was used most frequently as a seasoning ingredient in cooked and uncooked sauces, which in turn were presented as part of, or as an accompaniment to, dishes ready for the table. This being so it would be reasonable to assume that a prime function of verjuice was as an agent of flavour. But do the recipes themselves infer that this was in fact so? An aside in *Du Fait de Cuisine* gives one pointer, Maître Chiquart, aware that even at great banquets there will be one or two sickly guests requiring specially prepared foods, suggests the addition of a little verjuice to one such dish "pour lui donner aucun pou de goust"[29]. In another of his recipes, Chiquart adds verjuice to a broth to ensure it has "un bon goust". It is impossible to measure what "good flavour" meant in medieval terms, but there are definite indications that there was a penchant for a certain acid taste in some dishes at least. For example, Chiquart insists that in those dishes where sugar is present with verjuice, the quantity of sugar is never such that it abates the tartiness of the verjuice so that the resulting flavour is excessively sweet[30]. The author of the MP too recommends "ung petit de vertjus pour aguiser"[31] a capon stew. This liking for a certain piquancy is significant in that it emphasises, once again, the possible influence of Arabic cookery on that of the medieval West. Undoubtedly the Arabs were very fond of sharp flavours for they created many dishes seasoned with sour fruit juices and also with

> "vinegar that smarts the nostrils
> till they snuffle and they run"[32].

Though its use as a flavouring agent may have been of prime importance, it is possible that verjuice did perform other, additional functions. I should like to suggest the following hypotheses. The first relates to verjuice used as a marinade. In this context it would serve not only to impart flavour but also to soften the fibres of some meats and enable meat and fish to be kept for a longer period of time than is usually possible. Certainly veal, kid, and poultry, some of the meats most frequently associated with verjuice in medieval recipes are considered to be meats which are not easily digested and whose fibres need to be broken down in some way either by a tenderising agent or by the cooking process to render them more digestible. As for the matter of preservation, medieval ways of keeping cooked foods were limited and since it was customary for uneaten foods left over from one meal to re-appear repeatedly at subsequent ones it does not seem unreasonable to suggest that methods were sought which would help stay the natural process of putrifaction. Medieval cooks may have believed the addition of verjuice to be one such method. There are one or two indications that it did have some preservative qualities since it was used for pickling and in a non-culinary way for prolonging the life of cut flowers[33].

The second hypothesis relates to medieval medical practice. It is evident that chief cooks had to work closely with physicians who exerted their influence on the medieval diet for the *Forme of Cury* states in its introduction that it was

> "compiled of the chef Maister Cokes of Kyng Richard
> the Secunde Kyng of Englond aftir the Conquest...by
> assent and avyssement of Maisters of phisik and of
> philosophie that dwellid in his court"[34].

Briefly, since I can only touch on what is a vast subject, worthy of much more research than I can give time to here, medical advice[35], particularly that appertaining to dietary matters in medieval Europe was based on the teaching of the Ancient Greek physicians, especially Galen, about the cardinal humours and the importance of establishing an equilibrium between them for the maintainance of good health. Galenic doctrine also attributed humoural qualities to all foods. Each food had its own mixture of these qualities and was classified according to its degree of heat, cold, moisture or dryness. Master cooks had to be aware of the temperament of each foodstuff and know how to recognise which combination of ingredients in a dish would help maintain, rather then disturb, the humoural balance within the person who was to eat it. In addition, humoural qualities were also given to the seasons of the year and these too could affect the human temperament. Thus to maintain equilibrium in the body it was advisable in winter (which was considered to be cold and moist) to eat those dishes recognised as warmer and drier, whereas in summer (which was seen to be hot and dry) foods of a cooler temperament were to be preferred. It is these medical theories which may have determined the presence of verjuice in recipes. Verjuice is recognised to be cool and dry[36] and therefore may have been used to counteract the hotter, more moist nature of other ingredients in order to achieve the required balance of moderately warm and moderately moist, which was considered to be the ideal one for good health. It is perhaps significant that ginger, the spice most frequently cited with verjuice in recipes, is both warm and moist unlike most other spices which are classified generally as warm and dry. The fact also that some recipes do contain two sauces, one which is heavily spiced and is to be served only in winter, and another seasoned with verjuice to be eaten specifically in summer[37] does suggest that contemporary medical theory relating to the seasons may have had some influence on the culinary practice of the day.

Verjuice then was a store-cupboard staple with possible culinary and medical functions in continuous use in France and England during the Middle Ages. It continued to be used in subsequent centuries, though with diminishing regularity, until the nineteenth. Reasons for its decline are not obvious, but changing culinary tastes and traditions coupled with economic considerations affecting its availability may all have helped to effect its disappearance from the cookery books. We have seen various modern substitutes proposed in its place but which of these, if indeed any of them, comes closest to verjuice, not only in its nature, but also in its ability to perform the functions I have suggested for verjuice, remains a question that I, not being a professional cook do not feel able to answer.

Notes and References

1. *Le Menagier de Paris* c.1393 ed. G. E. Brereton and J. M. Ferrier. (Oxford: Clarendon Press 1981) (hereafter referred to as MP).
2. Godefroy, *Dictionnaire de l'Ancienne Langue Française*, Grand Robert; Littré; Grand Larousse; Oxford English Dictionary.
3. David, Elizabeth, 'Agresto'. *Petits Propos Culinaires* 7 (1981) p. 30.
4. I have not seen Durante's *Herbario Novo* but Elizabeth David writes "the entry in the book for Agresto is quite separate from that of Uva (grape)".

5. 'A Baghdad Cookery Book' (1226) trans. and ed. A. J. Arberry *Islamic Culture* 13 (1939) pp. 21-47; 189-214 (hereafter referred to as BCB).
6. BCB p. 37, see recipe headed *Tuffahiya*'.
7. BCB p, 37, see recipe headed *Hisrimiya*'.
8. Hartley, Dorothy, *Food in England* (London; Futura Books, 1985) p. 422. Since France and England shared a common cuisine in the Middle Ages the reason why the English preferred verjuice of crab-apples is curious. Perhaps it was a question of availability, though viticulture was practised here during this period and historical accounts do record the production of grape verjuice: see Wilson C. A. *Food and Drink in Britain* (London: Penguin 1984) p. 337.
9. Sass, Lorna, *To the King's Taste* (London: Murray, 1976) p. 132. The last part of this statement is somewhat puzzling. Medieval recipes do, generally, single out those occasions where verjuice is needed from those demanding vinegar. Only very occasionally is verjuice suggested as an alternative for vinegar. See Austin, T. ed. *Two Fifteenth Century Cookery Books*. (Early English Text Society: London, 1888); MS Harl 4016, pp. 72, 102, 106.
10. Bourin, Jeanne, *Les Recettes de Mathilde Brunel* (Paris: Flammarion, 1983) p. 10.
11. Hieatt, C. ed. *An Ordinance of Pottage*. Fifteenth Century Culinary Recipes in Yale University's MS Beinecke 163 with adaptations (London: Prospect Books,1988) p.117.
12. MP p. 271 No. 320. "se vous voulez avoir vertjus a Noël sur la treille, quant vous verrez que la grappe a son commencement se descouvera, et avant qu'elle soit en fleur, couppez la grappe par la queue et la iiie foiz laissiez la revenir jusques a Noël. Maistre Jehan de Hantecourt dit que l'en doit coupper le cep au dessoubz de la grappe et l'autre bourgun de dessoubz gecteroit grappe nouvelle".
13. MP p. 229. No. 156 "Item perdriz...reffaire...en eaue rose les trois pars, jus de pommes de orenges et le vin le quart."
14. *Du Fait de Cuisine*. compiled in 1420 by Maitre Chiquart Amizco, chief cook to the Duke of Savoy, ed. T. Scully *Vallesia* 40 (1985) pp. 103-231.
15. *Du Fait de Cuisine*. p. 164.
16. I refer to the following books written by and for cooks in noble households.
    (a) *Le Viandier*. ed. J. Pichon and G. Vicaire (Paris, 1892). This book was said, until recently, to have been written in the fourteenth century by Guillaume Tirel, also known as Taillevent, chief cook to the French King Charles V, but a recent editor, Paul Aebischer, in *Vallesia* 8 (1953) pp. 73-100 considers an early version may be dated to the thirteenth century and that the fourteenth and fifteenth century versions are *remaniements*.
    (b) The short anonymous *Traité de Cuisine écrit vers 1300*, printed in *Le Viandier* (ed. Pichon and Vicaire).
    (c) *Du Fait de Cuisine*, cited above.
    (d) *The Forme of Cury*, ed. C. B. Hieatt and S. Butler. *Curye on Inglysch*, E.E.T.S. (London 1985). This cookery book was compiled about 1390 by the chief cooks of King Richard II of England. An unedited manuscript, Crawford English manuscript 7, is housed in the John Rylands Library, University of Manchester, (see C. Hieatt 'Further Notes on The Forme of Cury et al: Additions and Corrections'. *Bulletin of JRUL* vol 70 No 1. Manchester (88) pp. 45-52.
17. MP p. 235 No. 185
18. MP p. 260 No. 279
19. MP p. 258 No. 270
20. *Le Viandier* p. 194.
21. (a) MP pp. 205, 214,219, 226, 283; *Le Viandier* pp. 80,85,153,185,225.
    (b) *Le Viandier* pp. 5, 10, 96.
    (c) MP p. 214; *Le Viandier* p. 177.
    (d) MP p. 263.
22. MP p. 214 no.97; *Le Viandier* pp. 5,10.
23. MP p. 219 no.113; *Le Viandier* pp. 10, 153. The juice of redcurrants is an ingredient included in

a recipe for 'Ribasiya' classified under 'Sour Dishes' in BCB p. 38 and indicates, once again the possible Arabic influence on French medieval cookery.

24. To specify, the following cookery collections were consulted, (details of the editions are in the notes above):
   (a) *Curye on Inglysch* (including the 'Forme of Cury' and other manuscripts).
   (b) *An Ordinance of Pottage*.
   (c) *Forme of Cury* (John Rylands manuscript).
   (d) *Two Fifteenth Century Cookery Books*.
   (e) *Traité de Cuisine écrit vers 1300*.
   (d) *Le Viandier* (ed. Pichon and Vicaire).
   (f) *Le Menagier de Paris*.
   (g) *Du Fait de Cuisine*. This collection is different from the others in that it concerns itself primarily with the details and preparation of those dishes a cook in a royal or aristocratic household may have been called upon to produce for his master's honoured guests at very special banquets.

25. *An Ordinance of Pottage* pp. 47, 48, 50, 51, 101. It seems a function of ginger was to abate the bitterness of grape verjuice. (cf. *Curye on Inglysch* p. 49. no. 31).

26. *Two Fifteenth Century Cookery Books*, p. 74; J. Rylands MS; *Forme of Cury*; 'Crustardes of Flessh', *Curye on Inglysch* p. 133. no. 161; p. 134 no. 164.

27. *Two Fifteenth Century Cookery Books* pp. 109, 110.

28. *Le Viandier* pp.152-156:160-169:207-208.

29. *Du Fait de Cuisine*. p.196. no.76.

30. Ibid p. 149. no. 12.

31. MP p. 207. no. 73.

32. BCB p. 22.

33. MP p. 272. no. 325.

34. *Curye on Inglysch* p.20.

35. All the information given here on medieval medical theories and their influence on culinary practice is heavily indebted to Terence Scully's masterly and very comprehensive introduction in *'On Cookery'* the English translation of his edition of *Du Fait de Cuisine*.

36. *On Cookery* p. xxiii note 16.

37. MP p. 214. no. 97; p. 238. no. 199.

Wheat, Staple Food for the Dead

By Rena Salaman

It is not only the living that require nourishment; the dead have been getting their fair share throughout the ages.

Food for the soul? Food for the dead? God's food? Or a combination of all three?

Grain of one kind or another has mostly been the staple on which communities relied throughout the ages. Barley - *krithari* - was the most popular grain in ancient Greece until around 6,000 BC when wheat made its appearance, most probably from the Near East and became a popular favourite. There is historical evidence that the Greeks had a special aptitude and preference for bread. (They still have a preference but not the aptitude any more.) Originally this was made with flour from barley mixed with wine or water and olive oil into a kind of dried biscuit which was called *maza*. Aristophanes often mentions it in the *Frogs* and also compares *maza* to an excited crowd in his *Peace*. In modern Greece we still use the same term in a slightly contemptuous tone to describe an amorphous and diverse crowd. And even after wheat flour had arrived on the scene and bread made from it was for sale in the special bakeries - *artopia* - barley bread or *maza*, which was much cheaper, was also readily available and its sales were much higher than those of ordinary bread.

Bread was given different names either according to its shape or the way it was cooked. For instance *Esharitis* was a thin flat bread that was cooked on a grid; *Krivanitis* on the other hand was the bread that was cooked in a *Krivanos* - which was a kind of primitive oven either built from mud or metal and which gave the bread a particularly attractive and identifiable flavour. Aristophanes praises explicitly this kind of bread. *Apopyrias* was bread cooked over an open fire as the word suggests.

According to shape there was a square loaf which was called *Blomiaios*, *Imiartion* which was a half-moon shape, *Plakitis* which was a small flat loaf, *Mystili* which was the full hearted pregnant round loaf, etc.

Needless to say there also existed the equivalent of our wholemeal or granary which was called *Aftopyros* or *Aftopyritis* which in real terms had a high fibre content, a term much used nowadays. Its opposite was a pure flour content bread called *Semidalitis*. Bread from Attica was much superior by any standards.

Coarsely ground wheat was also boiled with milk and made into *Tragos* - modern day *Trahanas* - which once cooked and dried is kept as a winter staple. All these delicacies though were meant for the living.

But wheat did not only provide nourishment for the living. The dead were also to benefit from it.

Food for the soul? Food for the dead? God's food? Or a combination of all three? Food has always been associated with death as it has been with life. The link although elusive at first with a little probing becomes quite tangible.

People were buried with their favourite foods which quite often was a sample of whatever counted as staple food for the living. Alternatively offerings for the dead consisted of the most familiar type of ingredients. In ancient Greece, according to Aristophanes' *Acharniai*, they cooked a special dish called *Panspermia* which as the word implies was a selection of grains and seeds. This was offered to the dead as well as the living members of the family. Moreover it was offered to the god of the Underworld Hades. A similar dish called *Kolyva* is specially made for the dead in modern Greece. This, although primarily made of wheat, contains all kinds of grains and nuts like its ancient equivalent perhaps. Although historical evidence about the ingredients of *Panspermia* is not explicit enough, the temptation to draw the continuity link is too great to resist. (So, I do it with caution and without passion and I am open to suggestions.)

What is the logic that justifies a staple food for the living and the dead alike though?

Is it because the staple - in this case wheat - is associated with the root of life? The point where everything begins but also everything ends? Is it because what comes from the earth - wheat - is destined for the the earth in a transcendental unison with the human body? Or is it an eternal circle where wheat transcends into another sphere and becomes a symbolic ingredient nourishing the soul? Or simply because an ingredient of such importance to life is regarded as the most sacred commodity which could be offered to the dead with the highest reverence attached to it?

When I got interested and wanted to know more about *Kolyva* I rang up my mother in Athens and she having assembled all her friends for a session, sent me a long and laborious recipe, the result of their combined wisdom. The recipe and its instructions make suitable reading for at least a piece out of *The Golden Bough* for its pagan and ritualistic element. For instance the impurities found in the wheat are not to be thrown into any ordinary dustbin. They are to be kept and thrown into the sea. Considering that the sea was not further away than ten minutes from our house this does not sound too eccentric. Water from rinsing and cooking the wheat is to be used for watering plants and thus providing further life rather than disposed of irreverently down a sink. While preparing it one has to prepare oneself spiritually and light at least one candle in the proximity of the table.

Kolyva is an extremely pleasing dish both in flavour and appearance - a kind of *muesli* - with different textures from all the nuts - flaked almonds, walnuts, sugared almonds, sesame seeds against the soothing smoothness of the boiled wheat. Different colours - the scarlet of the pomegranate seeds, the green of the parsley, the gold of the wheat and the nuts and different flavours and aromas of the spices - ground cinnamon, coriander and cummin. We used to love it as children. The addition of juicy crunchy pomegranate seeds made it almost into a festive dish for us and then there were the hot spices interspersed with the whole green leaves of fresh parsley which added an altogether different dimension. I will never forget the first time I tried the Indian dish of *Lamb Passanda* which contains a great quantity of ground coriander; I got a strong *déjà vu* feeling and *Kolyva* floated into my mind because this is where as a child I had encountered the aroma and flavour of ground coriander for the first time.

Kolyva is prepared for a funeral where it is distributed in handfuls to the participants at the end of the service. Forty days after a death it is prepared and distributed to the neighbourhood houses and the local church. But it is also prepared at prescribed times in the year. The equivalent of the Greek Orthodox All Souls day - *Ton Psihon* - which occurs during three consecutive Saturdays in the spring before and during Lent. And it is prepared yearly to commemorate somebody's passing away.

It is prepare on Fridays and the finished dish is taken away to the local church on Friday evening to remain overnight and to be blessed in a special service by the priest the following morning. A spare plate is prepared at the same time and distributed to at least three, six or nine households in the neighbourhood in order to remember the dead of the family and to contribute peace to their souls. Multiples of three are also important in the recipe, an allegory of the Holy Trinity. For instance the wheat should be rinsed three times.

Kolyva that is prepared on the first Friday of Lent is not sweetened with either sugar or nuts and has a very special significance for young girls. They attend the Saturday morning service and after the *Kolyva* has been blessed it is distributed to them by the priest. That night unmarried girls place three grains of *Kolyva* wheat under their pillow and according to popular belief they will then dream of the man that is to become their husband.

In this way *Kolyva* or its main ingredient wheat perpetuates life in an infinite cycle. Staple for the living, the dead and the dreaming.

Kolyva

Ingredients

    500g (1lb) whole wheat
    3 pinches of salt
    120g (4oz) small currants
    120g (4oz) raisins
    120g (4oz) coarsely chopped walnuts
    150g (5oz) shelled and flaked almonds
    1 pomegranate, shelled and seeded
    1tsp ground cinnamon
    1tsp ground coriander
    1tsp ground cummin
    2tbs leaves of fresh flat-leaved parsley

Topping

    500g (1lb) plain flour
    500g (1lb) caster sugar
    10 sugared almonds
    Handful of raisins

Method

Pick any impurities from the wheat. *Repeat three times.* (A touch of *The Golden Bough* here.) Keep any impurities and dispose of in the sea. Rinse wheat *three times* and dispose of water by watering a tree or plant. Soak wheat overnight. (With modern varieties this is not necessary.) Strain, cover with fresh water, add *three* pinches of salt and boil until tender. Strain, keeping the water for the plants as previously. Cover a table with lots of newspaper (I suppose it could be a sheet of plastic) and a clean tablecloth on top. Spread the wheat and let it dry for 5-6 hours or overnight.

Before the assembly starts you have to prepare yourself *spiritually*. For this purpose you light a candle and think of the dead that the Kolyva is to commemorate, recounting their names.

Mix the dry wheat with remaining ingredients gently, keeping some pomegranate seeds for decoration. Spread on one or two platters or flat trays evenly. Roast flour dry in a frying pan on gentle heat, turning continuously with a spatula until light golden and let it cool. Spread a thin layer of flour all over the surface of the wheat, pressing down lightly. This will absorb any remaining moisture. Spread a thin layer of sugar on top (using as much sugar or as little as you like) and press it down with a piece of paper so that it becomes quite solid.

Now comes the important task of decorating. Anything goes particularly crosses and flowers. Use sugared almonds and raisins, pomegranate seeds and/or ground cinnamon to make attractive patterns. Some people use silvered almonds or slivered peanuts but that is pure bad taste.

Ancient Vegetarianism -
Staple Foods and Customs in Azerbaijan

by Dr. Emil Salmanov, edited by Robert Chenciner

Caucasian cooking is more stern and simple in the highlands and more varied and tasty in the plains - especially in Azerbaijan. Azerbaijan, situated on the crossroads between the Caucasus and Iran, accumulated cooking and ingredients from both regions as well as from Turkic nomads.

To understand authentic Azerbaijan staple foods, it is necessary to study both the history and folklore of cooking local produce. The delicacies of kings differed from staple foods eaten by normal people, such as agricultural products and meat from domesticated animals in contrast to the hunt.

History

A brief look at Azerbaijan cooking shows that the national diet was linked to the development of civilisation. We start in the Nakhtchiwan region where the Azykh cave culture (250-300,000 BC)[1] has preserved the lower jaw of a humanoid - electron microscopy will show if he was a vegetarian (like *Sivapitecus indicus* from Potwar in Pakistan[2]). This is also likely because of current dietary habits. Firstly, contemporary meat dishes are always accompanied by fresh vegetables - not by a separate salad course. Secondly, some dishes have semi vegetal-meat forms - such as rice PILAW with MUSAMMA seasoning of meat and egg-plants. Thirdly, an ancient health reason - fruits eaten between the courses provided glucose, giving psychological and physiological satisfaction.

The 1st Period of staple food in Azerbaijan dates roughly from prehistory to the 2nd millennium BC, shifting from a vegetarian to milk-vegetarian diet. The 2nd Period, until 600 AD, diet was milk food supplemented by vegetables and meat from sacrificed animals. The 3rd Period, until today, diet is a mixture of meat, milk and vegetables, differing from the city to the country. Each period has its different dietary customs.

1st Period - vegetarian

The 1st Period resulted from the geography of Azerbaijan - a vast irrigated area between the Kura and Araks rivers, producing a large yield of cereals. Proof of the cultivation of cereals then is the aboriginal spelt, PARINDJ *(Triticum araraticum)*, from the highlands of several regions[3], which will only interbreed with another Caucasian aboriginal wheat from western Georgia - *Triticum timopheevi*[4]. Spelt was the preferred grain for PILAW before the introduction of rice to Azerbaijan, and continues to be used until today in northern Azerbaijan and southern Dagestan.

Archaeological excavations in Nakhtchiwan region revealed four cultivated wheats from the Neolitholic age (4th millennium BC): mild pellicle wheat *(T savitum L)*, hard wheat *(T compactum Dest)*, dwarf wheat *(T compactum Hest)* and spherical grain wheat *(T sterio cocum)*. There were also two barleys: *Hordeum savitum* and *H lagunculiforme Bacht*[5], and a millet, of which the grains were depicted on the handle of a jug found in a tomb of the 2nd millenium BC at Hodjaly[6]. The diversity of cereals together with the survival of several contemporary flour dishes made by ancient simple cooking techniques, supports the argument for a vegetarian diet.

The earliest flour meal is KHASHIL - a thin mash of flour and butter, seasoned with DOSHAB or BAKMAZ - an extract boiled from mulberry or grape juice. A variant is HORRA - a flour and milk mash without seasoning. After the introduction of rice during the Middle Ages YAYMA - a rice and milk porridge - entered the diet of the settled population.

UMADJ - a more sophisticated flour soup, represents the next level of mash technology. Flour, splashed with water is rolled into small balls between the palms. The balls are dropped into boiling water, which after adding sheep fat or butter is eaten as soup. The dry variant, GUYMAG - flour fried in butter then mixed with a little water into a porridge, seasoned with yellow ginger, saffron and salt - was eaten as an everyday meal. When made with rice it is called FIRNI. The rice is moistened in water for 2-3 hours, then dried and pounded into flour. The porridge is seasoned with salt, sugar and cinnamon. Later on when rice starch was separated, SUDJUK - a starchy jelly - was similarly prepared, seasoned with butter, salt, sugar, saffron and sometimes walnut.

The dumpling was a further development of the flour dishes. ARISHTA ISTISI are thin noodles, eaten with onions, fried in fat or butter - or with a sauce of sour milk and garlic, diluted with water. KHANGAL (which we described as Caucasian KHINGAL in OFS 1987) is a dough cut in small squares or rhombs and boiled in water. In Azerbaijan there are two types: GASHYG (spoon) KHANGAL - a soup, seasoned with onion fried in fat or butter, and AL (hand) KHANGAL - sometimes seasoned with force - meat and sour milk with garlic.

Ritual Dishes

There are some ritual meals with cereal dishes. GUYMAG, described above, and PISHI - a cookie made from wheat flour mixed with butter and milk, with a sweet filling, which are usually given to a woman two or three days after confinement. The sweet and spicy taste is different from ordinary meals, in honour of the importance (PISHI Pers) of the woman's condition.

FASALI - a cookie made from puff leavened pasta, fried in butter, looking like a pancake, was offered to the PIR or ODJAG spirits residing in sacred places such as a great stone, rock, tree, spring or holy ground. This gift symbolised a sacrifice of wheat-meal grown by the sun - so it took the form of a sun disk. Old wooden stamps with the solar design were also used to print on bread in Azerbaijan. The sun was the symbol of the god Mithra - a personification of sun, light and heaven. In Azerbaijan archeologists have found this symbol from 2nd millennium BC. FASALI is baked on a SADJ - an iron disk - which demands great skill because FASALI is very buttery and will not adhere to the inside of the TANDIR - a conical stove with an open fire inside. So a dexterous woman is called FASALI TANDIRA YAPAR - she who bakes FASALI in a TANDIR.

Other breads baked by the KOZDAMA[7] heating process were YUKHA - a thin bread, LAWASH - a paper-thin bread, both from unleavened dough, and SADJ TCHORAYI - baked on a SADJ over a fire in contrast to the TANDIR TCHORAYI - baked inside a TANDIR by strong radiation.

Around 1000 BC this cereal diet was combined with vegetables and fruits, both dried and fresh, to form the staple diet. In the HAFT SIN custom these vital foodstuffs were set out on the table during the most ancient holiday in Azerbaijan and Iran - NOW-RUZ BAYRAM, marking the start of the new year about 21st March, during the month of FARWARDIN (FRAVASHS - the spirits of nature), in the old Iranian solar calendar. The festival marks the end of winter and the start of spring, so the tablecloth was spread with seven staple plants beginning with the letter 'S,' which had been kept through the winter - SOFRE (or KHANTCHE) E HAFT SIN: SEPAND - ruta, SIB - apple, SIYAH DANE - black seeds, SENDJED - djidda berry, SAMANU - malt halva, SOMAGH - sumakh seeds, SIR - garlic. Each staple makes a symbolic contribution to the HAFT SIN festival.

1 SEPAND or in Azeri ADRASPAN *(Harmala ruta)* is the plant ruta, anciently used in fumigation for ritual purification of objects or houses before NOW-RUZ, (both by cleaning and throwing out worn or broken things, like European spring cleaning). In Iran today children are fumigated by ruta against the devil's eye. In Azerbaijan there is a kind of very fragrant ruta called SADO in Azeri *(Ruta gravelens)*, which is very poisonous and it is precisely this which explains its power against evil.

2   SIB, the apple was an important ancient vegetal staple, full of vitamins, and some varieties could be kept through the winter. Dried milled apple oat flour was added to milk or eaten separately[8]. This must have been widespread because it already had a name - AMARNA PISTA - in ancient Iranian[9]. The traditional eastern saying "Apples eaten at supper give a calm light sleep and slight diarrhoea" is similar to "An apple a day keeps the doctor away". In Azerbaijan an apple called ALMA in Azeri, grows everywhere, but the most famous and delicious are the QUBA apples, grown in the north of the country in the Quba region.

3   SIYAH DANE are buttercup *(Ranunculi generis)* seeds. SIYAH DANE - black seeds - are the seeds of a buttercup of the Ranunculus family *(Nigella orientalis)*, which grows on the grassy lower slopes of the small Caucasian chain[10]. The small black seeds, called in Azeri GARATCHORAK - black bread - were used as a spice or also sprinkled on to bread before baking.

4   SENDJED, a djidda berry *(Elaeagnus orientalis, Zizyphus generis)* is called in Azeri IYDA, UNNAB or UNNABI. Its flour is like pulp, tasting of sweet bread. It is high in calories with over 60% saccharose. This dry berry was popular on the caravan roads as it was dry, sweet, nourishing, durable, portable and lightweight. As the Pandjir (Afghan) riddle asks "what is a small red skin full of flour?"[11]

5   SAMANU, in Azeri SAMANI, a malt halva, prepared from moistened mature grains of wheat. It is thought that a vital force is released at the moment the grains start to grow, when the sprouts become yellow[12]. The SATTVA - energy in Sanskrit - is activated and when eaten transfers its force to the body. The grains are then crushed and sieved into a malt. Wheat flour dough is mixed with the malt and butter and cooked on a frying pan on a low fire for several hours. It is then mixed with spice and formed into small round loafs or balls - SAMANU. SAMANI is a simple soup made from the malt and wheat flour. So SAMANU has a symbolic meaning related to the annual renewal of nature. The Azerbaijan proverb "SAMANI, SAMANI HAR IL GOYARDARAM SANI" means "SAMANI, SAMANI, I try to make you grow every year".

6   SOMAGH (SUMAKH in Azeri) seeds, have been used as spice, tannin and dye from ancient times. Tanning sumakh *(Rhus coriaria)* grows on the lower dry rocky slopes. Milled sumakh is now used as spice for soups and meat - especially grilled kebab. Its deep purple colour gave its name to all similar colours in antiquity.

7   SIR or SARYMSAG in Azeri, a garlic *(Allium sativum)* was used from ancient times as a herb and universal medicine - also recommended by Ibn Sina (11th century). Garlic can be kept over winter as a natural vessel full of nitrous compounds, mineral salts, vitamins B & C and so on. So it was popular in both cooking and folk medicine, especially in mountain regions, where the diet was poorer.

On the NOW-RUZ table there are two other staples beginning with 'S' - SERKE and SABZI.

SERKE (SIRKA in Azeri) was a vinegar used in many vegetarian dishes. Most choice being SERKEBA - a vinegar soup and SERKESHIRE - a drink of a mixture of vinegar and grapejuice. Vinegar was also a universal remedy in folk medecine.

SABZI is a salad made from the first vegetables of spring: NA'NA - mint (Mentha), TARRE (TARHUN in Azeri) - leek *(Allium porrum)*, TOROBTCHE (GYRMYZY TURP in Azeri) - garden radish *(Raphanus sativus)*. SABZI was a later addition to the table after the cultivation of kitchen gardens.

Although fish and meat are now also found on the NOW-RUZ table, the observation of the HAFT SIN custom confirms the identity of the staple foods of the Indo-aryan diet adopted by Turkic tribes. Today too, in Azerbaijan, some HAFT SIN staples are substituted, but in the centre of the table there is always a saucer with green sprouting wheat grains.

2nd Period - Milk

Ceramic churns, dated from the 2nd millennium BC, are evidence of the new milk age in staple foods, when butter and other dairy products were introduced into the diet. Vegetal oil from FYSTYG - nuts from a beech, *Fagus orientalis* - was introduced after 1000 BC[13]. Although bones of domesticated animals - goats, cattle and sheep - show they were eaten, their main use was to produce milk. There appears to be evidence that about this time domestic animals were sacrificed and their meat eaten, from the Iranian epic poem by Firdousi, written during the 11th century, but based on earlier sources. The story symbolised the beginning of the meat eating era:

There was a struggle between the chief of the Iranian tribes, Feridun, and the chief of the Semitic tribes, Zohhak, who wished to seize his throne and sacrifice and eat his sacred cow Bermaye, whose milk had nourished Feridun. So one of Feridun's giant heroes made an iron mace in the shape of a cow's head, foretelling that he would smite Zohhak with it[14].

In antiquity only bulls were sacrificed and even today Caucasian mountaineers prefer to eat dried bulls' meat. The Zoroastrian holy book *Avesta* written before 700 BC, comments, and so provides evidence, on people starting to eat bull's meat:

> So to the priest the bull tells:
> You will be without heirs,
> You will be the target of evil,
> If you give to the people,
> To your wife and your sons
> Me tastily prepared.
>
> (*Avesta*, Y. 11, 1-2 tr Salmanov)[15]

During the 2nd millennium BC, when domesticated animals formed herds and flocks, the most simple dairy product was natural sour clotted milk - SHEYTAN GATYGHY in Azeri, literally 'devil's sour milk'. Today Azeris prefer GATYGH, real sour milk, fermented by PENDIR MAYASY - a rennet for cheese. Preparing rennet was a magic secret, kept by every family. An Azerbaijan lactophagic tribe, the Airums, demanded complete physical and spiritual purity of the rennet maker - either a post menopausal woman, or a pre-pubescent girl or a woman who was not menstruating or had made ritual ablution after intercourse or had completed 40 days parturition and been purified by water boiled with 40 small stones from a crystal clear spring[16].

Some of the ingredients of rennet had magical purposes: a dry rennet bag, sour dough, 5-10 rice grains, a little wheat boiled in water, a piece of sugar and alum. The egg-sized sour dough showed that in the past fermentation was possible without the rennet bag made from milk-fed calf's or lamb's intestines. So the transition from cereal to milk diet could have happened without regular meat eating. By the way, the old men of the Airums said that there is a mountain herb - DALAMA OTU - which causes fermentation souring warm fresh cow's milk[17].

There are two sorts of AGHARTY, Azeri for a white meal, a milk product made by separation or fermentation. The former are: SUD - pure milk of cow, sheep, goat and buffalo, GAYMAG - cream of cow and buffalo, KHAMA - sour cream of cow, KARA - butter. The latter are: GATYGH - sour milk, SUZMA - sour milk strained in a canvas bag, AYRAN - sour whey, KASMIK - curds, SHOR - cottage cheese, PENDIR - cheese, NOR - a whey byproduct left after cheese fermentation, boiled and strained in a canvas bag, GURUD - a dry cheese from sour milk, first strained then dried in small balls, YAGH-SHOR - a mixture of butter and SHOR, AKHTARMA - from a dense, cheese-like, fermented milk called DALAMA, mixed with its own secreted whey, then salted and kept for 7-10 days in a MOTAL skin, when it is eaten with bread.

Three other local dishes can be included in the AGHARTY group.

1　　SULAKH - a special dish with a cheesy taste, made of beestings - milk of newly calved cow or sheep - put into the placenta and grilled in hot ashes and coal.

2　　BULAMA - boiled mixture of normal milk and beestings.

3　　AGSOLAKH or AKHSAGULAKH - mixed boiled butter and sour milk, seasoned with garlic, diluted with water.

There are two Azerbaijan cheeses: TAZA PENDIR, kept in brine and the extraordinary MOTAL PENDIRI, made of fermented sour milk, strained through canvas or course calico under pressure of heavy stones. Next pieces of raw cheese are salted and put into the MOTAL, an inside out sheepskin (wool on the inside). After 6 months it turns into a cheese like amber butter. There is an earthenware pot in the form of a filled MOTAL hanging between two trees, from about 100 BC in the Museum of History of Azerbaijan, Baku, which gives a date before which the cheese must have been invented. Only someone who has eaten this cheese with bread and the crystal clear water of a spring sitting under a plane tree after coming a long way on foot, can appreciate this frugal and delicious country food.

3rd Period - Meat

The Arab invasion of Azerbaijan during the 7th century brought the nomad culture of the east, with a diet of sheep and camel, recalled by the saying: The three best things on earth are to sit on meat (ride), to eat meat and to put meat into meat (make love). The Mongol attacks destroyed the old irrigation system, changing agricultural land into pastures. Later Turkic tribes invaded and became assimilated, introducing Central Asian nomad culture. Eating mutton became the symbol of eating well. According to ADAT, local customary Muslim law, only mutton and lamb as the purest animals, were permitted meats. Recent research suggests that eating meat was an important stimulus for the development of the human brain, providing the necessary zinc trace element[18]. On the other hand, for instance, the Indus civilisation was developed by vegetarians.

In Azerbaijan sheep meat is cooked in two ways: KABAB - pieces of mutton roasted on a spit, and GOWURMA - mutton pieces fried in their own grease with melted fat tail of sheep. Cooled and stored in the prepared stomach of a sheep, it could be kept for a very long time.

Common people could not eat meat every day because families' flocks were not large enough. So the words GOYUN KASMAK - to slaughter a sheep means a sacrifice connected with a festive occasion, usually QURBAN BAYRAMY, the holiday of sacrifice (ID AL-ADHA in Arabic), marking the end of the pilgrimage to Mecca in memory of Ibrahim's (Abraham) willingness to sacrifice his son Ismail to Allah, who sent a ram in his place. The great holiday lasts for 3-4 days with entertainments, feasting, visits to relatives and friends and giving gifts. By Muslim tradition, with the help of offerings of sacrificed sheep a deceased Muslim can easily cross the hair-thin bridge SIRAT which leads to paradise. In Azerbaijan during the holiday sheep were also slaughtered in holy places - PIRS, which were usually only connected with pre-Islamic pagan beliefs in the power of stones, trees, rivers and mountains. Sometimes too sheep are sacrificed in return for a promise, asking deities to grant a large harvest or herd, to help recovery from or to have a pregnancy.

Notes

1 Gadzhiyev D B, Guseinov M M. Pervaya dlya SSSR nakhoda ashel'skova cheloveka. Yuyileinii sbornik *Ucheniye zapiski Azgosmedinstituta*, t XXXI, Baku, 1970.
2 Pilbeam D, Hominoid evaluation: Harvard program & field research in Pakistan, *Symbol* (Fall): 15.
3 Mustafezhev I D, *Azerbaijanin bugda kolleski asi*. B, 1962 - in the highlands of Latchin, Kalbadjar, Leric, Yardymly, Ismailly, Nakhichiwan and Karabakh regions.
4 Zhukovskii P M, *Kul'turiye rassteniya i ikh sorodichi*, Leningrad, 1971. p 27.
5 Abdullayev O A. *Zneolit i bronza territorii Nakhichevanskoi ASSR*, B, 1982, p 211.
6 Minkevich-Mustafayeva N V, 'Pamyatniki trekh osnovnikh grupp KHodzhali-Kedabekskoi kul'turi na territorii Azerbaijanin SSR i ikh datirovka'. v sb. *Material'naya kul'tura Azerbaijanin*. v. 4, B, 1962.
7 Lifshifts V A, 'Zoroastriiskii kalendar', v kn. Bikerman E. *KHronologiya Drevnevo mira*. Moscow, 1975, p 325.
8 Baranov P, Raikova I. *Darvaz i evo kul'turanaya rastitelnost*. Tashkent, 1928, p 78.
9 Morgenstierne G. 'Etymological vocabulary of the Shunghi group' - *Beitrage zur Iranstik*. B. 6. Wiesbaden, 1974, p 44.
10 *Flora Azerbaijanin IV* - Platanaceae. B, 1953.
11 Andreyev M S. 'Po etnologii Afganistana'. *Dolina Pandzhir*. Tashkent, 1927, p 87.
12 *Azerbaijanin kulinarizhasi*. B, 1986, p 172.
13 Beruni, Abu Reikhan. 'Farmakolnoziya v meditsine'. *Izbranniye proizvedeniya*. t 4. Tashkent, 1974.
14 Firdousi. *Shakhname*. t I. M, 1957, p 55.
15 Salmanov E. 'Ateshperestlizhin ile'. *Gobustan*, 1985, #I, p 29.
16 Karakashli K T. *Material'naya kul'tura Azerbaijanin*., B, 1964, p 240.
17 ibid, p 242.
18 *Nauka i zhizn'*. M, 1988, #9, p 120.

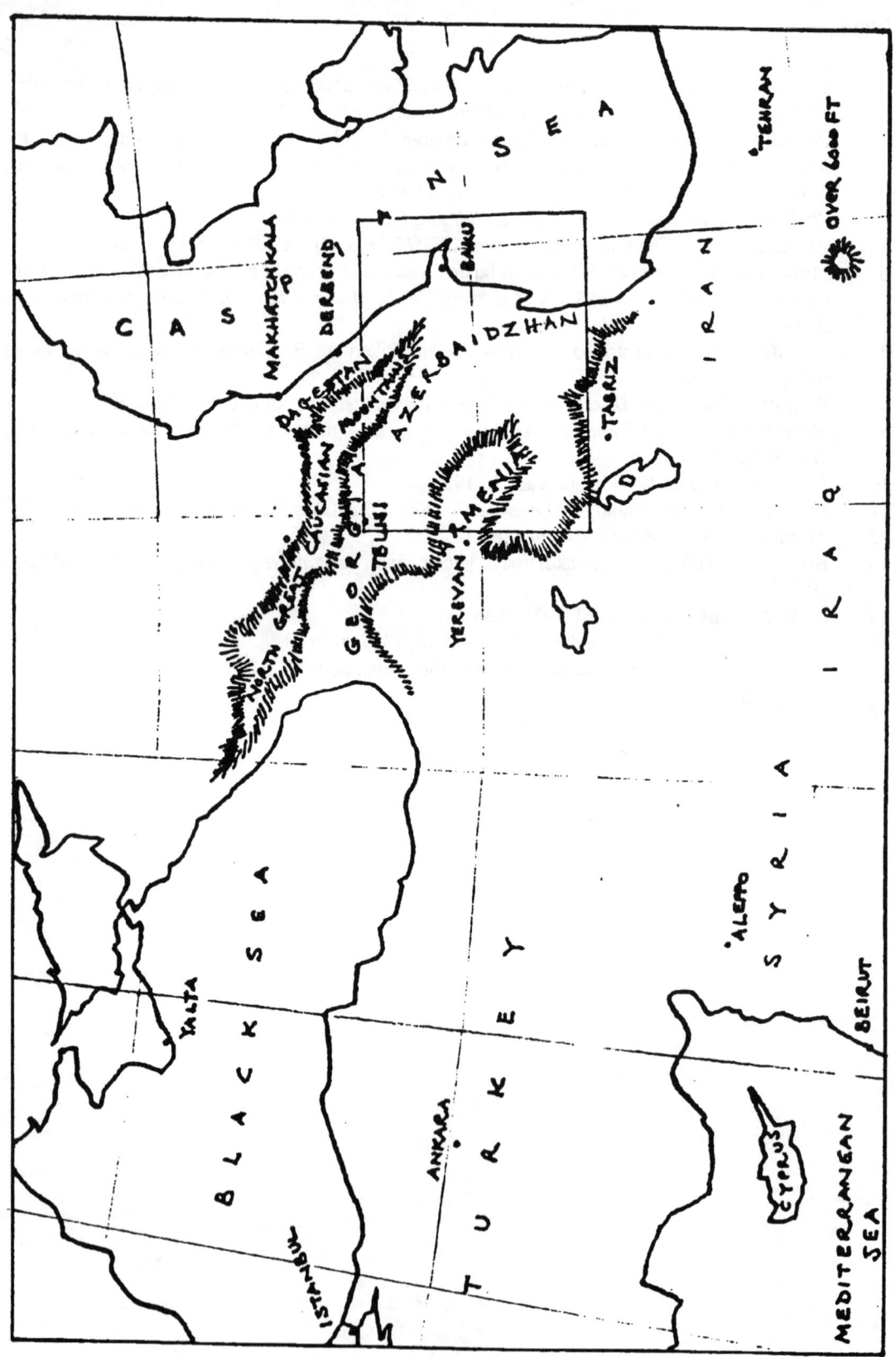

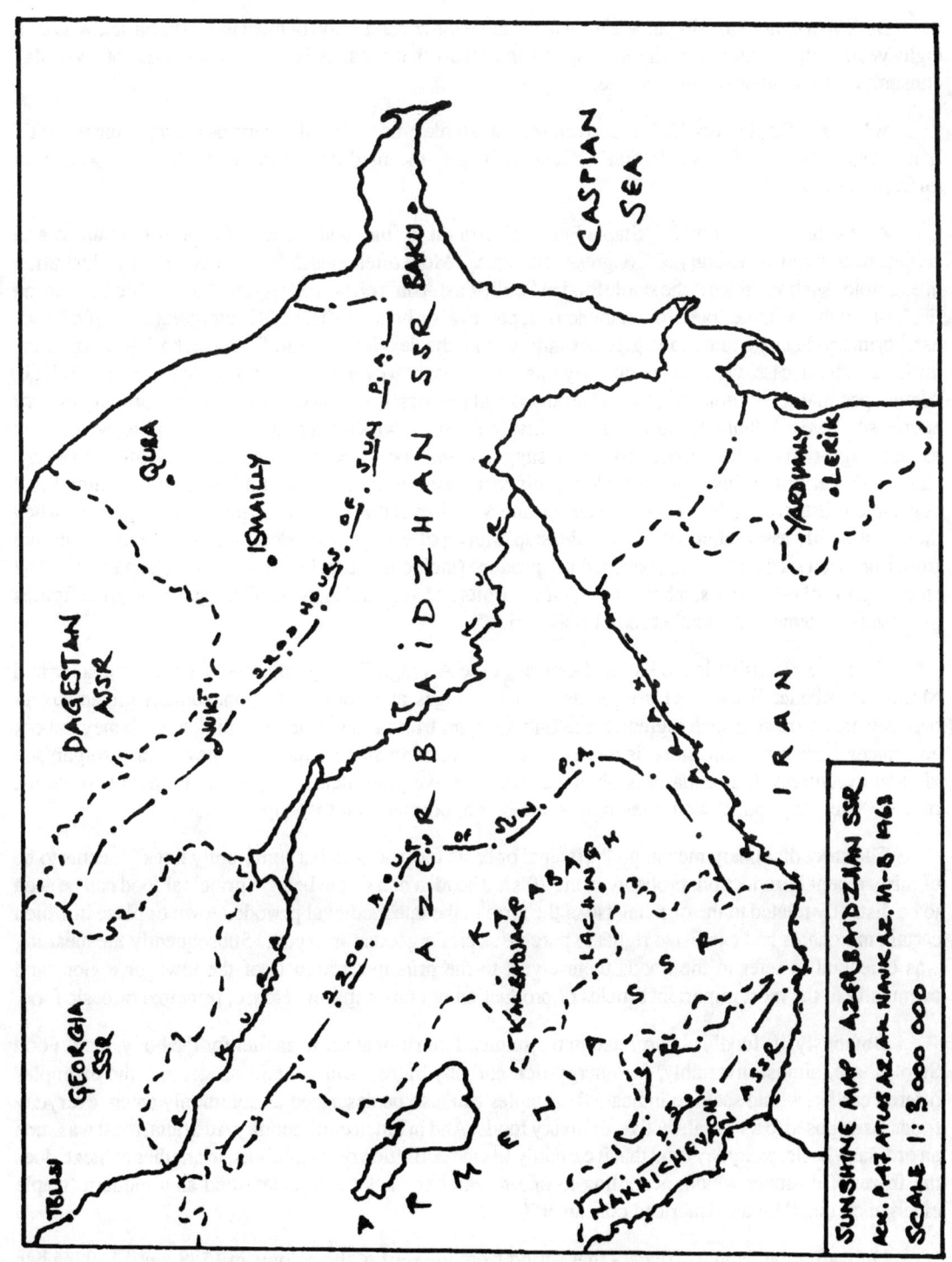

Feed The Man Meat!

by Barbara Santich

Believe it or not, this slogan was used to promote the consumption of meat in Australia, and less than eight years ago! Underlying the message is the credo that meat is basic, fundamental, of everyday consumption - in other words, a staple.

What is a 'Staple Food'? Is it derived from a 'staple crop'? Is it the same as a 'staple ingredient'? Is it necessarily part of a 'staple diet'? Clearly, 'staple' means different things to different people, in different contexts.

An explicit definition of a 'Staple Food' is difficult to find, and the term is not differentiated as a subject heading in the Library of Congress catalogues. More often, it is defined implicitly, by illustration or example - such as, "Rice is the staple food in South-East Asia"; or, "bread [is] the staple of Mediterranean life"; or again, "a staple food, for example rice, cassava, or bread"[1]. The FAO interprets a 'Staple Food' as a "primary dietary source" and gives examples from the developing world - wheat, barley, maize, rice, millet, sorghum, cassava, sweet potatoes, yams, plantains; however, it is also noted that since 1969-71, in "developed market economies", animal products had practically overtaken cereals as the primary dietary energy sources[2]. William A. Stini's idea of 'Staple Foods' is wider-ranging again, extending beyond the usual energy-rich and farinaceous foods; he suggests "manioc, sweet potatoes, bananas (more properly, plantains), various grains, milk and blood, reindeer flesh, insect grubs and a host of other animal and vegetable substances"[3]. Nor did Olga Hartley and Mrs. Leyel think only in terms of carbohydrates when they confidently asserted, in 1925, that "the staple foods of Europe, America, and the nations that derive from European races, are wheat, bacon, dairy produce (including milk, butter and cheese), and sugar. The consumption of other foods, other cereals, meats, fruits and vegetables, is small compared with the figures of these four items. The staple food of Asia is rice"[4].

Far more extensive is the list of 'Staples for the Average Family' offered by Irma Rombauer and Marion Rombauer Becker, which begins with beverages and ends with monosodium glutamate, in between encompassing such ingredients as butter, sugar, fruits and vegetables, milk, eggs, honey, sauces and spices[5]. Such an itemisation is more of a stock-list, or an aide-memoire for novice or disorganised kitchen executives, but its basis is obvious: these are everyday items, or potentially everyday items. Indeed, 'everyday foods' would seem to be the best approximation to 'Staple Foods'.

The strict dictionary meaning - "principal or basic food on which a community lives" - seems to be of fairly recent introduction, probably in the 70's[6]. The idea of a staple being a principal food can be seen to be distantly related to the original use of the term in the late medieval period: a town or place in which certain merchants had exclusive rights to purchase certain goods for export. Subsequently the meaning was extended to refer to the goods themselves, to the principal products of the town or region, and eventually to the most important articles of production or consumption. Hence, principal or basic food.

Obviously, if 'food' is interpreted in the standard nutritional sense as fuel for the body, then 'Food Staples' are, almost invariably, the energy-rich carbohydrates. But general usage, and the examples offered earlier, would seem to indicate that staples can also be described as commonly eaten, everyday foods, as opposed to a special, festive, or luxury foods. And in nineteenth-century Australia, meat was such an ordinary, basic, everyday food that it certainly fits the definition of 'staple' - as, in another context, does the frozen TV dinner which, according to one researcher, "has been reclassified as a modern 'staple emergency meal' by the American consumer"[7].

Only after the 'Hungry Years' - that period from the start of the colony, in 1788, until 1792, when the colonists survived on imported rations such as salt pork, flour, maize, beans and rice - did fresh meat enter into the Australian diet in any quantities; yet by 1831 the minimum convict ration was set at 12 lbs. of wheat, 7 lbs. of fresh meat and 4 oz. of salt per week. In the subsequent era of pastoral expansion, the typical weekly allowances for farm workers was known as 'Ten, Ten, Two & a Quarter' - 10 lbs. meat, 10 lbs. flour, 2 lbs. sugar, a quarter lb. tea, plus salt. Some farm labourers received even more than 10 lbs. meat each week, and 'Meat Three Times a Day' was the rallying catch-cry used by Mrs. Caroline Chisolm in 1847 to attract more emigrants to the Promised Land.

Ten pounds of meat per week may well have tempted many a new settler, but it was excessive only by modern nutritional standards; today's recommendation is no more than 125g of meat per day, or the weekly equivalent of just under 2 lbs. The 'common' or 'full' diets offered in London hospitals around the middle of the nineteenth century included fairly liberal rations of meat - and the physician could always prescribe more, presumably to those more affluent patients who could afford to order meals from outside[8]. Sir Henry Thompson favoured 7-8 oz of meat daily for the average English male - and that in addition to a couple of eggs and a portion of fish![9] And Samuel Pizey, addressing the Adelaide Philosophical Society in 1868, lamented the waste of good meat which, if properly preserved, might be exported to England to supplement English production and supply the 'necessary quota' of 8 oz. meat per person per day[10].

Any food that is eaten in relatively large quantities and on a daily basis must be cheap - and this is where Australia had the advantage. In the Sydney market in 1833, beef cost only 1½d per lb, if bought by the quarter, or 2½d per lb by the joint. The wholesale price of the 'best quality' beef and mutton was only 1d per lb in Sydney in 1851, while a 2lb loaf of bread cost 4½d[11]. In Melbourne in 1867, a four-pound loaf of bread cost 8-9d, while meat (beef and mutton) cost 1½-5d per lb[12]. At the end of the nineteenth century a fictional English visitor, at her first 'colonial meal', "was lost in admiration at the Colonial capacity...a huge hindquarter of mutton disappeared...and it was only afterwards that she learnt that this quarter sheep cost less than the meagre joint of a workman would have done in England."[13]

Cheaper again was the meat that hopped about on two legs - and there is ample evidence that it was greatly appreciated. For the early surveyors, explorers and pioneer farmers, game offered the only source of fresh meat and, by all accounts, was abundant. During the first years of settlement in Van Diemen's Land, at the start of the nineteenth century, convicts were issued a ration of 8 lb. kangaroo meat per week, and it is estimated that in six months the settlers (including convicts) ate 15,000 lb. dressed meat from haunches and tails[14]. Even in the 1840's, kangaroo meat was sold in Hobart when supplies of other fresh meats were scarce. Lieutenant Breton reported in 1834 that "the flesh of this animal is excellent eating, and is usually made into soup, or steamer... A steamer consists of pork and kangaroo minced together (the latter being without fat, the pork is substituted), and, when well cooked, is a dish fit for an alderman."[15] Ultimate praise, indeed! (Edward Abbott gives a recipe for Kangaroo Steamer in *The English and Australian Cookery Book* (1864), and a somewhat refined version of the same dish was still current in 1899, according to *The Goulburn Cookery Book*; many nineteenth-century writers speak very highly of the 'steamer'.)

Another way of cooking kangaroo was over an open fire of gum boughs - the 'sticker-up', described by Louisa Ann Meredith: "The orthodox material here is of course kangaroo, a piece of which is divided nicely into cutlets, two or three inches broad and a third of an inch thick. The next requisite is a straight clean stick, about four feet long, sharpened at both ends. On the narrow part of this, for the space of a foot or more, the cutlets are spitted at intervals, and on the end is placed a piece of delicately rosy fat bacon. The strong end of the stick-spit is now stuck fast and erect in the ground, close by the fire, to leeward, care being taken that it does not burn. Then the bacon on the summit of the spit, speedily softening in the genial blaze, drops a lubricating shower of rich and savoury tears upon the leaner kangaroo cutlets below, which forthwith frizzle and steam and sputter with as much ado as if they were illustrious Christmas beef grilling in some London chop-house under the gratified nose of the expectant consumer."[16] Lieutenant-Colonel Mundy's recipe is more succinct: "Skewer, or skiver (to use my informant's stronger word), skiver alternate slices of lean and fat on your ramrod, roast at a fire that any native will make with two sticks, or you yourself with a flash of gunpowder, (if you have no matchbox;) and if you happen to be hungry you will not require knife or fork, salt, pepper, or pressing."[17]

There can be no doubt that meat was a Staple Food for colonial Australians: it was a food of everyday consumption - indeed, a principal food; it was cheap; and if meat had not been available, in such quantities, diets may well have been deficient. Further, it was viewed as a staple, acquiring something of the character of a 'cultural superfood' (a staple which, because of its dietary pre-eminence, has special cultural or supernatural properties) and came to be considered a god-given birthright[18]. The advertising slogan, 'Feed the Man Meat', thus becomes an affirmation of cultural identity.

On the other hand, there is a (documented) colloquial usage of such terms as 'meat', 'mutton' and 'beef'[19]. It may not be exclusively Australian, but in Australia these words have definite sexual connotations - which places 'Feed the Man Meat' in an entirely different perspective!

> He had left the dead man's penis in a women's toilet at Melbourne's Flinders St Railway Station "to impress the ladies, show them a bit of meat", the court was told.

(From a report in "The Advertiser", 7/3/89)

REFERENCES

1. Rosemary Brissenden, *South East Asian Food* (Harmondsworth: Penguin, 1969), p. 20.
   Paula Wolfert, *Mediterranean Cooking* (London: Pan, 1980), p. 115.
   A.N.Rao (ed), *Food, Agriculture and Education* (Oxford: Pergamon, 1987), p. 75
2. FAO, *The Fifth World Food Survey* (Rome: FAO, 1987)
3. William A. Stini, 'Human Adaptability to Nutritional Stress', in *Nutrition, Food and Man*. ed. Paul B. Pearson & J. Richard Greenwell (Tucson: University of Arizona Press, 1980) p. 126.
4. Olga Hartley & Mrs. F. Leyel, *Lucullus: The Food of the Future* (London: Kegan Paul, 1925) p. 23.
5. Irma S. Rombauer & Marion Rombauer Becker, *The Joy of Cooking*, vol. 2 (New York: Signet, 1974), pp. 528-529.
6. Supplement to OED, 1986.
7. Norge W. Jerome, 'Frozen (TV) dinners - the staple emergency meals of a changing modern society', in *Food in Perspective*, ed. Alexander Fenton & Trevor M. Owen (Edinburgh: John Donald Publishers, 1981), p. 145.
8. Arthur J. Bellows, *The Philosophy of Eating* (New York: Hurd & Houghton, 1867), pp. 274-279.
9. Sir Henry Thompson, *Food and Feeding* (London: Frederick Warne, 1901), p. 60.
10. Samuel Pizey, Our Food Supply, paper read to the Adelaide Philosophical Society, 29.9.1868
11. G. C. Mundy, *Our Antipodes*, vol. 1 (London: Bentley, 1862), p. 417.

12 Rev. John Milner & Oswald Brierly, *Cruise of HMS Galatea (1867-68)* (London: W. H. Allen & Co, 1869).
13 Hume Nisbet, *In Sheep's Clothing* (London: F. V. White, 1900), p. 123
14 Robert Hughes, *The Fatal Shore* (London: Pan, 1988), p. 126.
15 Lieut. Breton, R.N., *Excursions in New South Wales*. 2nd ed., revised (London: Bentley, 1934), p. 221.
16 Mrs. Charles Meridith, *My Home in Tasmania. During a residence of nine years* (London: John Murray, 1852), pp. 54-55.
17 G. C. Mundy, pp. 338-339.
18 Paul Fieldhouse, Food & Nutrition, Customs & Culture (London: Croom Helm, 1986), p. 56.
19 Edward E. Morris, *A Dictionary of Austral English* (Sydney: Sydney University Press impr., 1972) A. Delbridge (ed), *Aussie Talk: The Macquarie Dictionary of Australian Colloquialisms* (Sydney: Macquarie Library, 1984).

Roots and Other Garden Vegetables in the Diet of Londoners, c. 1550-1650, and Some Responses to Harvest Failures in the 1590's.

By Malcolm Thick

The predominance of one staple in the food of a community will lead to famine and dearth when supply of that staple is interrupted. When grain harvests failed in the mid-1590s London, and much of the rest of England, faced such a problem. This paper first examines some immediate reactions to the dearth in pamphlets of advice published at, or soon after the crisis. Secondly, it will be argued that the grain shortages encouraged the already expanding market gardening around the capital and hastened the acceptance by Londoners of more garden vegetables in their daily diet.

I

W. G. Hoskins described the years 1595 to 1597 as 'The Great Famine [which] extended over nearly all Europe, lasting for some three years. In Hungary it was said that the Tartar women ate their own children. In Italy and Germany poor people ate whatever was edible - fungi, cats, dogs, and even snakes'. In Sweden people were 'so weak and their bodies so swollen that innumerable...died'[1].

England's worst year was 1596 when, despite a 'temperate winter' which 'made mens hearts to leape for joy, and the Barnes, as it were, to enlarge themselves for the receipt of this promised plentie', unseasonable weather, especially in harvest time, with 'never ceasing raine' and 'tempestuous winds which choake out the corne when it would have been shorne', destroyed early hopes and left the corn 'utterlie rotted and corrupted'. The shock of dearth in England was intensified by contrast with earlier abundance: the grain harvests of 1591, 1592 and 1593 were all good. These bountiful harvests led the government to repeal the Anti-Enclosure Act of 1563, and to allow limited export of wheat[2].

The bad English weather in these years affected all major bread grains: wheat, barley, oats and rye prices were very high, and alternatives such as peas, beans and dairy produce also became much more expensive. Cattle and sheep prices did not rise above average but, as such meat was normally beyond the pockets of the poor, one would not expect a close relationship between meat and grain prices in dearth years. That the dearth caused distress in London is clear from comments recorded at the time. Analysis of burials in selected London parishes shows that mortality rose in 1597, following the bad harvest of the previous year. Unless some unrelated outbreak of disease occurred, people in London in 1597 died from hunger or illnesses caused by malnutrition[3].

Government response to the crisis came from the Privy Council which issued *'The Book of Orders'*, (a comprehensive set of instructions on action to be taken in times of dearth), proclamations and exhortations to provincial justices and councils. The Council was particularly shrill against speculators who were largely blamed for keeping grain prices high. In 1597 it complained that, although the harvest was better that year, 'yet there are seene and fownde a number of wycked people in condicions more lyke to wolves or cormerants then to naturall men, that doe moste covetusly seeke to holde up the late great pryces of corne and all other victuells by ingrossing the same into their private hands'. The government tried to even out shortages by persuading the authorities in areas where grain was more plentiful to allow grain to be sent to places worse off[4].

In order to maximise grain available for bread the Council sought to restrict its non-food uses, regulating malting, brewing, and starch-making. Grain exports were banned, imports encouraged and even hijacking of foreign grain ships in the Channel authorised. Lastly, the Council tried moral suasion. The Archbishops were instructed to tell the clergy to preach against the gluttony of the rich and to encourage charity to the poor. The Lord Mayor was ordered to prevent excessive feasting in the City of London. To avoid civil unrest preachers were to urge the poor to be patient and accept their lot[5].

London, a city of some 200, 000 inhabitants and by far the largest urban area in England, posed the government particular problems in this time of dearth. The poor (and most of the rich) in London had no gardens and were totally dependent on food brought into the city. The Privy Council understood the problem of maintaining the flow of food despite shortages in areas which supplied the capital: 'wee have founde exceeding great difficulty to reconcile the wantes of the citty and countrie, the one requiring great supply, the other not so able in these as in other tymes to affoarde such stoare.' Nevertheless, the Council had frequently to order provincial authorities to release grain for London. In October 1596 Norfolk justices were told by the Council to hand over 1520 quarters of grain to London bakers 'because the cittie of London, being the chief place and resort of this realme, may in no wise be left to[o] much unprovided, and cannot be other wise sufficiently healped but by supplie out of other counties, amonge the which we suppose that the countie of Norfolk may best spare some reasonable porcion'. The London authorities, led by the Lord Mayor, tried to control the day to day supply and price of food in the City and suburbs and encouraged the rich to give food and alms to the poor[6].

II

The dearth produced some pamphlets in which individuals suggested what could be done to overcome the problem. Many laid emphasis on religious and moral failings as the causes of dearth. The Oxford preacher George Abbot wrote in late 1596 'that God is angry with us... Behold what a famine he hath brought into our land'. A Wakefield gentleman, Henry Arthington, also thought that 'the hand of God is heavy upon us, in most places in this Realme of England'. Arthington did, however, include some practical proposals in his pamphlet; abolition of customs duties on corn imports, official encouragement of imports, the rich abstaining from one meal a week and giving it to the poor, and action against 'covetous corne mongers who engross and hold back corne'[7]. Two works provide many interesting and practical ideas on the problem of feeding the poor when grain harvests failed and we will look at each in some detail.

The first to appear, in 1596, was; *'Sundrie new artificial remedies against famine, written by H. P. uppon thoccasion of this present dearth'*. The author was Sir Hugh Platt, a gentleman of means who, like many others of his class, chose to live in and near London. Son of a wealthy brewer, he was educated at Cambridge and Lincoln's Inn before devoting himself first to the study of classical literature and then to the natural sciences, agriculture and gardening. In 1605 James I recognised his work as an inventor with a knighthood. He was an eclectic writer: his interests ranged over distilling, chemistry, mnemonics, metallurgy, fishing, food preservation, cookery and general housekeeping, husbandry, gardening and many other miscellaneous matters. Much of his work on agriculture and gardening was of high quality[8].

'Sundrie new and artificial remedies' like many of Platt's works, was a hasty production, a quick response, as the title suggests to 'this present dearth'. It is written in short paragraphs, with occasional asides and suggestions for further research, and was to 'only break the yce, for those that shall follow in this kind'. The pamphlet begins conventionally with advice to pray for forgiveness of sins 'in these threatening daies of sword and famine'. Platt supports government measures to alleviate distress, calling on 'all inferior officers' to be diligent 'in the execution of those orders, which have lately beene penned'. Traders who charge more in times of scarcity are condemned and the rich are asked to give up a meal a week to the poor. Platt however, wished to go further than the sermons and government proclamations and 'by new, and artificial discoveries of strange bread, drinke, and food, in matter and preparation so full of variety, to worke some alteration and change in this great and dangerous dearth'.

After the introduction, the pamphlet falls into two parts, Platt's own thoughts on famine relief and a critical translation of a short, anonymous Roman treatise entitled *'Archora famis et sitis'* which 'raunged over all manner of trees, plants, roots, greene pulse and herbes, out of which [the author] might by any probabilitie draw any kind of sustenance for the relief of man'.

The forty suggestions in the Roman treatise, some with additional notes by Platt range from the sensible to the bizarre. The information that Roman post riders baked bread under their saddles, (Platt laconically comments 'I think that our climate will prove too cold'), or that the smell of new baked bread nourishes the body were of little use to anyone. There are, however, many useful ideas on substitutes for flour in breadmaking, such as; ground rice, Turkish wheat, lentils, beans, peas, vetch, panick, millet. Dough can be mixed with milk, wine, eggs and cheese to make it go further.

Substitutes for bread suggested include a fruit stew, parched corn, and a mixture of bean flour and cummin seed. Many ideas for alternative drinks are given. In extreme famine, bread can be made from ground leaves, and green wheat roasted and eaten. The Roman author sensibly observes that 'Men must be brought by degrees, and not too sodainlie from their usual and natural food and drinke, into these artificiall diets'.

Platt's own ideas were written to complement the Roman treatise, concentrating on 'some of those particulars which are most plentifull in their quantitie, least offensive in their nature, and most familiar with our soile and bodies'. To dull hunger he recommends liquorice root or ground tobacco chewed slowly. Platt has ideas on alternative ways of making non-food grain products: starch made from cuckoopint roots, a 'beer' made with dried ling flowers. He suggests passing bran back through the mill with a second batch of corn to produce more flour.

Platt includes several recipes for famine meals. One can, by boiling in two or three changes of water, 'take awaie a great part of that ranke and unsavorie tast of Beanes, Peas, Beechmast, Chestnuttes, Acornes, Vetches, and such like'. When dried they can be ground and mixed with $1/3$ or $1/4$ part of wheat flour and baked. 'If this may in some measure bee performed, then I doubt not but that the bulke and body of our meale and flower will be increased and multiplied at the least for poore mannes table'. Beechmast could also be pounded to a paste and eaten flavoured with herbs. To 'make an excellent bread of the rootes of Aaron called cuckowpit', (cuckoopint) the roots were to be cleaned, sliced, boiled until sweet, dried on canvas, and ground in handmills to 'make a most white and pure meale, which either of itselfe, or by the mixture of one thirde of wheate meale with it, maketh a most faire and savory bread'.

Here, in full is Platt's recipe for parsnip cakes ('Sweete and delicate cakes made without spice or sugar'):

> 'Slice great and sweete parsnip rootes, (such as are not seeded) into thin slices, and having washed and scraped them cleane, dry them and beat them into powder, (here a mil would make a great dispatch) searcing the same through a fine searce, then knead two partes of fine flower with one part of this powder and make the same into cakes, and you shall find them to tast very daintily. I have eaten of these cakes divers times in mine owne house, *Quare*, what may be done in carots, turneps and such like rootes after this manner'.

(Platt also thought turnips could be used instead of cuckoopint to make bread as 'there are both good store and the price of them likewise very reasonable').

Lastly, Platt recommends, not so much for famine relief as for long sea voyages, 'a certaine victuall in the forme of hollow pipes, or wafers wherewith...I furnished Sir Francis Drake in his laste voyage'. This food was pasta, probably macaroni. He gives ten reasons for using the 'certaine victual': it - will keep for 3 years; is light to carry; is easy to prepare; is fresh [i.e. not salted]; is cheap (2d per man per meal); substitutes for both bread and meat; can be used in the voyage if surplus to requirements; can be 'made as delicate as you please' with 'oyle butter, sugar, and such like'; can be readily made as 'There is sufficient matter to bee hadde al the yeare long'; can be supplied by Platt in good condition. Significantly, he gives no hint of how it is made - did he acquire the secret from Italy and wish to keep it for himself, or was he able to import pasta?

The second pamphlet, *'Profitable instructions for the manuring, sowing and planting of kitchen gardens. Very profitable for the common wealth and greatly for the helpe and comfort of poore people'*, was published first in 1599. Its author, Richard Gardiner, was a wealthy and philanthropic dyer of Shrewsbury who also ran a market garden of at least four acres and drew on his considerable experience of gardening to write a 'short and simple penning of this my practise and experience in Gardening'. Gardiner, in a preface 'to his loving neighbours and friends, within the towne of Shrewsburie' explains that in his 'olde age, or last daies,' he 'would willingly take my last farewell with some good instructions to pleasure the general number'[9].

The work is full of piety, from the preface with frequent references to God and His mercy, to the concluding prayer preceded by 'An exhortation to love, wherby all good works do effectually proceed'. But it is also an extremely practical book. The first twenty or so pages provide advice on vegetable growing equal to many modern gardening manuals. Gardiner begins by explaining how to raise and save the seeds of carrots, cabbages, parsnips, turnips, lettuce, beans, onions, cucumbers, artichokes, radishes, porrets and leeks. This first section may have been written before the dearth years of the 1590s, possibly as part of a projected larger book, for only in the concluding sentence is the dearth noticed when Gardiner hastily excuses further general discussion of vegetables with:

'I could yet heerein take occasion to write of divers rootes and hearbs, for sallets, to bee planted and sowed in gardens, which do not serve my purpose, for I rather desire to provide sufficient victuals for the poor and greatest number of people, to relieve their hungrie stomackes, then to picke dainty sallets, to provok appetite to those that doe live in excesse, the which God amend.'

He then begins what is almost a second pamphlet, specifically on the famine years, vividly describing his part in helping the poor 'in the great dearth and scarcitie last past in the Countie of Salop and elsewhere, for with lesse garden ground then foure ackers planted with Carrets, and above seaven hundreth close cabbedges, there were many hundreds of people well refreshed thereby, for the space of twenty daies, when bread was wanting amongst the poore in the pinch or fewe daies before harvest. And many of the poore said to me they had nothing to eate but onely carrets and Cabedges, which they had of me for many daies, and but onelie water to drinke. They had commonly sixe ware poundes of small close Cabedges for a penny to the poore. And in this manner did I serve them, and they were wonderfull glad to have them, most humbly praising God for them all.'

He sold cheaply to the poor in the dearth: normally large close cabbages were 2lb for 1 penny and small ones 5 lb for a penny. Other vegetables he usually sold as follows:

Large artichokes	1d each
Small artichokes	1d for 2 or 3
Green beans	1d per quart
Large turnips	2d per stone
Yellow carrots	2d per stone, 3d per stone January to March.

Gardiner thought carrots the most important vegetable for the relief of hunger and he gave a number of recipes for them, noting that, 'this last dearth and scarcitie hath somewhat urged the people to proove many waies for their better reliefe whereby I hope the benefit of Carret rootes are profitable'. He begins with the 'use of them amongst the better sort by the Cookes' [ie carrots for the rich], recommending that carrots be cut up and boiled to season broth, boiled with powdered [ie salt] beef and pork and with any meat the poor can afford. Red carrots make 'daintie sallets' with vinegar and pepper, to go with roast meat. Mixed into potages of any kind 'they effectually make those pottage good, for the use of the common sort ... Carrets well boyled and buttered is a good dish for hungrie or good stomackes'

Addressing himself specifically to dearth he wrote 'Carrets in necessitie and dearth, are eaten of the poore people, after they be well boyled, instead of bread and meate. Many people will eate Carrets raw, and doe digest well in hungry stomackes: they give good nourishment to all people, and not hurtful to any, whatsoever infirmities they be diseased of, as by experience doth proove by many to be true'[10].

III

While the pamphleteers wrote about the dearth, the people of London made what provision they could to replace bread in their diet. Both Platt and Gardiner recognised the potential part roots and other garden vegetables could play in feeding the poor and in doing so they reflected increased consumption of these foods in London and elsewhere in England in the late sixteenth century. Not that the production and sale of vegetables in London was a novelty: gardeners employed by rich citizens sold surplus vegetables at a regular market by St Paul's in 1345. John Stow recalled a commercial gardener name Cawsway who held land in Houndsditch until 1553 and was 'one that served the market with herbs and roots' and maps of London published in the 1560s and 70s depict market gardens in the suburbs[11].

In the 1590s Gerard wrote in his Herbal, 'The small Turnep groweth by Hackney, in a sandy ground, and those that are brought to Cheap-side market from that Village are the best that ever I tasted.' In 1593 concern was expressed about a recent increase in commercial gardening because 'it is to be feared that a number of poor people, living by roots, turnips, herbs, and such like are infected by the evil juice of such roots and herbs as are sown upon those corrupt laystalls and grounds which many gardeners and others of late have practised to sow, before they have lain a convenient time to rot and be fit for manuring'[12].

The expansion of market gardening around London in the second half of the sixteenth century was reflected in the acceptance by the poor of roots as part of their diet. In 1577 William Harrison found the poor now eating 'melons, pompions, gourds, cucumbers, radishes, skirrets, parsneps, carrets, cabbages, naevewes, turneps, and all kinds of salad herbes' whilst Thomas Cogan in the 1590s wrote of parsnips and carrots, 'The rootes are used to be eaten of both first sodden, then buttered, but especially Parseneppes: for they are common meate among the common people all the time of Autumne, and chiefly uppon fish daies'. The poor could regularly buy roots in London markets, although quality could not be assured for on 1593 a proposal was made ' for restraint of those that let out cellars and sheds under stalls where herbs, roots, fruits, bread, and victuals are noisomely kept till they be stale and unwholesome for man's body, and then, mingled with fresh wares of the same kind, are brought forth into the markets and there sold to the great deceit and hurt of the people'[13].

In the terrible dearth of grain in the 1590s roots played a larger part than before in feeding London's poor. As well as local produce, London received roots, mainly carrots, from the East Anglian ports of Norwich, Yarmouth, and Colchester. Most were shipped through Yarmouth; between October and March 1593-4 281 tons and 600 bushels went to London, in 1597-8 600 tons and 600 bushels were sent, and 1598-9 639 tons and 1 last were shipped. These totals may be small in comparison with the 111,075 quarters of grain which came to the port of London in the seven months ended May 26th 1597, but they would certainly have been a welcome addition to total food supplies[14].

The suppliers of roots from East Anglia were mainly Dutch and Flemish refugees who had settled at Norwich, Colchester, and Yarmouth in the second half of the sixteenth century to escape religious persecution. They began market gardening and, in particular the intensive production of roots. At Norwich in 1575 they were said to 'digge and delve a grete quantitie of grounde for rootes which is a grete succor and sustenaunce for the pore'[15].

The potential of roots as food in times of dearth not only impressed shrewd observers such as Platt and Gardiner. Following the shock of near famine in the 1590s, practical men saw profit in increasing the supply of roots to London markets to feed the ever growing population of the capital. In about 1600 many

Dutch gardeners moved to the Surrey bank of the Thames near London and 50 years later old men in Surrey could still recall 'the first Gardiners that came into those parts, to plant Cabages, Colleflowers, and to sowe Turneps, Carrets, and Parsnips, to sowe Raith (or early ripe) Peas'[16].

Across the Thames, husbandmen and gardeners in suburban Middlesex turned to roots and other vegetables. John Norden in 1607 mentioned Fulham in a list of carrot-growing areas and by 1616 the husbandmen and gardeners there were in legal conflict with the recently formed Gardeners Company of London because of their root production, which the Company sought to regulate. The dispute continued until 1633 by which time root growers in neighbouring Kensington and Chelsea were also involved. The Company was alarmed at the large scale of production undertaken by the Middlesex men and sought to bring them within its regulations on size of garden ground and numbers of employees. The Company lost the case, for the pragmatic reason that London could not afford restrictions on this source of food for the poor. These Middlesex producers 'by this manner of husbandry and ymployment of their grounds [furnish] the Cittys of London Westminster and places adjacent…with above fower and twenty Thousand loads yearly of Rootes'[17].

Because the expansion of market gardening increased supply, vegetables were more commonly eaten in London after 1600. A London schoolboy's diet in the early seventeenth century reflects this change: 'Our breakfast in the morning, is, a little piece of bread not buttered, but with all the bran in it, and a little butter, or some friute, according to the season of the yeare. To dinner we have herbes, or everyone a messe of porridge. Sometimes turneppes, colewarts, wheat and barley in porridge, a kind of delicat meate made of fine wheat flower and eggs. Upon fishe dayes, fleeted milke, in deepe porrengers (whereout the butter is taken) with some bread put in it. Some fresh fishe, if in Fish street can be had at a reasonable price. If not, salt fish, well wattered. After pease, or fitches, or beans, or lupins'[18].

In crude and satirical comment on London's eating habits and sanitary arrangements Ben Jonson wrote, in 1616, of a summer voyage up the Fleet Ditch, a minor tributary of the Thames between Westminster and City,

> ………………………………… 'How dare,
> Your daintie nostrills (in so hot a season)
> When every clerke eates artichokes and peason,
> (Laxative lettus and such windie meate)
> Tempt such a passage? When each privies seate
> Is fill'd with buttock? And the walls doe sweate
> Urine and plaisters? When the noise doth beate
> Upon your ears, of discomforts so unsweet?'[19].

Thomas Fuller, in 1662, when an estimated 10,000 acres of commercial gardens surrounded London, found it 'incredible how many poor people in London live thereon, so that in some seasons the gardens feed more people than the field'[20].

CONCLUSION

This paper has sought to show that, despite the preoccupation of the authorities during the 1590s dearth with grain supply, many ordinary citizens in London derived considerable nourishment from roots and other garden produce, either grown locally or shipped from East Anglia. Many were accustomed to eat garden vegetables in normal times: the dearth merely led them to eat more of such food instead of bread. Platt and Gardiner saw the roots as cheap and filling food, and the poor in London, both during and after the dearth, appear to have taken quite happily to them. Some may have consumed so many vegetables that they developed hunger-edoema, a condition characterised by waterlogged tissues and swollen limbs: the 'moist and loose flesh' which a London apothecary noted as a consequence of turnip eating in 1629[21].

The increasing popularity of roots is no wonder: they were inexpensive, palatable, and easy to prepare. Bread could not be produced by the urban poor - they had no facilities for grinding corn and baking bread, and had to rely on millers and bakers. Roots were easily stewed or baked and could, as Gardiner noted above, be eaten raw, an advantage not shared even by the later seventeenth century solution to hunger, the potato. In better times, roots, as they do today, filled out a meal (Gardiner thought carrots with boiled beef 'do save one quarter of beefe in the eating')[22].

Discounting problems of harvest failure, the sheer size and rapid expansion of London compared with other towns in England dictated a growing need for more and more food to supply the daily food markets of the capital. Latest estimates of London's population are:

1550	120,000
1600	200,000
1650	375,000
1700	490,000

In 1600 London contained about 5% of England's population, 7% in 1650, and 10% by 1700[23].

Expanding root and vegetable production was one response to London's growth. Roots produce, in terms of weight, very high yields per acre, (up to 40 tons on modern farms). Gardeners in the suburbs, with easy access by cart or water could send load after load of roots to market. The poor bought them because they were cheap but, because of the yield, returns to husbandmen and gardeners were large. Well might Thomas Fuller exclaim in 1662, 'Oh the incredible profit by digging of Ground!'[24].

Footnotes

1. W. G. Hoskins 'Harvest fluctuations and English economic history, 1480-1619' *Ag Hist Rev*, xvi, 1, 1968, p38; Gustaf Uttterstrom 'Climatic Fluctuations and Population Problems in Early Modern History', *The Scandanavian Economic History Review*, 3, 1955, pp27-28.
2. W. Barlow, trans., *Three Christian sermons made by Ludovike Lavatore, Minister of Zuricke in Helvetia, of Famine and Dearth of Victualls*, 1596; Henry Arthington, *Provision for the poore, now in penurie*, 1597; W. G. Hoskins, *op.cit.*, pp37-38.
3. A. B. Appleby, 'Nutrition and Disease: The case of London, 1550-1750', *Journal of Interdisciplinary History* VI, 1, 1975, p5; A. B. Appleby, *Famine in Tudor and Stuart England*, 1978, pp138-9.
4. *The Agrarian History of England &Wales*, vol IV, 1967, pp581-2; Appleby, *Famine in Tudor and Stuart England*, pp140-5.
5. Appleby, Ibid.
6. *Acts of the Privy Council*, 1596, p269, 1597-98, p291-92.
7. Arthington, *op.cit*; George Abbot, *An exposition upon the Prophet Jonah*, 1600, pp365-6, (quoted in Appleby, *Famine in Tudor and Stuart England*, p141).
8. Sir Hugh Platt, *Sundrie new and artificial remedies against famine, Written by H. P. upon thoccasion of this present dearth*, n. p. 1596; DNB, Compact ed. 1975, pp1293-1295; Blanche Henrey, *British Botanical and Horticultural Literature before 1800*, 1975, vol I, pp155-6; G. E. Fussell, *Old English Farming Books*, 1523 to 1730, 1957, pp14-16.
9. 'Richard Gardiner's "Profitable Instructions", 1603, ed. Dr. Calvert', *Shrops Arch Nat Hist Soc*, Series II, vol 4, 1892, pp241-2; Richard Gardiner, *Profitable instructions for the manuring, sowing and planting of kitchen gardens*, n.p. 1599.
10. Gardiner, *op.cit.*

11. H. T. Riley, *Memorials of London and London Life, I* 1868, pp288-9; John Stow, *A Survey of London* ed. H. Morley, n.d., p151.; *A Collection of Early Maps of London, 1553-1667*, intr. John Fisher, 1981. pp3-10. For a survey of market gardening in England from the mid-sixteenth century to 1750 see, Malcolm Thick, 'Market Gardening in England and Wales', chap. 18, *The Agrarian History of England and Wales*, vol V.II, 1640-1750, 1985.
12. John Gerard, *The Herbal, or General Historie of Plantes*, 1633, p232; B. L. Landsdowne MS 74, ff755-6 (I thank Joan Thirst for this reference).
13. William Harrison, *The Description of England*, ed. Georges Edden, New York, 1968, pp263-4; Thomas Cogan, *Haven of Health*, 1596, p63; B. L. Lansdowne MS 74, ff75-76.
14. Appleby, *Famine in Tudor and Stuart England*, p139.
15. SPD. Eliz, vol 20 no 49; The role of Dutch and Flemish refugees in stimulating root production is explored further in a forthcoming article.
16. Samuel Hartlib, *His Legacie*, 1651, pp8-9.
17. John Norden, *The Surveyor's Dialogue*, 1607, p207; Corporation of London, *City Reportories*, 33, f74, recto; 49, ff261-3.
18. G. E. Fussell, *The English Rural Labourer*, 1949, p29.
19. Ben Jonson, *Epigrammes* 133, (1616) 'On the famous voyage'.
20. T. Fuller, *The Worthies of England*, 1662, p7; Richard Bradley, *A General Treatise of Husbandry and Gardening*, II, 1726, p273.
21. John Parkinson, *Paradisi in sole*, 1629, p508; J. C. Drummond & A. Wilbraham, *The Englishman's Food*, 1939, pp106-9.
22. Gardiner, *op.cit.*; John Foster, *Englands happiness increased, ...by a plantation of roots called potatoes*, 1664.
23. R. Finlay and B. Shearer, 'Population growth and suburban expansion', *The Making of the metropolis, London 1500-1700*, ed. A. L. Beier & R. Finlay, 1986, pp38, 49.
24. P. McConnell, *The Agricultural Notebook*, 1976, p168; T. Fuller, *The Worthies of England*, 1662, p77.

Wheat and Rice Recipes of India

By Kathie Webber

This is a personal journey of discovery of the rice and bread dishes of India, rather than a solemn treatise intended to tidily demarcate areas of either bread- or rice-eating, for I do not claim to be expert after only a few months travel around the country. An account, then, of the preponderance for one or other of the staple foods noted in the different areas in which I stayed together with some recipes of favoured breads.

The simplest definition is that the rice-growing southern states ate rice as their staple, while the colder northern areas which could support wheat, grew it, ate it, and, depending on terrain, supplied their immediate neighbour states. My experience is that nowadays both rice and some kind of bread are eaten in many areas. Certainly, I was offered both but, as one would expect, this was mostly in the more expensive hotels. I was always pleased to discover that despite the hotel chains efforts at standardisation, the regional differences in ways of preparation remain largely untravelled, so that while I was often keen to find a bread or rice dish in the next state, I occasionally regretted leaving behind one I particularly liked as I crossed a state border.

Rice plants cover a quarter of the cultivated land in India and rice is the staple food in the south and east of the country; wheat in the north and centre. Prior to the railways, goods moved from village to village by bullock cart, and the rice/wheat dividing line must have seemed remarkably solid to travellers who crossed it and quickly found no evidence of the other staple eaten, until then, exclusively. Once the way was open for transporting goods quickly and efficiently, they began to move hundreds of miles, and since travelled goods are always more expensive than home-produced stuff they are more desirable. Those who could afford it would try the new foods, some of which would become part of the household repertory. On a food map of India, the bread/rice dividing line began to blur and was eventually almost indistinct. But for devotees of tradition there are still some outlying places which remain true to the local staple, preparing it in truly local ways.

My journey begins in Bombay for some Parsi food. The Parsis fled Persia in the 7th and 8th centuries to escape religious persecution, settling in Gujerat and Bombay and bringing with them their foods. During the intervening centuries their dishes have blended with the Indian way of life, but Parsis remain uniquely different in their extensive use of eggs in their dishes, and they also retain the Persian magical combinations of meat with dried fruits, particularly apricots. In the limited time I had I could only find a small cafe, and so had no chance to try a dhansak, the famed dish consisting of meat with vegetables and different varieties of split peas or dal, subtly spiced then finished with fried onions. Instead, I ate an egg chutney patty, a cheese tart and a potato pasty - Persian, Indian and British influences at work in these recipes. No rice in evidence here, but chapatis - unleavened bread the size of tea plates made from atta (wheat) flour, shaped and cooked on a tawa - a slightly curved heavy iron pan.

Bombay is cosmopolitan to the point that it is difficult to discover in a short time indigenous foods. Once the capital of Gujerat, today the capital of Maharashtra, earlier Bombay was part of the Portuguese territories, then given to the English in Catherine of Braganza's dowry when she married Charles II. It is a thriving multi-national city of film-making, jewellery and clothes factories with restaurants serving foods from all over India, gathered neatly and conveniently into one city. A good place to start a tour of India.

In largely vegetarian Gujarat, the state just north and west of Bombay, both rice and a bread are served. Puris seem to be the most popular; these are smaller versions of the chapati, made in exactly the same way but deep fried so that they puff. And highly nutritious, like all the brown breads, because the flour is ground from the whole wheat grains.

Often I had to remind myself I was in India when I travelled in Goa, where the Portuguese influence is apparent in the cathedrals, religion, saints and in the houses built in the fields and stringing the roadsides. But as a student of faces and their structure, it was here that I 'collected' my first face when I glanced across to the next seat in the local bus and discovered, in the shape of a local workman, Vasco da Gama, an almost perfect genetic throwback, even to the European skin colour. In Goa, crops of all kinds are grown and rice is important, the fields busy with women pushing the small plants into the soft mud of fields yet to be flooded. Now that Goa has become a major holiday resort, foods again are cosmopolitan with most dishes that one wishes to try available somewhere. Inevitably there are cafes which cater more for one nationality than any other. Here I began seriously collecting recipes, one local lady restaurateur prepared to trade Goanese fish recipes for a demonstration and recipe of a good vinaigrette dressing. A good bargain.

Rice is served mostly but of the breads, chapatis predominate. European breads, referred to as double bread because of the rising times, are used to make toast for breakfast. Indian breads either use no yeast (chapatis, puris, parathas) or rely on a single rising of yeast (naan) or on a natural fermentation to produce a light texture. Dosas and idlis, though made of rice flour and found further south, are a prime example of the natural fermentation process.

And it was while researching dosas and idlis in the state of Karnataka that by accident I discovered the chula or traditional stove. I had enquired about it earlier in my trip but was told that it was unlikely I should find a working example. The chula is a hollow brick cube with holes in the top to serve as burners for pans and a hole near the floor for fuel. Inside and out, the chula, in the traditional manner, had been plastered with mud then given a final coating of liquid cow dung. To one side, a low earthen platform had been levelled which was deeply covered with ash (from the cow dung cakes widely used in this area for fuel) on which pans were stood to keep them hot. Baking is done on this platform by standing a covered pan in the ash and heaping more hot ash on the lid. Finally on the chula was a flat section on which dosas were baked direct - the slight indentation the forerunner of the metal tawa. Because I had the wrong speed of film with me when I discovered this kitchen in the street of lorry-menders in Hassan on my way to the temples of Belur and Halebid, my photograph had to be taken with flash. For real, the kitchen was as black and hot as hell and lit only by the light escaping from around the pans on the burners or occasionally, but far more dramatically, by red flames showing when a pan was removed from its hole. Here I ate my first dosas, before I discovered they were cooked direct on a cow dung coating.

Dosas are a large dinner plate sized pancake, made from a batter of ground cooked rice, fermented overnight, then shaped by pouring the batter in concentric circles on the hot griddle. Cooked well on one side, briefly on the other, each dosa becomes crisp and golden with a pale, slightly soft-textured top, its rings apparent. These are served for breakfast, to be torn and dipped into a spicy sauce. Rice cakes called idlis, (also fermented overnight but steamed in small moulds which shape them into flat rounds or ovals) served with a thin vegetable curry sauce, also appear for breakfast. Like dosas, they are made from a batter of ground cooked rice, which is allowed to ferment naturally overnight, but there are variations which include split peas, and garlic. In Tamil Nadu at the temple complex of Mahabalipuram, south of Madras, I chose a snack lunch of idlis with a spicy yoghurt sauce and was served another variation of these rice cakes. They were flavoured with coconut milk.

Chapatis can be found in these two states, but other breads have been left behind.

Kerala in the south-east is a rice-growing state and rice is served at all meals, here eaten with the whole hand including the palm to the disgust of the finger-tips only northern peoples of India. Although one doesn't find the rich rice dishes of the north, the pilaus and birianis, the different uses to which the rice seed is put are numerous. Appams, pancakes with soft spongy centres and golden, crisp and lacy edges are a good example of a mopping-up ingredient in a meal made from rice flour rather than wheat flour. Again this is a fermented mixture, thinned with coconut milk just before they are baked. In Kerala I found part-cooked rice, a method which has been used here for centuries. Rice is par-boiled, then dried, a process that takes place in most homes, so that always the cook has the choice of raw rice or boiled rice. No wheat breads are evident except, as usual, in the larger hotels, and then it is the chapati, the simplest bread to make, that is offered.

In the northern states it was the breads which sometimes defied my attempts to define exact differences in preparation, just as the different ways with rice occasionally frustrated my note-taking in the south. Chapatis are baked flat on a tawa or griddle, the same mixture (an unleavened wheat dough) when deep-fried until it puffs is a puri. The same mixture again when spread with ghee and folded like our puff pastry and then deep fried becomes a flaky rich paratha. In the tandoor oven, a method of cooking which was brought into northern India by the Moghuls, naan is baked, as are the splendid variations of naan, keema (a minced meat mixture with which it is stuffed before baking) and the Peshwari naan stuffed with ground almonds and sultanas, a sumptuous bread indeed. In the grounds of one hotel in the south where an extension was being built, I spotted a northern Indian standing over an oil drum perforated at the bottom, and filled with coals. I was curious.

The temperature was over 100 deg F, he wasn't huddled over the oil drum for warmth. He was the cook to the building crew who came for the most part from Rajasthan, the desert state in the north west of India. He'd fashioned the oil drum into a tandoor and when he took off the lid to let me see, there stuck to the sides at the top were the tear drop shapes of naan. He'd stuffed his naan with a fresh chutney mixture. In the fierce heat of his makeshift tandoor the naan cooked quickly leaving the filling still slightly cool. Rotis are made in a similar way to chapatis though a mixture of flours is usual and they are often finished with a brief flash on either side over the flames to puff them. The same bread is sometimes stuffed with fenugreek leaves and then called kachoris.

Kashmir has its own breads more reminiscent of Middle Eastern and Persian breads than the familiar chapatis and naans. Poppy and sesame seeds are used to crust breads, there's the crumbly sour-dough kulchas, scone-like breads, and buns of all kinds. I was fascinated to watch a roti being made in the manner of pizza makers who fling the dough around their heads to achieve an even thinness. This was the rumali roti or handkerchief roti, named for its thinness and size.

Even rice dishes are sumptuous for when rice is served it is rarely plain, being dressed up as the pilau - a variation of the rice dishes pilaf and polo from Turkey and Persia respectively. This is a legacy of the Moghuls whose sumptuous foods, included almonds and raisins, meat cooked with the rice, saffron, dried apricots and oranges, embellished with beaten silver and gold on feast days. Biriani, is another example of the splendid way with rice. Here the main ingredient is studded with pieces of meat, vegetables, spices and nuts. And finally a khichdi, a dish which Persians at a feast long ago would recognise, it hasn't changed fundamentally. This is rice with minced meat and spices layered in the pan, cooked slowly and then finally dressed with spiced ghee and tossed very gently so as not to squash the rice.

And as a footnote, I went around India and never was fortunate enough to even see a dish garnished with beaten silver, let alone taste it. That experience came much nearer home, at L'Ortolan in Shinfield near Reading where John Burton-Race finishes one of his stunning chocolate puddings with this precious metal.

DOSAS

Simplest when prepared by the volume method.

> 2 cups long grain rice
> ½ cup skinned white dal
> salt
> ½ cup coconut milk
> vegetable oil

Rinse the rice before soaking it for several hours. Similarly rinse, pick over and soak the dal for the same amount of time. Drain both well. Put the rice through a processor until it forms small semolina-like grains, then gradually add about ¾ cup of water to make a batter. Pour this batter into a bowl, then very finely grind the dal, adding about ½ cup of water a little at a time and about 1 teaspoon salt. Continue blending until this mixture is frothy. Pour it into the rice bowl and mix well. Cover with a cloth and stand the bowl in a warm place (overnight in summer, but perhaps as long as 24 hours in winter; I haven't been able to check the cold weather timings yet) until the batter becomes a mass of tiny bubbles. Just before cooking, thin it to a crepe batter consistency with coconut milk.

To cook the dosas, you need a heavy pan or griddle and a ladle. Heat a very little oil in the pan, stir the batter and ladle about 4 tablespoons into the pan and, with the bottom of the ladle moving in ever increasing and continuous spirals, thin it to a large circle. Sprinkle another teaspoon of oil over the top, cover the pan and cook the dosa for about 2 minutes until it is golden on the bottom, then flip it over and uncovered briefly cook the underside. The cooks in Hassan where I first ate dosas poured the batter in a spiral from an old mineral water bottle with a finger over the neck to control the flow.

BHATURA

Sour dough bread from the Punjab, which like Dosas I find easiest to make by the volume method.

> ¾ cup plain unset yoghurt
> 2 teaspoons sugar
> ½ teaspoon bicarbonate of soda
> 1 cup plain flour
> 1 tablespoon ghee
> 2 cups atta or chapati flour
> 2 teaspoons salt
> ½ cup tepid water
> oil for deep frying

Mix the yoghurt with the sugar, soda and plain flour and leave covered with a cloth overnight for the mixture to ferment naturally. In cold weather this might take as long as 24 hours unless you make use of an airing cupboard. Rub the ghee into the atta or chapati flour mixed with the salt, add the fermented mixture and as much tepid water as you need to make a bread dough. Knead it well then cover it and leave it in a warm place for 2 or 3 hours before dividing it into portions. Roll each one thinly to the size of a small tea plate and fry them one at a time in hot deep oil, spooning oil over them as they cook. Bhaturas puff as they cook and change colour when done. Then they should be lifted out and drained well. Serve warm.

Rye, a Daily Bread and a Daily Treat

by Joop Witteveen

To all of us it is so evident that bread is baked with wheat that we can hardly imagine that this has not always been the case. Right up to this century the daily bread of the majority of the population of the northern half of Europe was made of rye, not of wheat. Throughout the ages wheat was perceived as healthier, more nutritious and easier digestible than rye, but wheat was rather scarce and therefore expensive. So wheat was for the rich and rye for the common people. The irony of history is that nowadays, modern well-to-do people again have turned to eating bread made with rye or a mixture of rye and wheat because this is supposed to be healthy.

Wheat needs a fertile soil and much heat to ripen, while rye can flourish in poor soil and stand a cold climate. Therefore, wheat was grown in the southern half of Europe and rye in the northern one, the dividing-line running across Germany. After World War II, due to fertilizer, wheat could be grown north of the dividing-line and in 1957, for the first time in the history of West Germany, the yield of wheat surpassed that of rye, while in East Germany that happened in 1966[1].

The wild rye is assumed to originate in Central Asia, in the Afghan and Turkmen regions and to have spread westwards from there. In the second millennium BC it reached the Baltic countries and in the next millennium Germany. About 400 BC it was actually grown there. In Roman times rye was also raised in Central Europe, Hungary and England[2]. On the British Isles it was never generally grown with the exception of Norfolk, with its sandy grounds, where rye was cultivated on a larger scale.

In the northern half of the European continent, rye was the only cereal produced at a large scale until the middle of this century. The Netherlands were the most western part of the rye belt which extended far into Russia. In this paper I will deal with the use of rye in the Netherlands.

Growing rye was only possible on the higher and drier - but also poorer - grounds in the eastern half of the Netherlands. The yields were small and often hardly sufficient to nourish the population[3]. In the 14th century, when the cloth industry in the Netherlands was well under way and the cities were expanding, rye was imported by the Hansa organization from the Baltic countries[4]. In the 16th century Amsterdam monopolized the corn-trade and in the following centuries this 'Mother-trade' would become the basis for the economic and cultural prosperity of Amsterdam. In the middle of the 19th century, in the agrarian provinces of the northern, eastern and southern parts of the Netherlands, 80% of the population ate only one kind of bread, the rye bread. In Utrecht this percentage was 50%, and in the urbanized province North-Holland 47%. In the wheat growing provinces South-Holland and Zealand these figures were 34% and 15%[5].

In the middle of the 19th century, of the rye grown in the Netherlands the Frisian rye was thought to be the best and the most palatable, followed by that grown in Overijsel and Guelders, provided it was grown on heavy soil. From foreign countries the strongly dried rye from St Petersburg and the Baltic countries was most suitable for baking bread, followed by the undried rye from Prussia, Dantzig and Königsberg. Of much lesser quality was the rye imported from Odessa and the Black Sea[6].

After 1900 the consumption of rye bread sharply declined. Wheaten bread had taken its place. But up till now many people eat at their lunch at least one slice of rye bread, a lunch which usually consists of open sandwiches with savoury and sweet fillings.

1. THE BAKING OF RYE BREAD

There are big differences between rye breads from the north and the south of the Netherlands. The difference is caused by the way of milling the grains. The southern breads are made from finely milled rye flour, the northern from crushed grains. In his book *'On Food and Cooking'* Harold McGee very lucidly explains which physical processes play a role in the baking of wheaten bread but he does not comment on the baking of rye bread. A few dispersed remarks on the difference between wheat and rye give some clues as to how rye bread is made in a way so different from wheaten bread:

a. Rye proteins form a very weak gluten opposed to wheat proteins which form a strong enough gluten to produce raised breads.

b. Rye flour contains an unusual amount of pentosans, or long chains of 5-carbon sugars. These compounds have a very high water-binding capacity, which means that rye bread retains moisture better than wheat[7].

These characteristics determine the way rye bread is made.

Frisian rye bread

The black Frisian rye bread, which resembles the German Pumpernickel, is baked from crushed grains and not from flour. The native Frisian rye is generally considered to be the most palatable of Dutch rye species, however, its ability to bind water is far less than those from abroad, which gives the loaves a less solid texture. Hence the bread is made from a mixture of native and foreign rye. The bread is made thus:

> 8 or 9 old rye breads are soaked overnight in 6 or 7 litre water. The next morning the breads are crumbled and cooked in the soaking liquid for some time and stirred continuously. The cooked breads are mixed with 65 kilogram crushed rye grains, 1.25 kilogram salt and 37 litre water. The dough is left to cool and to stiffen. Pieces of dough, weighing about 1 kilogram, are dipped into rye bran and shaped in a mould into flat square loaves. They are then coated with a pap of rye flour, water and a dark molasses and placed upright on the floor of the oven. After having baked for about two hours the loaves are covered with planks on which wet sacks are spread, to prevent the crusts from hardening. The baking time is 20 hours at a temperature of 190°C.

In some parts of Friesland a firmer and lighter coloured type of bread is preferred. A greater quantity of foreign rye is used and the baking time is much shorter, 6 to 8 hours.

Groningen rye bread

This bread looks very similar to the Frisian kind, as it has the same shape and colour, but it tastes much sweeter.

> At night 70 kilograms of crushed rye grains, 2/3 part native and 1/3 foreign, and 70 litres cold water are well mixed. When the mixture has somewhat stiffened it is put into oblong pans, called *'zoetpannen'* (pans for sweetening), and baked in a slow oven till the next morning. During the slow baking the ferment diastase converts some of the starch of the rye into sugar, but when a temperature of 68°C. is reached, the conversion ceases. At a higher temperature a part of the starch converts into dextrine and fructose. Slow baking for a long time makes the bread taste sweet.
>
> The next morning the contents of the sweetening pans are soaked in boiling water and then kneaded. 1.5 kilograms salt is added along with some rye flour, to give

the dough the right stiffness. The further processing of the rye bread is the same as for Frisian bread. The baking time is 12 to 14 hours.

Guelders rye bread

In the province of Guelders rye flour is used to make a sour and a sweet kind of rye bread.

> To make sour bread, very finely milled rye meal is mixed with tepid water into a very thin dough 24 hours in advance. This dough is left for 24 hours in a warm place till it has become sour. The next morning, 1 hectolitre rye meal - half native and half foreign - with 1 kilogram salt and 66 litres cold water are mixed with 6 litres of the sour dough into a firm dough. From this dough square flat loaves are shaped and covered with bran. The baking time is 13 hours. After having baked for half an hour the loaves are covered with planks to prevent the crusts from cracking.

For the sweet Guelders rye bread fine milled native flour is used, which is mixed with cold water, salt (1.25 kilogram per 100 kilograms flour), and a little yeast. The handling is the same as for sour bread.

Limburg rye bread

In this southern province finely milled rye flour is used as well. In the northern parts of the Netherlands rye bread is baked at a moderate temperature for a very long time in a moist atmosphere. On the other hand, being rather dry, the Limburg bread is baked at a higher temperature and at a much shorter time. The dough is made with a sourdough to give the bread some lightness, but with a small quantity only to prevent the bread from acquiring a sour taste. The Limburg bread is made this way:

> A piece of dough is left to ferment until late in the afternoon, mixed with some rye flour and tepid water into a firm dough at a temperature of 30°C. This dough is kept overnight in the rye flour that will be used the next morning to bake the breads with. A third part of the flour consists of native rye and 2/3 part is of foreign origin. The next morning a dough is made of the sour dough, the rye flour and half that quantity of tepid water.

> Pieces of the dough, weighing about 5 pounds, are shaped into balls, rolled out lengthwise and then brushed with lard. They are left to rise in a draughty place and are covered with a cloth lest they should develop a crust. After having risen, the pieces of dough are pressed into a frame to give them a square flat shape and then brushed with a pap of rye flour. They are baked at a temperature of at least 250°C for two hours[8].

All these types of rye bread have hardly risen and are solid and compact. They do taste very well, but are rather filling and soon give a feeling of satisfaction.

In my youth, around 1940, we ate at least one and often two slices of rye bread every day at breakfast and at supper, along with slices of wheaten bread. The favourite fillings were bacon or cheese. In winter we ate it with lard, pepper and salt, and when available, with greaves or cracklings.

The recipes quoted above are taken from two well-known handbooks for bakers, published just before and just after World War II. At that time the ovens were heated with faggots and peat. The kneading of the dough was done mechanically. In earlier times the baker had to knead the stiff and tough dough with his bare feet[9]

II. CAKES MADE OF RYE

In the 14th century sweet cakes were also baked from rye. A mixture of honey and water was cooked and then mixed with rye flour and spices. A complication was that sourdough or yeast are not active in such a mixture and the cakes did not rise. In Deventer, an important Hansa trading centre in the east of the Netherlands, one had learned in the course of the 14th century, that cake dough would rise when potash and some buttermilk were added. In an acid environment potash produces carbon dioxide when heated. Potash was imported from the Baltic countries.* It is not clear whether the use of potash as a leaven was invented in Deventer or originated in the Baltics. Honey cakes were made in all the countries of the rye belt in the northern half of Europe. From the household accounts of the Dukes of Guelders we learn that *'pepper cakes'* were eaten at the festivities of New Year's Eve 1398[11]. The honey- or pepper-cakes got their name from the spices they were flavoured with. Pepper was the name of a mixture of spices usually including pepper as well as the name of the spice itself. In a cake recipe from the middle of the 15th century the spice mixture for pepper-cake was composed of equal parts of ground pepper, ginger powder, ground cloves and cinnamon powder, half a part of nutmeg and a fourth part of mace[12].

Deventer was the first city in the Netherlands to export cakes to other cities, especially to Amsterdam. Soon, bakers outside of Deventer produced cakes too and each city had its own type of cake. Well known became those from Gorkum, Schoonhoven, Venlo, 's Hertogenbosch and Groningen. Bread and cakes were produced and sold by the Bakers Guild, but the city-government fixed the prices. When in times of scarcity the prices of wheat, rye or honey were high, the price of the bread and cakes remained the same, but their weight was lowered by the municipality. Every week or every fortnight the weight of the bread and cakes were adapted to the price.

In Amsterdam the cake bakers, together with the pie bakers and rusk bakers, detached themselves from the age old Bread Bakers Guild in 1694 and founded a guild organization of their own[13]. Their independence from the Bakers Guild gave the cake bakers the opportunity to modernize their trade and to develop themselves into modern pastry bakers or pâtissiers. At its foundation the bakers guild had decided that the most common kind of cake its members would bake was *Deventer cake*[14]. For this cake equal amounts of honey and water were used[15]. Cakes of better quality must contain more honey. They were not only flavoured with a spice mixture but also with candied peel (orange and citron), candied ginger and pieces of sugar-candy. When the number of sugar refineries in the Netherlands increased during the 17th century, the cake bakers adapted the use of molasses for sweetening.

Groningen citron cake

As an example for how to bake a cake I will give the recipe of a Groningen candied peel cake. It dates back from the first half of this century.

> 3 kilograms honey, 2 kilograms molasses and 2 litres water are boiled together and then poured through a strainer into 10 kilograms rye flour. This must be well mixed. The dough will become very firm, stiff and tough. Leave it at a cool place for several days. To work up the dough, one should bring it into the bakery a day in advance, so it can get warm and easy to handle. To knead the dough a wooden instrument must be used, called a *'koekbraak'* (cake-brake). An illustration of this instrument is on the cover of *PPC 21* [16]. While braking the dough, 3 kilograms honey is mixed into it, along with 150 grams of mixed spices. These spices are usually pepper, cloves, cinnamon and ginger. A rising agent is added too. This

---

* Potash was used in the cloth-industry to fix colours[10]

agent is chemically pure potash and hartshornsalt. When well mixed, pieces of dough, weighing one kilogram or half a kilogram each, are taken and moulded into an oblong square shape. The sides and upper side are covered with finely cut candied citron peel and pieces of sugar-candy and lightly brushed with oil. Next the cakes are dusted with fine flour and placed on an oiled baking sheet. When the cakes are put in the oven they are brushed with milk. They are baked in a moderate oven at 190°C. for one hour [17].

In the later Middle Ages cake was considered to be a treat for the rich, in the course of time it became a treat for the burghers and the lower orders. When drinking tea came into vogue at the end of the 17th century - and the drinking of coffee somewhat later - serving a slice of cake with the tea and the coffee soon became a matter of course. In the northern provinces of the Netherlands this custom lasted until well after World War II. On weekdays the cake was plain but on Sundays and holidays it had a finer and more complex taste.

In the past decades most bakers in the countryside have vanished and rye breads and cakes are hardly baked anymore in the traditional way. But one type of cake - mostly factory made - has survived and is widely sold : the *'ontbijtkoek'* (breakfast cake). Buttered slices of this cake are eaten with coffee or tea and at lunch, on top of an open sandwich. Breakfast cake is a common filling for open sandwiches and in most hotels it is served at breakfast.

CAKE AND PASTRY

The first printed professional book for pastry bakers in the Netherlands is by Gerrit van den Brenk, pastry baker and confectioner in Amsterdam: *'tZaamen-Spraaken tusschen een Mevrouw, Banketbakker en Confiturier...in 4 delen'* (Dialogues between a Lady and a Pastrycook-Confectioner...in 4 parts), published about 1750. The book is exceptionally rare, only two complete copies are known, one of which is misbound. It is a book for the professional and it describes the art of pastry making, an art not yet a hundred years old and now in its heydays. The upper class were served pastry with their tea and coffee in beautiful china. Or, in the elaborate last course of a banquet, the dessert. For this course the table was elegantly decorated with allegorical representations made of sugar and sometimes from Saxon porcelain. The pastrymaker made the pastry and confectionery and the decorations too[18]. Van den Brenk, who was a member of the Amsterdam cake bakers guild, did not see himself as a cake baker but as a modern craftsman, a pastrybaker. But nevertheless, his book contains a chapter called: *'Complete Instructions for Cake Bakers and their Pupils'*, in which recipes of the old craft of cake baking, recipes which are age-old, are published for the first time. Here for the first time one finds recipes for the since long famous cakes from Venlo, 's Hertogenbosch, Amsterdam and Groningen. In the centuries to come these cakes would be inseparable from the cup of tea or coffee of the lower classes, the so-called *'bakkie troost'* (cup of comfort).

REFERENCES

1. Harold McGee, *On Food and Cooking. The Science and Lore of the Kitchen*, London-Sydney, Unwin Hyman, 1987, p 325.
2. Don and Patricia Brothwell, *Food in Antiquity, A survey of the diet of early peoples*, New York-Washington, Frederick Praeger, 1969, p 100.
3. W. Jappe Alberts & H. P H. Jansen, *Welvaart in Wording, Sociaal- Economische Geschiedenis van Nederland van de vroegste tijden tot het einde van de middeleeuwen*, 's Gravenhage, Martinus Nijhoff, 1964, p 150-70.
4. Ibid, p 234.
5. Tentoonstelling: *Brood, de geschiedenis van het brood en het brood- gebruik*, Rotterdam, Museum Boymans-van Beuningen, 1983, p 63.
6. H. J. Wilson, *Het Brood, zijne Geschiedenis, Bereiding en Vervalschingen*, Kampen, K.van Hulst, 1864, p 16; G.Buisman, *Practisch Handboek voor den Brood- en Banketbakker*, 3rd edn, Amsterdam, Uitgevers-Maatschappij v/h van Mantgem & de Does, 1946, p 77.
7. McGee, op cit, p 291, 235-6.
8. H. Menkveld, A. C. Spil, K.v.Til, *Voor den Bakker*, 3rd edn, Purmerend, J. Muusses, n.d. (1946), p 68; Buisman, op cit, p 77.
9. Noël Chomel & J.A.Chalmot, *Huishoudelijk Woordenboek*, Leiden & Leeuwarden, Joh.le Mair & H.A.de Chalmot, 1769, vol I p 330-4.
10. Jappe Alberts & Jansen, op cit, pp 141, 234, 242.
11. G. van Hasselt, *Bydragen voor d' oude Geldersche Maaltyden*, Arnhem, 1805, p 133.
12. G. van Hasselt, *Arnhemsche Oudheden*, Arnhem, 1804, vol 4 p 243.
13. *Ordonnantie van het Koekebakkers, Beschuitbakkers en Pasteibakkers Gild*, Amsterdam, Pieter Mortier, n. d. (1773), p 16.
14. Ibid, p 15.
15. 18th century manuscript recipe, in: Gerrit van den Brenk, *Het Tweede Deel der 'tZaamen-spraak tusschen een Mevrouw en Confiturier*, Amsterdam, Erven de Weduwe Jacobus van Egmont, n. d. (about 1760). Amsterdam University Library, sign. 1080E50 1, 2.
16. J. Witteveen, 'Potash and Hartshornsalt', London, *PPC 21* 1985, p 66-8.
17. Buisman, op cit, p 73-4.
18. van den Brenk, op cit (ref 15), p 15-25. See also: Barbara Wheaton, *Savouring the Past*, London, Chatto and Windus, 1983, p 186ff.

Rice and Wheat in Middle Eastern Cultures

by Sami Zubaida

Rice and wheat, both staples in different parts of the region, constitute a rich field of cultural representations and symbols. The two staples, and different qualities and modes of preparation of each, mark differences and boundaries between groups, social statuses and social occasions.

The predominant staple is wheat or its (inferior) equivalents, barley, millet and corn. In a mostly arid zone, rice growing is limited. Anatolia, Greater Syria and Northern Iraq, including Kurdistan, may be seen to share common elements of a culinary culture. In this culture wheat is predominant, and rice, historically, a scarce luxury, largely confined to the towns, the rich and special occasions. This was also true of parts of Iran. Rice was more common in southern Iraq, grown in the marshy region at the intersection of the Tigris and the Euphrates. It is also grown in Egypt, but has never constituted an important element of the Egyptian diet.

However, even in rice growing areas, rice has rarely supplanted wheat and barley, but coexisted with them. To my knowledge, it was only in small pockets in Northern Iran that rice was exclusive of wheat (see below).

Of course, there are different qualities of rice and wheat, and different methods of processing and preparation of each. In particular, wheat is most commonly known as some form of bread, but in Anatolia, Greater Syria and Northern Iraq it is also known as *burghul*, crushed wheat. This is made by boiling the grain, then draining and drying it in the sun (or in an industrial oven), then cracking. The product, no doubt familiar to most of you, is a pliable, easily softened grain, most commonly cooked with water, seasonal vegetables and some fat or oil. In this form it is a common and important element in the diet of country people and the urban popular classes. It can also be added to salads, as in *tabbouleh*, or to soups as a grainy thickener. *Fireek* is another, less common, wheat product, made by burning and cracking the green wheat in the fields. It can be boiled like *burghul*, but for a much longer time, or can be added to soups, stews and stuffings. A lovely nutty taste all its own.

Rice is variable in its qualities, especially between round and long grain, and the cooked texture of the grain. Rice, usually long grained, which can be cooked in separate and distinct grains, and the cooking methods and skills which can achieve this effect, are generally the most highly valued. In Egypt, where long grain rice is uncommon, cooks try to achieve the same effect with the local round grains. Rice can also be ground into flour and cooked as bread, or pancakes, usually mixed with milk and fat, sometimes mixed with wheat flour. These foods are uncommon, and, to my knowledge, confined to parts of Iran and Southern Iraq. Rice, ground or whole, can also be combined with meat or fish in dumplings known as *kebbeh*, parallel to the more commonly known preparations made with *burghul* in the wheat areas.

These distinctions of grains, qualities and methods of processing and cooking, constitute a cultural vocabulary of social boundaries, statuses and occasions. In the predominantly wheat region of Turkey, Syria and Northern Iraq, as already noted, rice was/is luxury food (I use the past tense because the modern international food economy has changed many things in recent years). In Ottoman times a rice dish constituted a separate item of a full dinner at the tables of the rich, cooked in butter and eaten by itself, usually at the end of the meal. Edward Lane described such a meal in Egyptian rich households (which partook of Ottoman culture) of the nineteenth century (Lane, 1895-1978, pp 151-2). He tells us that the boiled rice is called *ruz mufelfel*, adding 'the *pilav* of the Turks'. *Ruz mufelfel* remains the term for boiled rice in Egypt and Greater Syria, and indicates separate grains (was it in distinction from rice puddings or gruels?). Rice (more often round grained) is also used in stuffings of vegetables and fowls (Lane lists these items as constituting part of the upper class table). For the common people, and especially in the rural areas, rice was a dish for special occasions, and for the prosperous provincials, a once a week indulgence.

An interesting phenomenon in this region is the substitution of *burghul* for rice. *Burghul* (Arabic; Turkish: *bulgar*) is cooked in imitation of the boiled rice, in water or stock till it is dry with separate grains, oil (more rarely butter) added, sometimes with vegetables and pulses, such as chick-peas, broad beans and tomatoes, more rarely, with meat. The terms used imitate rice terminology, *bulgar pilaf* in Turkey, and *burghul mufelfel* in the Arab parts. It (or *fireek*) is used for stuffing vegetables or fowl. It is as if the common people are imitating the high culinary traditions, but with an affordable ingredient. This is made explicit in the customary imperative to serve rice on wedding feasts and similar occasions. Failure to do so is a measure of poverty or meanness. This is indicated by the popular Arabic saying (in parts of Greater Syria) *el-izz bil-rizz wil-burghul yushnu' nafsu*, 'pride (or well-being) is in rice, and let the *burghul* hang itself'. Another indication of the status of rice is that it is almost always cooked in butter or meat fat, like meat, while *burghul* is usually cooked in oil, like vegetables.

These status distinctions apply only to *burghul* versus rice, not to bread, which has its own status hierarchy. White wheat bread is at the top of the hierarchy and barley bread at the bottom. Fine breads cooked with milk, butter, sugar, preserves or garnished with meat, are in another class, like pastry. In many parts of the region, barley bread is synonymous with poverty or meanness. The inhabitants of the two Iraqi cities of Baghdad and Mosul cultivate negative stereotypes of one another which include miserliness. I, a Baghdadi, recall strolling in the market in Mosul with members of my family and hearing the gibe shouted from one shopkeeper to another across the market: 'what do they have for supper?', answer 'barley bread and water melon'!

Iran is the Middle Eastern country with the most distinguished traditions of rice cookery. It gave the term *polo, pilaf, pulau* to its neighbours. The Persian *polo* refers to dishes of rice cooked with other ingredients, as distinct from *chelo* or plain boiled rice served as an accompaniment to *kebab* (at its best with pats of butter and a raw egg yolk mixed with the steaming fragrant rice on the serving plate, sometimes decorated with colourful streaks of saffron), or to *khoresht*, a generic name for stews of meat with vegetables, pulses or fruit. *Polo* is familiar to the British restaurant goer as the 'pulau rice' of Indian restaurants or the *pilafi* of the Greek. However, the methods of preparation do not always follow the Persian. For both *polo* and *chelo* the rice (long grain) is washed then soaked for at least an hour and preferably much longer, in cold water. It is then drained and added to a large quantity of boiling salted water. When the grains are just cooked (approximately three minutes) the rice is drained and washed under a cold tap. For *chelo* the drained rice is then returned to the pan over heated fat (traditionally) or oil, the pan covered with a towel (to absorb the steam) and a lid and left to steam for at least half an hour, preferably longer. For *polo*, some meat, vegetable or pulse is added to the fat, either in a separate bottom layer or mixed into the rice before it is allowed to steam in the same way. Some vegetables (green haricot or broad beans) are boiled with the rice.

In most Iranian regions and cities, rice, though a very important part of food culture and culinary aesthetics, was not an everyday meal for most people, but more occasional, depending on availability, price and the prosperity of particular households. But there are regions where rice is grown and constitutes a daily staple. However, in these instances there are different qualities of rice, and the daily staple for most people is provided by the cheaper varieties cooked by quicker and more economical methods. This is the case in the rice growing region of coastal Guilan, by the Caspian in the north. An ethnographic study of this region by Bazin and Bromberger (1979) documented the different uses of rice and bread. It is not widely known that there is in Iran a third method of preparing rice, *kate*, used for lower quality rice, sometimes round grained, for everyday meals of country people in certain regions, for whom the *polo* and *chelo* are foods for special occasions. The preparation of *kate* is quicker and simpler: the rice is washed but not soaked, placed in a pan and covered with one and a half time its volume of water, boiled until all the water is absorbed, a small amount of fat then added, and the whole lot allowed to steam. The result is sticky rice, which when cool forms a mass which can be cut with a knife, or made into balls and stored. It is served as an accompaniment to *khoresht*, boiled vegetables or fried fish. But it is hardly ever served with *kebab*, a discordant combination of humble rice and exalted grilled meat. This rice can be stored for future consumption, made into balls and taken on journeys, or eaten for breakfast the next morning. If this is the intention, then fat is not added in the cooking, but it may subsequently be heated up with fat.

Bazin and Bromberger note a division of the region into the humid coastal part where rice consumption predominated over wheat, and the arid mountainous zone where bread was eaten exclusively. It would seem that until recent decades, and before the spread of commercial bakeries, inhabitants of the first zone ate rice exclusively, three meals a day. Breakfast consisted of *kate* left over from the previous day, eaten with cheese and/or jam, washed down sometimes by hot sugared milk, but always with three glasses of sugared tea. We see here that the rice substitutes for bread in the breakfast of bread regions. But unlike the felt inferiority of *burghul* to rice in parts of Turkey and Syria, the inhabitants of coastal Guilan were militantly proud of their rice culture, which they considered superior to bread and healthier. Bazin and Bromberger quote earlier observers at the turn of the twentieth century who cite common sayings to that effect: an angry husband says to his wife, 'Go eat bread and burst!', and an angry parent would threaten his child with being sent to Arak (in the interior) where he would have to eat bread.

REFERENCES

M. Bazin and C. Bromberger (1979), *Documents pour l'etude de la repartition de quelques traits culturels dans le Guilan et l'Azerbayjan oriental*, CNRS, Paris.

E. W. Lane, *Manners and Customs of the Modern Egyptians*, (written in Egypt during the years 1833-35), London and The Hague 1895/1987.